T0299750

Large Outdoor Fire Dynamics

Large Outdoor Fire Dynamics provides the essential knowledge for the hazard evaluation of large outdoor fires, including wildland, WUI (wildland–urban interface), and urban fires. The spread of outdoor fires can be viewed as a successive occurrence of physical and chemical processes – solid fuel combustion, heat transfer to surrounding combustibles, and ignition of heated combustibles – which are explained herein. Engineering equations frequently used in practical hazard analyses are derived and then integrated to implement a computational code predicting fire spread among discretely distributed combustibles. This code facilitates learning the procedure of hazard evaluation for large outdoor fires.

Chapters cover underlying assumptions for analyzing fire spread behavior in large outdoor fires, namely, wind conditions near the ground surface and fundamentals of heat transfer; the physical mechanism of fire spread in and between combustibles, specifically focusing on fire plumes (both reacting and non-reacting) and firebrand dispersal; and the spatial modeling of 3D objects and developing the computational framework for predicting fire spread.

The book is ideal for engineers, researchers, and graduate students in fire safety as well as mechanical engineering, civil engineering, disaster management, safety engineering, and planning. Companion source codes are available online.

Keisuke Himoto, DrEng, is a senior researcher at the National Institute for Land and Infrastructure Management in Tsukuba, Japan. His research interests cover a broad range of fire safety issues in the built environment but with a special focus on large outdoor fires. He is the developer of various fire-related computational models, including one of the first physics-based computational models for fire spread in densely built urban areas.

Large Outdoor Fire Dynamics

Keisuke Himoto

CRC Press
Taylor & Francis Group
Boca Raton London New York

Front cover image: Trybex/Shutterstock

First edition published 2023
by CRC Press
6000 Broken Sound Parkway NW, Suite 300, Boca Raton, FL 33487-2742

and by CRC Press
4 Park Square, Milton Park, Abingdon, Oxon, OX14 4RN

CRC Press is an imprint of Taylor & Francis Group, LLC

Library of Congress Cataloging-in-Publication Data
Names: Himoto, Keisuke, author.
Title: Large outdoor fire dynamics / Keisuke Himoto.
Description: First edition. | Boca Raton : CRC Press, 2023. |
Includes bibliographical references and index.
Identifiers: LCCN 2022029396 | ISBN 9780367561680 (hbk) |
ISBN 9780367561697 (pbk) | ISBN 9781003096689 (ebk)
Subjects: LCSH: Wildfires–Mathematical models. |
Forest fires–Mathematical models. |
Heat transfer. | Fire protection engineering.
Classification: LCC SD421.45.M37 H56 2023 |
DDC 634.9/618–dc23/eng/20220928
LC record available at https://lccn.loc.gov/2022029396

ISBN: 978-0-367-56168-0 (hbk)
ISBN: 978-0-367-56169-7 (pbk)
ISBN: 978-1-003-09668-9 (ebk)

DOI: 10.1201/9781003096689

Typeset in Sabon
by Newgen Publishing UK

Access the Support Material: www.routledge.com/9780367561680

Contents

Preface

Large wildland fires with significant societal impacts have occurred frequently in recent years. This is believed to be partly associated with hotter and drier weather in some areas because of climate change. The frequency of such fires is expected to increase further in the future. The awareness of such risks is shared worldwide. Efforts are underway to adequately assess the risk of wildland fires and incorporate the results into the design of fire-resilient societies.

Urban areas in Japan have suffered from numerous conflagrations over the centuries. Cities outside of Japan have not necessarily been exempt from urban conflagrations, as evidenced by the occurrences of the Great Fire of London in 1666 and the San Francisco Earthquake and Fire in 1906. However, urban areas in Japan have historically been comprised of wooden buildings with low fire resistance performance maintaining small separations between each other. Given this background, the frequency of conflagrations in the cities of Japan has been outstanding. Although the situation has improved in recent years, the occurrence of conflagrations such as the 1995 Kobe Earthquake and Fire and the 2016 Itoigawa Fire exhibited the remaining risks in cities of Japan.

Wildland fires and urban conflagrations are both large outdoor fires where successive fire spread occurs among combustibles distributed outdoors. Exposure to fire plumes and dispersal of firebrands are their common mechanisms of fire spread. However, research on these two variants of outdoor fires has been developed under independent scientific systems. This book was inspired by the idea that the hazards of outdoor fires could be described in a unified manner. The growing concern for protecting structures and properties from wildland–urban interface (WUI) fires, another variant of outdoor fires, has also encouraged the writing of such a book.

This book aims to provide basic knowledge essential for the hazard evaluation of large outdoor fires. Fire spread in outdoor fires can be viewed as a successive occurrence of physical and chemical processes, which comprise combustion of a solid fuel object, heat transfer from burning fuel object to adjacent fuel objects, and ignition of fuel objects subject to external heating. It explains the basic mechanisms of the processes in the context of outdoor fires and derives engineering equations frequently used in practical hazard analyses. It further integrates engineering equations to implement a computational code predicting fire spread among discretely distributed 3D fuel objects. The computational code, together with its description, aims to facilitate readers to learn the procedure of hazard evaluation for large outdoor fires.

It is the author's sincere hope to contribute in any way to the development of related research and the enhancement of resiliency in the societies vulnerable to outdoor fires through the discussion of this book.

Nomenclature

ALPHABET

A	Area (m^2), Area of an opening (m^2), Pre-exponential factor (s^{-1})
A_B	Fuel burning area (m^2)
A_F	Fuel surface area (m^2)
A_P	Projected area of a firebrand (cm^2)
A_T	Total area of enclosure surfaces (m^2)
A_W	Wall surface area (m^2)
B	Width of an opening (m)
b	Plume width with respect to velocity (m)
B^*	Dimensionless plume width (–), Scaling dimensionless parameter for firebrand transport (–)
Bi^*	Biot number (–)
c	Heat capacity (kJ $\cdot$ kg^{-1} $\cdot$ K^{-1})
C_D	rag coefficient (–)
C_L	Lift coefficient (–)
C_M	Moment coefficient (–)
c_P	Specific heat capacity at constant pressure (kJ $\cdot$ kg^{-1} $\cdot$ K^{-1})
D	Reference length (m)
D_C	Thickness of a wood stick (m)
d	Depth (m), Distance (m), Representative width of fuel component (cm)
d_B	Width of flame front (m)
d_H	Flame heated length (m)
d_P	Firebrand diameter (cm)
$d\omega$	Solid angle (sr)
E	Activation energy (J $\cdot$ mol^{-1}), Emissive power (kW $\cdot$ m^{-2})
$E = (\alpha, \beta, \gamma)$	Euler angle
F	View factor (–), Force (F)
f	Coriolis parameter $\left(\text{rad} \cdot \text{s}^{-1} \right)$
$\mathbf{f}, \mathbf{f}_D, \mathbf{f}_G, \mathbf{f}_L$	Aerodynamic force, Drag, Gravity, Lift applied to a firebrand (N)

Fo^*	Fourier number (–)
Fr^*	Froude number (–)
g	Acceleration due to gravity ($\text{m} \cdot \text{s}^{-1}$)
Gr^*	Grashof number (–)
H	Height, Height of an opening (m)
H_C	Height of a wood crib (m)
H_{fl}	Mean flame height (m)
$\Delta H, \Delta H_C$	Heat of combustion ($\text{kJ} \cdot \text{kg}^{-1}$)
ΔH_A	Heat of combustion per unit mass of fresh air consumed ($\text{kJ} \cdot \text{kg}^{-1}$)
ΔH_{ig}	Heat required for ignition ($\text{kJ} \cdot \text{kg}^{-1}$)
ΔH_O	Heat of combustion per unit mass of oxygen consumed ($\text{kJ} \cdot \text{kg}^{-1}$)
h	Convective heat transfer coefficient ($\text{kW} \cdot \text{m}^{-2} \cdot \text{K}^{-1}$), Height (m)
h_k	Effective heat transfer coefficient ($\text{kW} \cdot \text{m}^{-2} \cdot \text{K}^{-1}$)
h_T	Total heat transfer coefficient ($\text{kW} \cdot \text{m}^{-2} \cdot \text{K}^{-1}$)
I	Radiation intensity ($\text{kW} \cdot \text{m}^{-2} \cdot \text{sr}^{-1}$)
i_P	Moment of inertia ($\text{kg} \cdot \text{m}^2$)
k	Thermal conductivity ($\text{kW} \cdot \text{m}^{-1} \cdot \text{K}^{-1}$)
$k\rho c$	Thermal inertia ($\text{kJ}^2 \cdot \text{s}^{-1} \cdot \text{m}^{-4} \cdot \text{K}^{-2}$)
L	Length, Representative length, Thickness of plane wall (m)
$\bar{L}$	Mean beam length (m)
L_{fl}	Mean flame length (m)
L_G	Latent heat of gasification ($\text{kJ} \cdot \text{g}^{-1}$)
L_V	Latent heat of vaporization of water ($\text{kJ} \cdot \text{g}^{-1}$)
$M(t)$	Mass of wood crib at time t (kg)
M_0	Initial mass of wood crib (kg)
M_W^*	Dimensionless mass loss rate (–)
m	Mass (kg)
$\mathbf{m}$	Moment applied to a firebrand ($\text{N} \cdot \text{m}$)
$\dot{m}_A$	Mass consumption rate of fresh air, Mass inflow rate by ventilation ($\text{kg} \cdot \text{s}^{-1}$)
$\dot{m}_B$	Mass burning rate ($\text{kg} \cdot \text{s}^{-1}$)
$\dot{m}_B''$	Mass burning rate per unit area ($\text{kg} \cdot \text{m}^{-2} \cdot \text{s}^{-1}$)
$\dot{m}_C''$	Surface oxidation rate per unit area ($\text{kg} \cdot \text{m}^{-2} \cdot \text{s}^{-1}$)
$\dot{m}_e$	Cumulative amount of air entrainment ($\text{kg} \cdot \text{s}^{-1}$)
$\dot{m}_O$	Mass consumption rate of oxygen ($\text{kg} \cdot \text{s}^{-1}$)
$\dot{m}_S$	Mass outflow rate by ventilation ($\text{kg} \cdot \text{s}^{-1}$)
N	Fractional contribution of cross-wind velocity on velocity along trajectory (–)

$\dot{N}''_{FB}$	Firebrand generation rate (number of firebrands generated per unit time) per unit floor area ($m^{-2} \cdot s^{-1}$)
N_k	Number of category-k firebrands generated from a fire source
n_k	Number of category-k firebrands landed on recipient combustible-i
Nu^*	Nusselt number (–)
$\overline{Nu}^*$	Mean Nusselt number (–)
P	Atmospheric pressure ($kg \cdot m^{-1} \cdot s^{-2}$) (Pa)
p	Pressure ($kg \cdot m^{-1} \cdot s^{-2}$) (Pa)
$p_{FS(i)}$	Probability of fire spread caused by firebrands to recipient combustible-i
$p_{FS(j \to i)}$	Probability of fire spread caused by firebrands from a fire source-j to recipient combustible-i
$p_{I,k}$	Probability of a category-k firebrand to cause ignition of recipient combustible-i
$p_{T,k}$	Probability of a category-k firebrand to be transported from a fire source to recipient combustible-i
$p_{T(j \to i),k}$	Probability of a category-k firebrand to be transported from a fire source-j to recipient combustible-i
Pr^*	Prandtl number (–)
$\dot{Q}, \dot{Q}_B$	Heat release rate (HRR) (kW)
$\dot{Q}', \dot{Q}'_B$	Heat release rate per unit length ($kW \cdot m^{-1}$)
$\dot{Q}'', \dot{Q}''_B$	Heat release rate per unit area ($kW \cdot m^{-2}$)
Q^*	Dimensionless heat release rate (–)
$\dot{Q}_D$	Rate of radiant heat loss through opening (kW)
$\dot{Q}_F$	Rate of heat loss to fuel surface (kW)
$\mathbf{Q}_P$	Attitude angle of a firebrand (rad)
Q_s^*, Q''_s	Dimensionless heat release rate with respect to distance along trajectory (–)
$\dot{Q}_W$	Rate of heat loss to compartment surfaces (kW)
Q_W^*	Dimensionless heat release rate (–)
$\dot{q}''$	Heat flux ($kW \cdot m^{-2}$)
$\dot{q}''_{cr}$	Critical heat flux for ignition ($kW \cdot m^{-2}$)
q''_{ex}	Incident heat flux from external heat source ($kW \cdot m^{-2}$)
q''_{fl}	Incident heat flux from flame ($kW \cdot m^{-2}$)
q''_{loss}	Surface re-radiation from fuel surface ($kW \cdot m^{-2}$)
$\dot{q}''_{net}$	Net incident heat flux at solid surface ($kW \cdot m^{-2}$)
$\dot{q}''_R$	Radiant heat flux received at solid surface ($kW \cdot m^{-2}$)
$\dot{q}'''$	Rate of heat generation per unit volume ($kW \cdot m^{-3}$)
R	Thermal resistance ($m^2 \cdot K \cdot kW^{-1}$), Universal gas constant $= 8.314 (J \cdot mol^{-1} \cdot K^{-1})$

r	Radius (m), Radial distance from centerline (m), Stoichiometric air–fuel ratio
R_C	Thermal contact resistance $(\mathrm{m^2 \cdot K \cdot kW^{-1}})$
Re^*	Reynolds number (–)
r_0	Equivalent radius of window flame (m)
r_{st}	Stoichiometric oxygen–fuel ratio
$r_{st,A}$	Stoichiometric air–fuel ratio
S	Spacing between wood cribs (m)
S_C	Spacing between wood sticks (m)
S^*	Dimensionless distance along trajectory (–)
s	Distance along trajectory, Distance between point source and heat receiving point (m)
T	Temperature (K)
T^*	Dimensionless temperature rise (–)
T_f	Film temperature (K)
T_{ig}	Surface temperature at ignition (K)
T_P	Pyrolysis temperature (K)
T_S	Surface temperature (K)
T_∞	Ambient temperature (K)
T_0	Initial temperature (K)
ΔT	Temperature rise above ambient (K)
t	Time (s)
t_C	Time for chemical reaction (s)
t_{fire}	Fire duration (s), (min)
t_{ig}	Time to ignition (s)
t_M	Time to form flammable gas–oxygen mixture (s)
t_P	Time to pyrolysis (s)
U,V,W	$x-,y-$, and $z-$ components of velocity $(\mathrm{m \cdot s^{-1}})$
U	Velocity, wind velocity, reference velocity $(\mathrm{m \cdot s^{-1}})$
$\mathbf{U}$	Airflow velocity $(\mathrm{m \cdot s^{-1}})$
U_g, V_g	$x-$ and $y-$ components of geostrophic wind velocity $(\mathrm{m \cdot s^{-1}})$
$\bar{U}$	Average wind velocity $(\mathrm{m \cdot s^{-1}})$
U^*	Friction velocity $(\mathrm{m \cdot s^{-1}})$
U_W^*	Dimensionless velocity (-)
$\mathbf{u}$	Airflow velocity $(\mathrm{m \cdot s^{-1}})$
$\mathbf{u}_P$	Firebrand velocity $(\mathrm{m \cdot s^{-1}})$
$\mathbf{u}_R$	Airflow velocity relative to firebrand velocity $(\mathrm{m \cdot s^{-1}})$
U_∞	Ambient wind velocity $(\mathrm{m \cdot s^{-1}})$
u, v, w	$x-,y-$, and $z-$ components of velocity $(\mathrm{m \cdot s^{-1}})$
u'	Fluctuation component of wind velocity $(\mathrm{m \cdot s^{-1}})$
u_r	Radial velocity $(\mathrm{m \cdot s^{-1}})$
u_1, u_2	$x-$ and $y-$ components of unit wind velocity vector $(\mathrm{m \cdot s^{-1}})$

V	Volume (m^3)
v_{fl}	Flame spread rate ($m \cdot s^{-1}$)
V_P	Firebrand volume (m^3)
v_P	Regression velocity of a wood stick ($m \cdot s^{-1}$)
W	Total mass of fuel (kg), Side length of a fire source (m), Upward velocity ($m \cdot s^{-1}$)
W^*	Dimensionless upward velocity
W_F	Burnable mass of fuel (kg)
W_0	Vertical ejection velocity of a firebrand ($m \cdot s^{-1}$)
x, y, z	Spatial coordinate (m)
X_P	Position of a firebrand (m)
Z^*	Dimensionless height (-)
Z_N	Height of neutral plane (m)
z_{ref}	Reference height (m)
z_0	Roughness length (m)
Δz	Adjustment for virtual origin (m)

GREEK LETTERS

α	Thermal diffusivity ($m^2 \cdot s^{-1}$), Mass flow rate coefficient (–), Absorptivity (–), Incidence angle (rad), Fire growth rate ($kW \cdot s^{-2}$)
β	Coefficient of volumetric expansion (K^{-1}), Coefficient for plume width b(–)
δ	Thickness (m), Thermal penetration depth (m)
ε	Emissivity (–)
θ	Azimuthal angle (rad), Flame tilt angle (rad)
κ	Absorption coefficient (m^{-1}), von Karman constant (–)
λ	Wavelength (m)
μ	Viscosity coefficient ($\mu Pa \cdot s^{-1}$), Mean
ν	Kinematic viscosity coefficient ($m^2 \cdot s^{-1}$)
π	Circumference ratio (–)
ρ	Density ($kg \cdot m^{-3}$), Reflectivity (–)
ρ_P	Firebrand density ($kg \cdot m^{-3}$)
ρ_∞	Ambient gas density ($kg \cdot m^{-3}$)
$\Delta\rho$	Density difference ($kg \cdot m^{-3}$)
σ	Stefan-Boltzmann constant($kW \cdot m^{-2} \cdot K^{-4}$), Standard deviation
τ	Transmissivity (–)
ϕ	Moisture content (–), Equivalence ratio(–), Incident angle, polar angle (rad)
χ	Burn type factor ($m^{1/2}$)
χ_R	Radiative fraction of heat release(–)

ψ	Latitude (°)
ω	Rotational speed of the earth $\left(\cong 7.292 \times 10^{-5}\right)(\text{rad}\cdot\text{s}^{-1})$
$\boldsymbol{\omega}$	Angular velocity of flow field $(\text{rad}\cdot\text{s}^{-1})$
$\boldsymbol{\omega}_P = \left(\xi_P, \eta_P, \zeta_P\right)$	Angular velocity of firebrand $(\text{rad}\cdot\text{s}^{-1})$
$\boldsymbol{\omega}_R$	Angular velocity of flow field relative to firebrand $(\text{rad}\cdot\text{s}^{-1})$

ACCENTS

$(\dot{\ })$	Per unit time
$(\bar{\ })$	Mean
$(\ )'$	Per unit length
$(\ )''$	Per unit area
$(\ )'''$	Per unit volume
$(\ )^*$	Dimensionless

SUBSCRIPTS

b	Black body
cr	Critical
fl	Flame
ig	Ignition
m	Centerline
R	Radiation
S	Surface
∞	Ambient
0	Initial

Chapter 1

Introduction

This book systematically describes the knowledge essential to quantitatively evaluate the hazard of fire spread in large outdoor fires. In this chapter, we first define a large outdoor fire, as discussed in this book. We then present the overall structure of the book and an overview of the discussions in each chapter.

1.1 LARGE OUTDOOR FIRES

1.1.1 Definition

To begin our discussion, we first define a large outdoor fire, which we refer to in this book, as follows:

> 'A non-explosive combustion that occurs outdoors under an external weather condition, spreading successively from one combustible to another resulting in a simultaneous combustion of many combustibles that cannot be easily suppressed by firefighting activities, and causing losses to lives and properties.'

By definition, 'fire' is generally included in but not fully consistent with the 'combustion phenomenon'. The term 'fire' has the connotation of being a phenomenon that can cause damage. A 'large-scale combustion phenomenon' is not necessarily a 'large outdoor fire', but it is a 'large outdoor fire' when hazardous to lives and properties.

According to the terminology given by ISO, examples of fire that fall under the above definition of a large outdoor fire are as listed below (ISO, 2021).

- Forest fire: unwanted fire burning forests and wildlands (this term is used not exclusively but primarily in Europe).
- Bushfire: unplanned fire in a vegetated area (this term is used not exclusively but primarily in Africa and Oceania).

DOI: 10.1201/9781003096689-1

- Wildland fire: fire occurring in forests, scrublands, grasslands, or rangelands, either of natural origin or caused by human intervention (this term is used not exclusively but primarily in North America).
- Urban fire: unwanted fire burning in an urbanized area.
- Wildland–urban interface (WUI) fire: wildland fire that has spread into the wildland–urban interface. It is also possible for a fire to start in the wildland–urban interface and spread into the wildland.

Fires in the first through third bullet points refer to those occurring in forests and wildlands. However, they are termed differently by region. The behavior of each fire may differ because the spatial distribution and physical structure of forest and wildland fuels vary by region. However, we will use the term 'wildland fires' to refer to all these fires that occur in forests and wildlands as their fundamental nature is common. Primary fuels in urban fires in the fourth bullet point are houses and structures. Although the characteristics of fire sources are different, the process of fire spread from one combustible to another follows a mechanism similar to wildland fires. WUI fire in the fifth bullet point has combined characteristics of both wildland and urban fires. WUI fires occur in spaces with a mixture of wildland and structural fuels. In this book, all the fires in the above list are collectively referred to as large outdoor fires. Note that oil and chemical tank fires may also be included in the list of large outdoor fires. However, they are not our primary focus as they are combustion phenomena of liquid fuels often involving explosive reactions.

1.1.2 Examples

Examples of large outdoor fires that fall into the above definition and have occurred since 1990 are listed in Table 1.1 (Pagni, 1993; JAFSE, 1996; Graham, 2003; Flat Top Complex Wildfire Review Committee, 2012; Ferreira-Leite et al., 2013; CalFire, 2013–2021; Ribeiro et al., 2014; Takeya et al., 2017; Tedim et al., 2020). However, they are not free from regional bias, nor are they exhaustive. As indicated by the size of the area burnt, and the number of fatalities and structures destroyed in the table, large outdoor fires often cause significant human and property losses. In addition, there are also cases involving impacts on the social and economic activities due to disfunction of business activities, the health issues of local residents due to exposure to smoke and other hazardous products, and the environmental influences both locally and globally.

What is the background causing such extensive losses in outdoor fires? There are several difficulties in taking countermeasures compared to the other types of fire, such as those in buildings and vehicles.

The first is the difficulty in achieving the key principle of fire-fighting: extinguishing a fire while still small. There are various ignition

Table 1.1 Examples of large outdoor fires since 1990 (Pagni, 1993; JAFSE, 1996; Graham, 2003; Flat Top Complex Wildfire Review Committee, 2012; Ferreira-Leite et al., 2013; CalFire, 2013–2021; Ribeiro et al., 2014; Takeya et al., 2017; Tedim et al., 2020).

Year	Name	Site	Area burnt (ha)	Structures destroyed	Fatalities
1991	Oakland Hills fire	California, US	620	3,276	25
1993	Laguna Beach fire	California, US	6,475	>400	0
1994	1994 Eastern Seaboard fire	Sydney, Australia	800,000	225	0
1995	Kobe fire	Kobe, Japan	83	>7,000	~500
2000	Cerro Grande fire	New Mexico, US	17,400	235	0
2002	Hayman fire	Colorado, US	55,893	600	6
2002	Rodeo-Chediski fire	Arizona, US	189,651	426	0
2003	2003 Canberra bushfires	Canberra, Australia	1,120,000	488	4
2003	Cedar fire	California, US	110,578	2,820	15
2003	Okanagan Mountain Park fire	British Columbia, Canada	25,912	238	3
2007	October 2007 California wildfires (including Witch Creek fire)	California, US	393,415	3,143	17
2007	Greek forest fires	Greece	270,000	>3,000	84
2008	2008 California wildfires	California, US	644,940	>1,000	32
2009	Black Saturday bushfires	Victoria, Australia	450,000	>3,500	173
2011	Flat Top Complex wildfires	Alberta, Canada	22,000	510	0
2013	Yarnell Hill fire	Arizona, US	3,400	129	19
2013	Picões fire	Portugal	14,000	NA	0
2013	Caramulo fire	Portugal	9,146	NA	4
2014	Valparaíso wildfire	Chile	800	>2,500	15
2014	Västmanland wildfire	Västmanland, Sweden	15,000	71	1
2015	North Complex fire	California, US	129,068	2,455	16
2016	Fort McMurray wildfire	Alberta, Canada	590,000	3,244	0
2016	Gatlinburg wildfires	Tennessee, US	7,240	2,460	14
2016	Itoigawa fire	Niigata, Japan	4	147	0
2017	2017 Chile wildfires	Chile	523,000	>1,000	11
2017	June 2017 Portugal wildfires	Leiria, Portugal	45,329	NA	66
2017	October 2017 Iberian wildfires	Portugal and Spain	54,000	NA	49

(continued)

Table 1.1 Cont.

Year	Name	Site	Area burnt (ha)	Structures destroyed	Fatalities
2017	October 2017 Northern California wildfires (including Atlas, Tubbs, Nuns, and Redwood valley fires)	California, US	99,100	8,900	44
2017	Thomas fire	California, US	114,078	1,063	2
2018	Camp fire	California, US	62,053	18,804	85
2018	Attica wildfires	Attica, Greece	1,431	>4,000	102
2019	2019–20 Australian bushfires	Australia	24,300,000-33,800,000	>9,352	34
2020	2020 California wildfires (including August Complex fire)	California, US	1,779,730	10,488	33
2020	2020 Oregon wildfires	Oregon, US	494,252	>3,000	11
2021	2021 California wildfires (including Dixie and Caldor fires)	California, US	1,039,641	3,629	3

causes of large outdoor fires. Triggers of wildland fires can be either natural, such as lightning strikes, or anthropogenic, such as arson or hot metal fragments/sparks generated from power lines (Fernandez-Pello, 2017). In fires that occur in forests and wildlands, detection and reporting could delay due to the absence of people in the vicinity. Firefighters are often incapable of promptly responding to fires after receiving a fire call, given that fire stations are only sparsely located in forests and wildlands. Consequently, the start of firefighting activities is delayed. Conflagrations in urban areas can be caused either by the spread originating from ordinary fires due to strong winds or the spread originating from multiple ignitions following earthquakes. Compared to fires in forests and wildlands, most fires under ordinary situations can be promptly extinguished, given that fire stations are more densely located in urban areas. However, in strong wind conditions, fires often gain momentum in a short time and become uncontrollable. In post-earthquake conditions, firefighters are often split into multiple units to respond to multiple fires. Consequently, the principle of early stage extinguishment also fails. In either case of wildland and urban fires, fires develop when their momentum exceeds the capability of firefighters to extinguish them. The momentum of a fire increases with time, whereas the ability of firefighters to extinguish a fire is essentially limited.

Once a fire is developed, the fire spread may continue until combustibles are spatially disconnected.

The second is the difficulty in controlling the safety condition of combustibles next to an ignition source. In a building, for example, the use of flammable materials as linings is generally avoided near a cooking stove to prevent an easy ignition. Although ignition causes vary, we can narrow down locations where fires are most likely to occur, and necessary preventive measures can be taken accordingly. If we can similarly specify locations where outdoor fires are likely to start, we may also prevent them from spreading by removing combustibles near the ignition source in advance. However, the situation in outdoor fires is somewhat different from that of fires in a building. Even though we can narrow down the range of concerned areas, it still remains vast for the safety control of combustibles next to an ignition source to an extent equivalent to the case in a building.

The third is the difficulty in confining the burn area due to a wide range and continuous distribution of combustibles. The upper bound of the burn area depends on the extent to which combustibles are spatially interconnected. For example, the maximum possible burn area due to a fire started in a building should correspond to its size unless there is only a small separation between adjacent buildings. In contrast, combustibles are continuously distributed in wildlands or urban areas where outdoor fires occur, allowing a fire to spread over a wide area. Generally, interior spaces in large buildings are divided into multiple fire compartments to disconnect the spatial continuity of combustibles. For a similar purpose, open spaces are often maintained as fire breaks to prevent fire spread in outdoor fires. However, it is not a measure as reliable as a fire compartment inside a building. Fire breaks have limited effectiveness in preventing spotting ignition by firebrands that can be dispersed over several hundred meters or even more downwind of a fire source.

In the past, fires in buildings frequently caused dozens of fatalities. Even today, such fires have not been fully eliminated, but the situation differs by country and region. However, the risk is becoming manageable in many countries and regions, as fire protection standards have been strengthened incrementally through the experience of past major fires. On the other hand, the risk of large outdoor fires remains unmanageable. The frequent occurrence of destructive fires, as summarized in Table 1.1, indicates that the risk of large outdoor fires is still not at a level acceptable for some societies. In order to establish a technological framework to maintain a safe environment against outdoor fires, we need to develop systematic methodologies to evaluate the hazard of fire spread.

1.2 SCOPE OF THIS BOOK

Issues can be raised from different viewpoints in discussing large outdoor fires.

- Influence of external factors such as the weather condition on the outbreak of large outdoor fires.
- Mechanism of fire spread in large outdoor fires.
- Human behavior related to large outdoor fires, including evacuation and firefighting activities.
- Impact of large outdoor fires on socioeconomic activities.
- Relationship between large outdoor fires and climate change.
- Measures to reduce the negative impact of large outdoor fires.

Although these are all important, it is beyond our ability to discuss all these issues in this book. We focus our discussion on the second bullet point, the mechanism of fire spread in large outdoor fires.

Large outdoor fires are caused by the successive occurrence of fire spread between multiple combustibles. Therefore, to consider measures to prevent a large outdoor fire is nothing but to consider measures to prevent fire spread between combustibles. This book provides a systematic overview of the knowledge necessary to quantitatively evaluate the process of fire spread in large outdoor fires.

Figure 1.1 schematically illustrates the process of fire spread in a large outdoor fire where multiple combustibles are discretely distributed. When one of these combustibles ignites due to external heating, it creates a fire plume forming a new fire source for combustibles further beyond. An extensive area downwind of a fire source is covered by its fire plume as tilted in a crosswind. Heated combustibles ignite when they thermally decompose and release flammable gas into the gas phase. Numerous firebrands released from a fire source are widely dispersed in the downwind. A new ignition point may be created at a combustible where firebrands that sustain a high

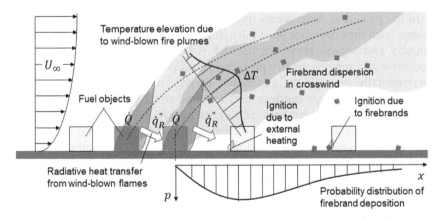

Figure 1.1 Schematic representation of the process of fire spread in a large outdoor fire.

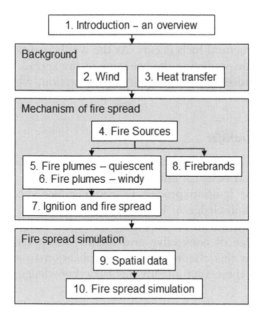

Figure 1.2 An overview of this book.

thermal energy state for a certain duration are received. The above process of fire spread is broken down into several sub-processes, each discussed in a separate chapter in this book.

An overview of this book is shown in Figure 1.2. The volume can be divided into three parts. The first two chapters of this book (Chapters 2 and 3) explain underlying assumptions for analyzing the behavior of fire spread in large outdoor fires, namely, the wind condition near the ground surface and the fundamentals of heat transfer. The following five chapters (Chapters 4, 5, 6, 7, and 8) explain the physical mechanism of fire spread between combustibles, specifically focusing on the combustion of solids, the behavior of fire plumes (with and without wind), the ignition of solids, and the firebrand dispersal. The last two chapters (Chapters 9 and 10) explain the spatial modeling of 3D objects and develop the computational framework for predicting the fire spread.

Chapter 2: Wind

Chapter 2 describes the characteristics of winds that can significantly affect fire behavior. Wind supplies fresh air to a burning combustible and tilts the fire plume formed above it. In addition, it causes a scatter of firebrands generated from the fire source in the downwind. If the fire size is large, the fire plume effect may reach a high altitude. However, we focus on the behavior of local winds near the ground surface, where fire spread between vegetative and structural fuels occurs. As the wind condition is generally a given condition for predicting the behavior of fire spread, we also discuss the statistical features of wind velocity and direction, which vary significantly by location.

Chapter 3: Heat transfer

Chapter 3 explains the fundamentals of heat transfer, which is a particularly important technical basis for analyzing the process of fire spread between combustibles. The ignition process of combustibles subject to external heating requires knowledge of heat conduction in solids. Analysis of the process of fire spread from a fire source to its surrounding combustibles requires knowledge of convective and radiative heat transfer. Although the description in this chapter assumes application to outdoor fires, it can be skipped if the reader already has basic knowledge of heat transfer engineering.

Chapter 4: Fire sources

Chapter 4 provides an overview of the combustion mechanism of solid fuel, the most common type of fuel in outdoor fires. We also discuss methods for estimating the heat of combustion of various fuels, necessary for determining the heat release rate due to combustion. The fire sources assumed in wildland fires are different from those in urban fires. In wildland fires, examples of combustibles are vegetative fuels such as horizontally layered fallen leaves, twigs, barks and branches, and shrubs and tree canopies. In urban fires, they are structural fuels comprising wooden structural members and materials. Combustion characteristics of fuels in each fire are the premise for the discussions in Chapters 5 and 6.

Chapter 5: Fire plumes – quiescent environment

Chapter 5 analyzes the characteristics of a fire plume formed above a fire source in a quiescent environment. Among several possible approaches, we invoke the classical plume theory, which explicitly treats the relationship between physical quantities. We derive an evaluation method for temperature, width, and velocity in different regimes of the fire plume (continuous flame, intermittent flame, and buoyant plume regimes) as a function of the height above the fire source. We derive another evaluation method for the flame height on a fuel of various geometries by further extending this concept. Characteristics of an external flame ejected from an opening of a building, which is an important factor of fire spread in urban fires, are also discussed.

Chapter 6: Fire plumes – windy environment

Chapter 6 analyzes the characteristics of a fire plume formed in the downwind of a fire source, with the theoretical framework following the one provided in Chapter 5. In contrast to the no-wind case, the thermal effects on combustibles downwind of a fire source cannot be evaluated without accounting for the tilt of the fire plume due to crosswind, in addition to the temperature rise along the central axis. Therefore, its trajectory is evaluated based on the velocity ratio of the wind-induced horizontal advection and buoyancy-induced updraft. Prediction equations for the flame geometry are also presented, combining dimensionless numbers representing the fire size and crosswind velocity.

Chapter 7: Ignition and fire spread processes

Applying the heat transfer theory presented in Chapter 3, Chapter 7 describes the ignition process of a solid combustible subject to radiative and convective heating. The time to ignition of a solid combustible can be regarded as the time the surface temperature reaches its critical temperature, accompanied by an increase in the internal temperature by heat conduction. By using analytical solutions for the temperature rise of a semi-infinite solid subject to heating, tractable equations are derived for the time to ignition. Furthermore, the rates of fire spread over continuously distributed combustibles and between discretely distributed combustibles are described using the obtained equations of the time to ignition. The critical distance to avoid fire spread for discretely distributed combustibles is formulated through deterministic and probabilistic approaches.

Chapter 8: Firebrands

Chapter 8 describes the spot ignition behavior due to the scatter of firebrands released from a fire source. The occurrence of spot ignition by an individual firebrand can be regarded as the successive occurrence of three sub-processes: generation, transport, and ignition. The behavior of a firebrand in each sub-process is affected by the conditions of its own, ambient wind, and recipient combustible, each involving significant uncertainty. Hence, probabilistic approaches are adopted in the formulation of each sub-process. Given that an enormous number of firebrands are generally released into the airflow from various fire sources, a framework for evaluating the occurrence probability of spot ignition is developed by aggregating the probabilistic behavior of individual firebrands.

Chapter 9: Spatial data modeling

Chapter 9 explains how to process 3D spatial data to simulate fire spread between irregularly arranged combustibles. Following the methods presented in the previous chapters, we can evaluate fire spread under a simple arrangement of combustibles, for example, when identical combustibles are aligned in a straight line at an equal spacing. However, such an idealistic arrangement of combustibles is not likely to be the case in actual outdoor fires. In the non-regular arrangement of combustibles, a numerical approach is required for calculating the heat transfer and spotting of firebrands based on the relative positional relationship of the fire source and the target combustible.

Chapter 10: Fire spread simulation

Chapter 10 describes the fire spread simulation model developed based on the concepts and methods presented in the previous sections. Fire spread can be divided into two types: fire spread in continuously distributed combustibles and between discretely distributed combustibles. We focus on the latter. The model evaluates fire spread between non-regularly and discretely distributed 3D fuel objects due to the radiative and convective heat transfer from fire plumes and the spotting of firebrands. The model is implemented with Python, and the code can be downloaded from the website (https://github.com/khimoto298/lofd/).

Chapter 2

Wind

Fuel, weather, and topography are often referred to as the three primary factors that affect the behavior of outdoor fires.

The fuel condition includes the type (combustion characteristics), shape (size, thickness), distribution (continuous, discrete), and state (moisture content), which respectively affect the ignition and subsequent burning behavior of combustibles.

The weather condition also has a wide range of factors that can affect the behavior of outdoor fires. Representative examples include wind, temperature, humidity, and precipitation. Wind often causes negative impacts on preventing fire spread. This is because it tilts the flame and scatters numerous firebrands to promote the ignition of combustibles on the leeward side. The effects of temperature and humidity are critical in changing the ignitability of combustible materials rather than on the behavior of flames formed above them.

The topography condition affects the behavior of outdoor fires by creating localized wind around combustibles. This is a factor that increases the complexity in evaluating the fire behavior because the prediction of local wind over uneven terrain often presents difficulty compared to that over flat terrain. On a smaller scale, obstacles such as structures and vegetation are also factors that can change the local wind behavior.

Given that many outdoor fires are anthropogenic and that firefighting activity by firefighters takes an important role in controlling them, human activities could be added as the fourth factor that affects the behavior of outdoor fires. Nevertheless, we cannot discuss all of these factors in this chapter because of the broad range of topics involved. Thus, we focus our discussion on the weather condition, especially the wind condition, which presumably has the greatest impact on the behavior of outdoor fires. Readers interested in a broader discussion of the weather condition are referred to other literature (Pyne et al., 1996; Johnson et al., 2001). They also provide comprehensive descriptions of the fuel condition, which has special importance in evaluating the behavior of wildland fires.

2.1 SURFACE WIND

The effect of a fire plume may reach a high altitude in outdoor fires depending on its size. If it rises as high as the altitude of the free atmosphere, it can cause the dispersion of combustion products over a wide area. However, if we focus on the hazard of fire spread between fuel objects, or the hazard of fire plumes on firefighters and evacuees in the downwind of fire sources, it is often sufficient to know the condition of natural wind near the ground surface (surface wind). Importantly, the condition of the surface wind is not determined solely by the atmospheric motion near the ground, but rather is closely related to the atmospheric motion at higher elevations. In this section, we overview the stratified structure of the atmosphere, followed by the important characteristics of the geostrophic and surface winds.

2.1.1 Stratified structure of the atmosphere

The thickness of the Earth's atmosphere is approximately 1,000 (km). However, 99% of its total mass is contained in the stratosphere and troposphere within 50 km above the ground. The 50 km is less than 1% of the Earth's radius, which is approximately 6400 km. Thus, the atmosphere is often analogized as the thin outer skin of the Earth (Kondo, 2000).

The lower atmosphere is kept at a fairly high temperature by gases such as water vapor. Based on the temperature distribution, the stratified structure of the atmosphere is classified into the troposphere, stratosphere, mesosphere, and thermosphere. Although the troposphere is thinner at high latitudes and thicker at low latitudes, its thickness is approximately 11 km (Stull, 1988). The atmosphere is mixed up and down in the troposphere by convective motion. Cloud formation associated with the movement of water vapor is also active.

The bottom layer of the troposphere is called the planetary boundary layer or atmospheric boundary layer, which is distinguished from the free atmosphere above it (Figure 2.1). The boundary layer is the layer of fluid in the immediate vicinity of a material surface, in which a significant exchange of momentum, heat, or mass occurs between the surface and the fluid (Arya, 2001). The thickness of the boundary layer varies with the surface temperature, wind speed, roughness of the surface, topography, and large-scale motion in the vertical direction. Thus, it can range from a few hundred meters to several kilometers, depending on location and time (Stull, 1988; Arya, 2001). Airflow in the atmospheric boundary layer is generally turbulent, involving random fluctuations in time and space. It is the meteorology within this atmospheric boundary layer that is of primary interest in the analysis of outdoor fires.

An important feature of the atmospheric boundary layer distinct from the other part of the atmosphere is that the average wind velocity is distributed in the vertical direction. The average wind velocity is zero at the surface and

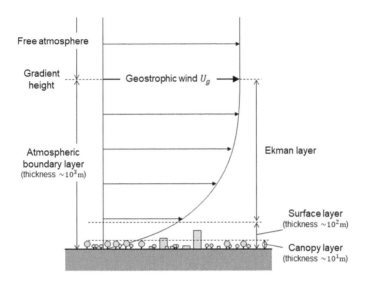

Figure 2.1 Schematic representation of the atmospheric boundary layer.

increases with height (Figure 2.1). The geostrophic wind in the free atmosphere above the atmospheric boundary layer has an approximately constant
wind velocity, U_g, because the frictional forces of the ground surface can be
neglected. In the Ekman layer, the upper layer of the atmospheric boundary
layer, major factors affecting the atmospheric motion are the pressure gradient force due to the atmospheric pressure, the Coriolis force due to the
Earth's rotation, and the frictional force of the ground surface. In the surface
boundary layer, whose thickness is approximately one-tenth of the atmospheric boundary layer, vertical variations in the wind velocity and temperature are significant due to the predominant effect of the surface roughness
(Kondo, 2000). The actual ground surface is covered by plant colonies such
as grasslands and forests (vegetation canopy), and groups of buildings of
various sizes (urban canopy). A canopy is a space where part or all of the
sky is covered with vegetation or structures. The canopy layer is a layer of
the atmosphere within the surface boundary layer involving large spatial
variations of wind speed due to the direct influence of individual land objects.

2.1.2 Geostrophic and surface winds

The motion of an object can be described by Newton's law of motion as
follows:

$$F = m\frac{dU}{dt} \tag{2.1}$$

where F is the force, U is the velocity, m is the mass, and t is the time. This law also governs the atmospheric motion. In the free atmosphere above the gradient height, which is the depth of the atmospheric boundary layer, the frictional force of the ground surface can be neglected. Thus, the equation of the atmospheric motion can be formulated by considering the pressure gradient force $(\partial P / \partial x, \partial P / \partial y, \partial P / \partial z)$, the Coriolis force $(-\rho f V, \rho f U, 0)$, and the gravity $(0, 0, \rho g)$, as follows:

$$\left. \begin{array}{l} \rho \left(\dfrac{\partial U}{\partial t} + U \dfrac{\partial U}{\partial x} + V \dfrac{\partial U}{\partial y} + W \dfrac{\partial U}{\partial z} \right) - \rho f V = -\dfrac{\partial P}{\partial x} \\[3mm] \rho \left(\dfrac{\partial U}{\partial t} + U \dfrac{\partial V}{\partial x} + V \dfrac{\partial V}{\partial y} + W \dfrac{\partial V}{\partial z} \right) + \rho f U = -\dfrac{\partial P}{\partial y} \\[3mm] \rho \left(\dfrac{\partial u}{\partial t} + U \dfrac{\partial w}{\partial x} + V \dfrac{\partial w}{\partial y} + E \dfrac{\partial w}{\partial z} \right) + \rho g = -\dfrac{\partial P}{\partial z} \end{array} \right\} \tag{2.2}$$

where (U, V, W) is the $x-$, $y-$, and $z-$ components of the wind velocity, p is the atmospheric pressure, f is the Coriolis parameter, ρ is the air density, and g is the acceleration due to gravity. Eq. (2.2) considers the diffusion term negligible compared to the convection term. Similarly, the z-axis component of the Coriolis force is considered negligible compared to gravity. The Coriolis force is a fictitious force acting on the atmosphere when viewed from the rotating Earth, which is proportional to the wind velocity and acts in the direction perpendicular to the wind direction. Using the angular velocity of the Earth's rotation, ω, and the latitude, ψ, the Coriolis parameter, f, is expressed as follows:

$$f = 2\omega \sin \psi \tag{2.3}$$

According to this equation, the Coriolis force acts perpendicular and to the right of the wind direction in the northern hemisphere $(f > 0)$, whereas it acts perpendicular and to the left of the wind direction in the southern hemisphere $(f < 0)$.

Under a steady flow condition, when the time and spatial derivatives of the velocity can be neglected, Eq. (2.2) can be simplified as follows:

$$\left. \begin{array}{l} -fV = -\dfrac{1}{\rho} \dfrac{\partial P}{\partial x} \\[3mm] fU = -\dfrac{1}{\rho} \dfrac{\partial P}{\partial y} \\[3mm] g = -\dfrac{1}{\rho} \dfrac{\partial P}{\partial z} \end{array} \right\} \tag{2.4}$$

The U and V determined by this equation are the $x-$ and $y-$ components of the geostrophic winds, U_g and V_g, respectively:

$$\left.\begin{aligned}
U_g &= -\frac{1}{\rho f}\frac{\partial p}{\partial y} \\[2mm]
V_g &= \frac{1}{\rho f}\frac{\partial p}{\partial x}
\end{aligned}\right\} \tag{2.5}$$

Weather maps describe the complex atmospheric motion using isobars representing atmospheric pressure distribution corrected to a hypothetical height, such as the sea level. They can be used to estimate approximate wind conditions. A mass of air is subject to the force induced by the unevenness of the pressure distribution. Figure 2.2(a) schematically shows the relationship between isobars and geostrophic wind in the northern hemisphere. As Eq. (2.5) implies, the geostrophic wind must be parallel to the isobars with low pressure to the left in the northern hemisphere and low pressure to the right in the southern hemisphere. However, this is true only in the free atmosphere. Figure 2.2(b) schematically shows the relationship between isobars and surface wind in the northern hemisphere. In addition to the pressure gradient force and the Coriolis force, the effect of frictional force needs to be considered near the ground surface. This is due to the viscous nature of the atmosphere, which creates frictional resistance to the airflow. When the airflow is steady, the three forces are balanced in the atmospheric boundary layer near the ground surface. The frictional force works opposite to the combined force of the pressure gradient and Coriolis forces. Consequently, the wind direction near the ground surface is non-parallel to the isobars, but is tilted counterclockwise in the northern hemisphere and clockwise in the southern hemisphere. Given that the frictional force changes due to the frictional drag of the ground surface, the tilt angle of the surface wind to

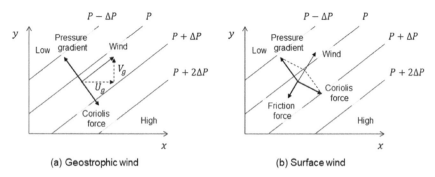

Figure 2.2 Geostrophic and surface winds associated with isobars in the northern hemisphere (Arya, 2001).

the isobars is smaller in areas where it is small (e.g., sea), and larger in areas where it is large (e.g., urban area).

2.1.3 Vertical wind velocity profile in the atmospheric boundary layer

The wind in the boundary layer is significantly affected by the frictional drag of the ground surface. The frictional drag slows down the airflow in contact with the ground surface. This results in a reduction in the average wind velocity in the horizontal direction near the ground surface. The rate and extent of the change in the average wind velocity depend on the geometric unevenness of the ground surface. Figure 2.3 schematically shows vertical profiles of the average wind velocity near the ground of different surface roughnesses. In Figure 2.3, the altitude, z_B, is the upper edge of the boundary layer, above which the average wind velocity can be assumed to be approximately uniform. As the roughness increases, the thickness of the boundary layer also increases. Thus, the vertical gradient of the average wind velocity is larger on smooth terrain and smaller on uneven terrain.

The profile of wind velocity in boundary layers has been extensively studied theoretically. The average wind velocity near the solid surface parallel to the mean flow, $\bar{U}(z)$, can be expressed by the following log law model (Tennekes et al., 1972; Arya, 2001):

$$\bar{U}(z) = \frac{U^*}{\kappa} \cdot \ln\left(\frac{z}{z_0}\right) \tag{2.6}$$

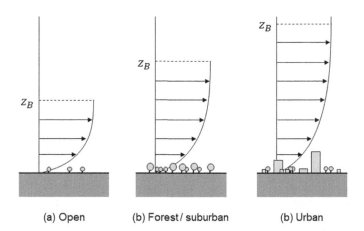

(a) Open   (b) Forest/ suburban   (b) Urban

Figure 2.3 Schematic representation of wind profiles near the ground surface (Oke, 1987).

Table 2.1 Roughness length of homogeneous surface types (Wieringa, 1993).

Surface type	Roughness length z_0(m)
Sea, loose sand, and snow	≈0.0002 (velocity dependent)
Concrete, flat desert, tidal flat	0.0002–0.0005
Flat snow field	0.0001–0.0007
Rough ice field	0.001–0.012
Fallow ground	0.001–0.004
Short grass and moss	0.008–0.03
Long grass and heather	0.02–0.06
Low mature agricultural crops	0.04–0.09
High mature crops (grain)	0.12–0.18
Continuous bushland	0.35–0.45
Mature pine forest	0.8–1.6
Tropical forest	1.7–2.3
Dense low buildings (suburb)	0.4–2.3
Regular-built large town	0.7–1.5

where U^* is the friction velocity, $\kappa(= 0.4)$ is the von Karman constant, z is the height, and z_0 is the roughness length. z_0 is the height where $\bar{U}(z) = 0$, which depends on the geometric roughness of the ground surface (the average height of the obstacles on the ground surface that obstruct the airflow) and its spatial distribution. According to the results of previous measurements, the values of z_0 can be summarized as shown in Table 2.1, depending on the type of ground surface (Wieringa, 1993). z_0 is approximately $10^{-4} - 10^{-3}$ m over lake and sea surfaces, $10^{-3} - 10^{-1}$ m over farmland and grassland, $10^{-1} - 10^0$ m over forest, and $\sim 10^0$ m over urban areas. The friction velocity, U^*, needs to be calculated from the vertical profile of observed wind velocities. However, observation data of vertical wind profile are not easily available. Thus, in practice, U^* is often calculated by substituting the value of z_0 as below (Arya, 2001):

$$U^* = \frac{\kappa \cdot \bar{U}\left(z_{ref}\right)}{\ln\left(z_{ref} / z_0\right)} \tag{2.7}$$

where $\bar{U}\left(z_{ref}\right)$ is the average wind velocity at a reference height, z_{ref}.

The power law model is an alternative to the log law model, empirically derived from observed wind velocity profiles. In the power law model, the average wind velocity, $\bar{U}(z)$, is expressed as follows:

$$\frac{\bar{U}(z)}{\bar{U}\left(z_{ref}\right)} = \left(\frac{z}{z_{ref}}\right)^{\alpha} \tag{2.8}$$

Table 2.2 Power indices of homogeneous surface types (Counihan, 1975).

Surface type	Power index α
Smooth (ice, mud, snow, sea)	0.08–0.12
Moderately rough (short grass, grass, crops, rural)	0.13–0.16
Rough (rural-woods, woods, suburb)	0.20–0.23
Very rough (urban)	0.25–0.40

where α is the power index determined by the surface roughness and z_{ref} is the reference height below which the model is applicable. Table 2.2 summarizes the values of the power index, α. α is approximately $0.1 - 0.15$ over a smooth flat surface such as the sea, $0.15 - 0.25$ over open flat land typical of low vegetated land and rural areas, $0.2 - 0.3$ over forests and low-rise urban areas, and $0.3 - 0.4$ over high-rise urban areas (Counihan, 1975; Wieringa, 1992). It is noteworthy that the log law model can reasonably generalize winds formed over relatively homogeneous surfaces, whereas the power law model has greater applicability to actual natural winds (Ohkuma et al., 1996).

The relationship between the surface roughness and the vertical profile of wind velocity shown in Tables 2.1 and 2.2 was established based on a wide range of observations. However, the observation data used to validate the models were not necessarily systematically organized in terms of the surface roughness and wind velocity conditions. In other words, the base data involved a large variability. In addition, as expressed in the log- and power-law models, the wind velocity profile in the canopy layer near the ground surface is generally significantly slowed down. However, the wind can be locally intensified due to the effect of land objects, which causes a deviation from the average wind velocity profile. Thus, in principle, both models can only approximate the average wind velocity profile above the canopy layer. However, in practice, the wind velocity evaluated by these models is often extrapolated into the canopy layer. As a simple alternative, the wind velocity at the upper end of the canopy layer could be regarded as the velocity uniformly distributed in the canopy layer (Ohkuma et al., 1996).

2.2 STATISTICAL FEATURES OF WIND

Wind information is critical for predicting the behavior of fire spread and thus evaluating its hazard. Features of wind are unique to each region. Strong winds frequently occur in some regions while rarely occur in others. The occurrence of large outdoor fires in the past is often relevant to the frequency of strong winds in the region. However, even in strong-wind-prone regions, strong winds rarely blow ceaselessly throughout the year. As wind condition is associated with the heating of the Earth's surface by solar radiation, it generally fluctuates cyclically on a daily or seasonal basis. Unless prescribed, the onset date and time of an outdoor fire are virtually unpredictable. Thus, the statistical features of wind in the target region are important for performing hazard evaluations.

2.2.1 Characteristic values at a specific location

Wind can be represented by a vector quantity that combines wind velocity and direction. The wind velocity and direction of natural winds change dynamically. The instantaneous wind velocity, U, can be decomposed into an average, $\bar{U}$, and a variation from the average, u', as follows:

$$U = \bar{U} + u' \tag{2.9}$$

In Eq. (2.9), U can be regarded either as a component of the wind velocity vector or as a scalar quantity that is a composite of the components. $\bar{U}$ can be defined as:

$$\bar{U} = \frac{1}{t}\int_0^t U(t)\,dt \tag{2.10}$$

where t is the averaging time, which is chosen in accordance with the purpose of analysis. Although wind direction is inherently 3D, it is often regarded as 2D, focusing on the components in the horizontal plane. As with the case of averaging the wind velocity in Eq. (2.10), the average of the wind direction can similarly be obtained.

The frequency distribution of wind velocity at a specific site, U, is typically asymmetric, with the overall distribution slightly skewed toward the origin. Among various probability distributions, the Weibull distribution is most commonly used for the approximation. When the recurrence interval of U follows the Weibull distribution, the probability density function, $f(U)$, and the cumulative distribution function, $F(U)$, can be expressed as:

$$f(U) = \frac{k}{c}\left(\frac{U}{c}\right)^{k-1} \exp\left[-\left(\frac{U}{c}\right)^k\right] \tag{2.11a}$$

$$F(U) = \int_0^U f(U)\,dU = 1 - \exp\left[-\left(\frac{U}{c}\right)^k\right] \tag{2.11b}$$

where c and k are the site-specific constants, and the scale and shape parameters, respectively. When $k = 1$, $f(U)$ corresponds to the exponential distribution, and when $k = 2$, $f(U)$ corresponds to the Rayleigh distribution.

If U follows the Weibull distribution, the following relationships for the average, $\bar{U}$, and variance, σ^2, are available:

$$\text{Average: } \bar{U} = \int_0^\infty Uf(U)dU = c\,\Gamma\left(1+\frac{1}{k}\right) \tag{2.12a}$$

$$\text{Variance: } \sigma^2 = \int_0^\infty (U-\bar{U})^2 f(U)dU = c^2\left[\Gamma\left(1+\frac{2}{k}\right)-\Gamma^2\left(1+\frac{1}{k}\right)\right] \tag{2.12b}$$

where Γ is the gamma distribution defined as:

$$\Gamma(x) = \int_0^\infty \xi^{x-1}\exp(-\xi)d\xi \tag{2.13}$$

Figure 2.4 shows the cumulative frequency distribution of wind velocity and the frequency distribution of wind direction (wind rose) using hourly observation records (averaged values of the last 10 minutes of each hour) at the Automated Meteorological Data Acquisition System (AMeDAS) station in Tsukuba, Japan, for the 10-year period from 2011 to 2020. As for the wind velocity, the histogram of the observed records is overlaid by a regression using the Weibull distribution in Eq. (2.11b). Although there is a slight discrepancy, the Weibull distribution reasonably regressed the observed records. As for the wind direction, winds from the northeast, south, and northwest are relatively frequent in Tsukuba, which conversely implies that no specific wind direction is dominant. However, the wind rose shows that relatively strong winds are likely to occur from two directions, the south or northwest. This could also be associated with the deviation of the observed

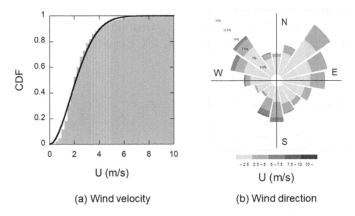

(a) Wind velocity   (b) Wind direction

Figure 2.4 Statistical features of the wind in Tsukuba, Japan, from 2011 to 2020.

wind velocity from the Weibull distribution, which is known to be a good regressor. In other words, the obtained wind velocity distribution in the left panel of Figure 2.4 is not that of any one dominant wind direction, but a mixture of multiple wind directions.

2.2.2 Time variation at a specific site

Figure 2.4 provides an overview of the 10-year observation records of the wind in Tsukuba, exhibiting its long-time statistical features. However, the behavior of fire spread in outdoor fires is affected by wind velocity and direction that vary over a relatively shorter period of time. The atmospheric boundary layer over the ground surface is generally unstable during the day and becomes stable during the night. Accordingly, even if the wind velocity in the free atmosphere (geostrophic wind velocity) does not change from day to night, the wind velocity in the surface layer becomes stronger during the day and weaker during the night. The diurnal variation in the wind velocity decreases as the altitude increases (Stull, 1988; Arya, 2001). Winds in coastal areas are characterized by sea breezes blowing from the sea to the land during the day and land breezes blowing from the land to the sea during the night. The change in the wind direction is due to the difference in the absorption rate of solar radiation and heat capacity between the sea and land surfaces, which causes a difference in temperature distribution between day and night (Oke, 1987).

Then, how can we describe the diurnal or seasonal variations in wind velocity and direction? One possible approach is to use a database of wind conditions observed in the target area of analysis. We randomly sample wind conditions over a continuous time range from the database without modification and use them as the assumed wind conditions. The procedure is simple, and the wind conditions are those of the target area itself, which obviously represents the site-specific features. This could be a viable option in practice as long as there is no constraint to using the observed data of a particular year for future extrapolation.

An alternative to this approach is to develop mathematical models to predict possible future wind conditions using a database of past observations. Among various mathematical models, time series models are an option suitable for modeling wind conditions (Hamilton, 1994; Box et al., 2015). The AutoRegressive Moving Average model (ARMA model) is one of the most basic time series models given by:

$$U_t = \sum_{i=1}^{m} a_i U_{t-i} + \sum_{j=1}^{n} b_j \varepsilon_{t-j} + \varepsilon_t \tag{2.14a}$$

$$\varepsilon_t \sim Normal\left(0, \sigma^2\right) \tag{2.14b}$$

where U_t is the wind velocity at time t, a_i and b_j are the coefficients, m and n are the numbers of order, and ε_t is the white noise that follows a normal distribution with a mean of zero and variance of σ^2. White noise is a noise that has the same intensity at all frequencies, which can be regarded as a highly irregular noise.

The ARMA model is a combination of the AutoRegressive model (AR model, $\sum_{i=1}^{m} a_i U_{t-i}$) and the Moving Average model (MA model, $\sum_{j=1}^{n} b_j \varepsilon_{t-j}$). By combining the two models of different features, the autocorrelation of time series data can be expressed in a flexible manner. The ARMA model is equivalent to the AR model when $n = 0$. Thus, Eq. (2.14a) takes the form:

$$U_t = \sum_{i=1}^{m} a_i U_{t-i} + \varepsilon_t \tag{2.15}$$

This indicates that U_t can be estimated based on the wind velocities at one or more preceding time points. The value of m determines the number of preceding time points to be referenced. The model obtained by adjusting the coefficients of m time points is called the m-th order AR model. On the other hand, the ARMA model is equivalent to the MA model when $m = 0$. Thus, Eq. (2.14a) takes the form:

$$U_t = \mu + \sum_{j=1}^{n} b_j \varepsilon_{t-j} + \varepsilon_t \tag{2.16}$$

where μ is the average of U_t. Eq. (2.16) indicates that U_t can be estimated based on the deviations from μ at one or more preceding time points. The value of n determines the number of preceding time points to be referenced. The model obtained by adjusting the coefficients of n time points is called the n-th order MA model. Another desirable feature of time series models in modeling wind conditions is that they are capable of not only considering the short-term autocorrelation, but also incorporating cyclical variations, such as diurnal and seasonal variations, into the model.

2.3 TOPOGRAPHIC EFFECTS

As described in the previous sections, region-scale winds near the ground surface are governed by the motion of the free atmosphere. Local winds are further affected by terrains and, on a smaller scale, obstacles such as structures and vegetation. In this section, we collectively refer to such effects on local winds as the topographic effects. The topographic effects on the behavior of fire spread can be classified into two categories. One is the case that accompanies the change in the burning behavior itself of combustibles when the natural wind passes around them. The other is the case that accompanies the change in the behavior of fire plumes formed above burning combustibles. In this section, we briefly overview the former of the two. The latter case that involves the deflection and diffusion of fire plumes is discussed in Chapter 6.

2.3.1 Effect of terrain

(1) Combined effect of terrain and solar radiation

Sea and land breeze and mountain and valley winds are typical examples of local winds formed under the combined effect of terrain and solar radiation (Figure 2.5).

Due to the difference in thermal properties between the ground and sea surfaces (or water surfaces in general), the diurnal temperature variation of the sea surface remains smaller than that of the ground surface. This forms a local circulation near the seashore, as described in Figure 2.5(a). The temperature of the ground surface is higher than that of the sea surface during the day. This creates a pressure difference between the land and the sea, causing winds from the sea to the land near the ground surface. The wind blows in the opposite direction in the upper level for its compensation. Such local circulation of the atmosphere is called the sea breeze. As opposed to the daytime, the wind blows from the land to the sea near the ground

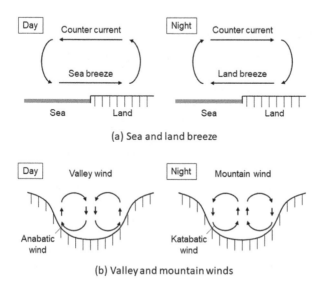

(a) Sea and land breeze

(b) Valley and mountain winds

Figure 2.5 Schematic representation of the local wind formation under the combined effect of terrain and solar radiation (Oke, 1987).

surface during the night. The wind blows in the opposite direction in the upper level for its compensation. Such local circulation of the atmosphere is called the land breeze. The sea and land breezes caused by surface temperature differences are most commonly developed in the summer season under high atmospheric pressure. This is due to the scarcity of clouds and the weakness of the wind system at the synoptic scale.

In a valley, the slope part receives more solar radiation than the bottom part. This causes a temperature difference between the two, forming a unique wind system in the valley, as described in Figure 2.5(b). During the day, the air near the slopes is heated from the ground surface and becomes warmer than the air in the upper level. Consequently, a shallow upward flow along the slope (anabatic wind) is created. A downward flow is formed in the center of the valley for its compensation. Such a local circulation is called the valley wind. However, as the exposure of a slope to solar radiation depends on its orientation, such a wind system is most likely to be created in a deep and straight valley that runs from north to south. Local winds formed in a valley with a different orientation or a complex shape tend to exhibit asymmetric or incomplete patterns (Oke, 1987). During the night, the thermal environment changes due to the radiative cooling from the ground surface, creating a downward wind along the slope (katabatic wind). An upward flow is formed in the center of the valley for its compensation. Such a local circulation is called mountain wind. Similar to sea and land breezes, valley and mountain winds are most commonly developed in the summer season under high atmospheric pressure.

(2) Mechanical effect of terrain

Generalizing the mechanical effect of terrain on the local wind is not straightforward as the types of landform are diverse. However, the major effects of typical landforms can be summarized below (Figure 2.6).

Over a gentle terrain, the airflow is generally formed along the ground surface without separation. On an upward slope, the airflow is accelerated as it is contracted in the vertical direction. On the contrary, it decelerates on a downward slope. For example, over a 2D hill, the wind velocity is highest near the ridge and lowest in the valley.

Over a steep terrain, as opposed to a gentle terrain, the airflow separates from the ground surface and creates circulating flows. For example, when approaching a steep hill, the bulk flow is headed toward the ridge. Similar to the case of a gentle terrain, the wind velocity reaches its maximum near the ridge as the airflow is contracted in the vertical direction. The remaining portion stays at the foot of the hill and forms a vortex flow. As the wind direction is opposite to that of the main flow, the wind velocity of this vortex flow is smaller. The airflow over the hill separates from the ground surface as it passes the ridge and forms a large circulation flow called a wake on the

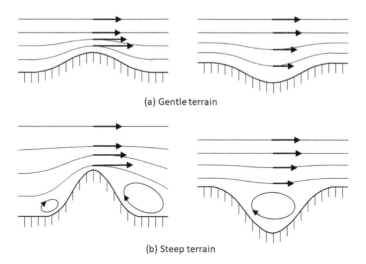

(a) Gentle terrain

(b) Steep terrain

Figure 2.6 Schematic representation of the local wind formation over a gentle and steep terrain (after Oke, 1987).

leeward side of the hill. Similar to the vortex flow on the windward side, the circulation flow on the leeward side is weak, or even unstable. This implies that such a location is prone to gusty winds, although the average wind velocity is low. In canyons (depressions), the bulk flow passes over and does not flow into the canyon. Inside the canyon, isolated from the bulk flow, a circulatory flow is formed due to the effect of air viscosity.

2.3.2 Effect of obstacles

The effect of obstacles such as buildings and vegetation on the local wind varies depending on their shape and size. Consider a rectangular obstacle with sharp edges, as shown in Figure 2.7. The flow pattern is similar to the airflow around the steep hill in Figure 2.6(b). The pressure is positive on the windward side and negative on the leeward side. Although the windward face of the obstacle is not shown in Figure 2.7, the airflow impinging on the face separates at about a point of approximately one-third of the height from the top (Hanna et al., 1982). This point of the airflow separation is called the stagnation point. The impingement of the airflow causes a small vortex flow near the ground surface on the windward side of the obstacle. The airflow separations also occur at both edges of the windward side of the obstacle. When this separation is viewed from the top, the airflow is contracted along the outer edge of the separated streamlines, creating zones of high wind velocity. In addition, vortex flows are generated inside

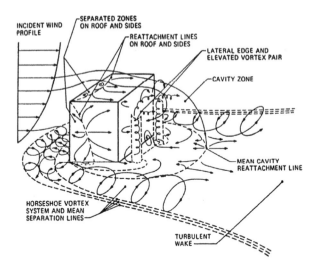

INCIDENT WIND
PROFILE

SEPARATED ZONES
ON ROOF AND SIDES

REATTACHMENT LINES
ON ROOF AND SIDES

LATERAL EDGE AND
ELEVATED VORTEX PAIR

CAVITY ZONE

MEAN CAVITY
REATTACHMENT LINE

HORSESHOE VORTEX
SYSTEM AND MEAN
SEPARATION LINES

TURBULENT
WAKE

Figure 2.7 Airflow near a sharp-edged rectangular obstacle in a boundary layer (Hanna et al., 1982).

the separated streamlines near the side faces of the obstacle. A large circulation flow called the wake is created on the leeward side. The flow pattern changes as the aspect ratio of the obstacle changes.

In general, obstacles are arranged continuously in wildland and urban areas, and the airflows formed around each obstacle interact with each other. Although the shape of actual obstacles is diverse and their arrangement is often uneven, we consider a row of rectangular obstacles, as shown in Figure 2.8. The height and width are both H, and the length perpendicular to the wind direction is L. The spacing between the obstacles is uniform at W. If substantial W is maintained, the wake flow on the leeward side of an obstacle and the vortex flow on the windward side of another obstacle next to it do not interfere with each other, as shown in Figure 2.8(a) (isolated roughness flow regime). However, if W decreases, the vortices formed in the cavity spaces start to interfere with each other, and the isolated roughness flow regime shifts to the wake interference regime, as shown in Figure 2.8(b). This is characterized by secondary flows in the cavity space where the leeward flow of the cavity eddy is reinforced by deflection down the windward face of the next obstacle (Oke, 1988). If W is further decreased, a stable circulating vortex is formed in the cavity spaces, as shown in Figure 2.8(c). This regime is called the skimming flow regime, where the bulk flow does not enter the cavity. This situation is identical to the circulation eddy formed in a steep depression, as shown in Figure 2.6. The bounds of the

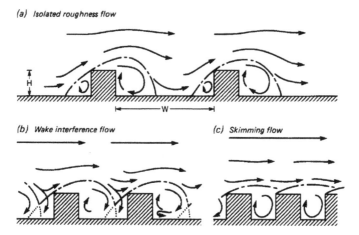

Figure 2.8 The flow regimes associated with airflow over obstacle arrays of increasing H / W (Oke, 1988).

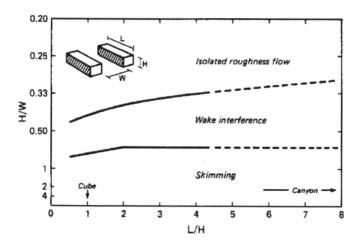

Figure 2.9 Threshold lines dividing the flow into three regimes as functions of the H / W and L / H (Oke, 1988).

three regimes are often identified by a combination of two ratios: the ratio of the cross-sectional dimensions of the obstacle to its spacing, H / W, and the ratio of the length of the obstacle perpendicular to the wind direction to its cross-sectional dimensions, L/H (Oke, 1988). The thresholds dividing the flow into three regimes are described in Figure 2.9.

Chapter 3

Heat transfer

Heat transfer is an important physical process that transfers thermal energy from a fire source to surrounding objects in outdoor fires. This causes the temperature rise and ignition of the surrounding objects, resulting in the expansion of the burning area.

There are two modes of heat transfer: heat conduction and thermal radiation. In heat conduction, atoms, molecules, or free electrons are responsible for heat transfer. It is the motion and collision of atoms and molecules that cause the heat conduction in gases and liquids. In contrast, it is the lattice vibration and motion of free electrons that cause the heat conduction in solids. Thermal radiation is the exchange of heat in the form of electromagnetic waves, which is associated with the vibration and rotation of molecules in thermal motion. When a solid surface is in contact with a fluid at a different temperature, heat transfer from the hotter side to the colder side occurs in conjunction with the fluid motion. This is convective heat transfer, another basic mode of heat transfer, often listed alongside heat conduction and thermal radiation. However, the actual mode of heat transfer in convective heat transfer is nothing but heat conduction. Convective heat transfer is the mode of heat transfer caused by the transport of the fluid after being heated from the solid surface by heat conduction.

Figure 3.1 shows a schematic representation of heat transfer from a fire source to an adjacent object at a certain separation. Heat transfer from a fire source to its adjacent object is caused by thermal radiation and convective heat transfer. The heat received by the object is transferred into its body by heat conduction. This may cause thermal decomposition of the object depending on the extent of temperature rise, which further leads to its ignition. In this chapter, we discuss the fundamentals of heat transfer necessary for analyzing such phenomena. However, as the contents of this chapter are also covered in general textbooks on heat transfer engineering, readers who are already acquainted with this subject may skip this chapter.

DOI: 10.1201/9781003096689-3

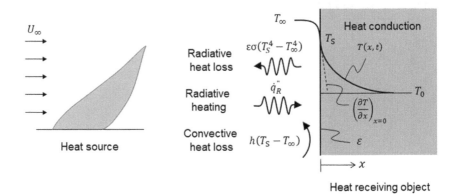

Figure 3.1 Form of heat transfer from a heat source to an adjacent object.

3.1 HEAT CONDUCTION

In this section, we describe the fundamentals of heat conduction in a solid material. The analysis approaches can be used to evaluate the internal temperature of a fuel object subject to external heating.

3.1.1 Heat conduction equation

Heat conduction is a mode of heat transfer when a temperature gradient is present in a solid material, as schematically described in Figure 3.2. In heat conduction inside a solid material, thermal energy is transferred from a high-temperature zone to a low-temperature zone through the interaction of its constituent molecules and electrons. According to Fourier's law, the amount of thermal energy transferred per unit area and unit time, which is called the heat flux, $\dot{q}''$, is given by:

$$\dot{q}'' = -k\frac{\partial T}{\partial x} \tag{3.1}$$

where k is the thermal conductivity, and T is the temperature. Fourier's law is an empirical law stating that the $\dot{q}''$ due to heat conduction is proportional to the temperature gradient in the solid material. The negative sign on the right-hand side of Eq. (3.1) indicates that the heat transfer occurs opposite to the temperature gradient. In other words, if the temperature gradient in the solid material is negative, as shown in Figure 3.2, the thermal energy is transferred in the positive direction.

Heat conduction problems can be formulated using the conservation law of thermal energy in conjunction with Fourier's law. For a 1D solid

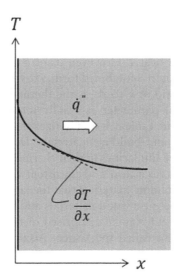

Figure 3.2 Temperature gradient and the corresponding heat flux in a solid material.

material as shown in Figure 3.2, the heat conduction equation governing the internal temperature change can be obtained by applying Fourier's law as a boundary condition to the conservation law of thermal energy as follows (Incropera et al., 2007):

$$\frac{\partial T}{\partial t} = \frac{k}{\rho c}\left(\frac{\partial^2 T}{\partial x^2}\right) + \frac{\dot{q}'''}{\rho c} \tag{3.2}$$

where ρ is the density, c is the specific heat capacity, and $\dot{q}'''$ is the heat release rate per unit volume. $k / \rho c = \alpha$ on the right-hand side of this equation is the thermal diffusivity. The spatial and temporal distributions of T in a solid material can be obtained by solving this equation. Once the distribution of T is obtained, the $\dot{q}''$ can be calculated from Fourier's law. In other words, the heat conduction equation can be viewed as a potential equation, and the thermal energy flow inside of a solid material can be calculated as its dependent.

Note that heat conduction in solids is a 3D physical process by nature. 1D heat conduction in Eq. (3.2) is only its partial representation. However, for evaluating the hazard of outdoor fires, where we focus on the temperature rise of externally heated fuel objects and the associated ignition behavior, the thickness relative to the heated surface area is generally small for many presumable fuel objects. In such a situation, the solution of 1D heat conduction equation is often applicable to analyze their internal temperature. Thus, we limit our discussion to 1D heat conduction in this chapter.

3.1.2 Thermo-physical properties

As indicative in Eqs. (3.1) and (3.2), the heat conduction of solid materials is controlled by several thermo-physical parameters, the thermal conductivity, k, density, ρ, and specific heat, c. However, ρ and c do not appear independently in the equations, but in a combined form, ρc, which is the heat capacity per unit volume. The ratio of k to ρc, is the thermal diffusivity, $\alpha = k / \rho c$, another thermo-physical parameter especially important in the transient phase of heat conduction. Table 3.1 summarizes the thermo-physical properties of selected wood and building materials. Despite the differences in type, these materials have similarities in their thermo-physical properties. As shown in Figure 3.3, there is a clear and consistent $k - \rho$ relationship among the selected wood and building materials that the larger the ρ, the larger the k (JSTP, 2000). This is because wood materials and most building materials are nonconductors. A solid may be comprised of free electrons and atoms bound in an arrangement called the lattice. Accordingly, the mechanism of heat conduction can be divided into two effects: the migration of free

Table 3.1 Thermo-physical properties of selected wood and building materials (Gronli, 1996; Incropera et al., 2006; JSTP, 2000; USDA, 2010).

Material	Thermal conductivity $k\,(\mathrm{kW \cdot m^{-1} \cdot K^{-1}})$	Density $\rho\,(\mathrm{kg \cdot m^{-3}})$	Specific heat $c\,(\mathrm{kJ \cdot kg^{-1} \cdot K^{-1}})$	Thermal diffusivity $\alpha\,(\mathrm{m^2 \cdot s^{-1}})$
Hardwoods (ambient temperature, 12% moisture content, k, c, and α are calculated)				
Beech	0.18×10^{-3}	680	1.62	1.67×10^{-7}
Birch	0.19×10^{-3}	710		1.66×10^{-7}
Chestnut	0.13×10^{-3}	450		1.76×10^{-7}
Maple	0.16×10^{-3}	560		1.71×10^{-7}
Oak	0.18×10^{-3}	660		1.68×10^{-7}
Softwoods (ambient temperature, 12% moisture content, k, c, and α are calculated)				
Balsa	0.067×10^{-3}	200	1.62	2.07×10^{-7}
Cedar	0.10×10^{-3}	340		1.83×10^{-7}
Douglas-fir	0.14×10^{-3}	510		1.73×10^{-7}
Pine	0.13×10^{-3}	450		1.76×10^{-7}
Spruce	0.12×10^{-3}	400		1.79×10^{-7}
Common building materials (ambient temperature)				
Brick	0.72×10^{-3}	1920	0.84	4.49×10^{-7}
Cement mortar	0.72×10^{-3}	1860	0.78	4.96×10^{-7}
Concrete	1.4×10^{-3}	2300	0.88	6.92×10^{-7}
Glass	1.0×10^{-3}	2500	0.75	7.47×10^{-7}
Glass wool	0.036×10^{-3}	32	0.84	13.4×10^{-7}
Gypsum board	0.14×10^{-3}	800	1.13	1.55×10^{-7}
Hardboard	0.094×10^{-3}	640	1.17	1.26×10^{-7}
Particle board	0.078×10^{-3}	590	1.30	1.02×10^{-7}
Plywood	0.12×10^{-3}	545	1.22	1.81×10^{-7}
Sheathing	0.055×10^{-3}	290	1.30	1.46×10^{-7}
Soil	0.52×10^{-3}	2050	1.84	1.38×10^{-7}
Steel	45×10^{-3}	7860	0.48	119×10^{-7}

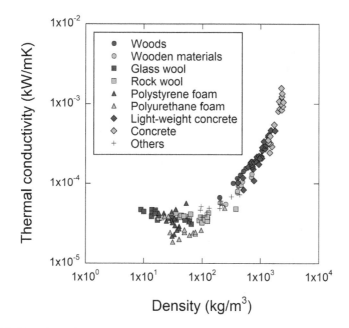

Figure 3.3 Correlation between the thermal conductivity and the density of various wood and building materials (JSTP, 2000).

electrons and lattice vibration waves. In pure metals, the electron contribution to heat conduction is dominant, whereas in nonconductors and semiconductors, the vibrational wave contribution is dominant (Incropera et al., 2006). However, in Figure 3.3, the k of low-density fiber insulation materials such as glass wool and rock wool decreases as ρ increases. This is because the convective flow through large voids within fibers enhances the effective thermal conductivity of the materials.

The thermo-physical properties of the materials in Table 3.1 were measured under the respective reference conditions; the wood materials were measured at room temperature and 12% moisture content, and the other materials were measured at the ambient temperature. However, thermo-physical properties of materials strongly depend on their temperature and moisture content. Thus, it is important to know their varying nature, especially in a fire environment where the materials are exposed to large temperature variations.

(1) Thermal conductivity of wood materials

Wood is an anisotropic, porous material. The combination of multiple modes of heat transfer, such as heat conduction through the cell walls, air and water movement through the pores, and radiative heat exchange between the pore surfaces, increases the behavioral complexity in porous materials. Thus, factors affecting the effective thermal conductivity, k, of wood materials can be represented by various parameters, including the density, moisture content, extractive content, grain orientation, structural irregularities, and temperature (USDA, 2010). Air and water movement through the pores is of special importance in the heat conduction behavior of wood materials. As the moisture content substantially differs between live and dead wood materials (the moisture content of green wood can range from about 30% to more than 200%) (USDA, 2010), there is variation in k even within the same species. Anisotropy of the material structure, often represented by the grain orientation, also causes a variation in the k. The k is similar in radial and tangential directions, whereas k along the grain ranges from 1.5 to 2.8 times greater than that across the grain (TenWolde et al., 1988). When considering the ignition of wood materials in outdoor fires, we generally focus on the heat conduction across the grain as it is the direction most materials are exposed to external heating in actual fires.

One of the first attempts at systematically investigating the thermal conductivity of wood in terms of density and moisture content was by MacLean (MacLean, 1941). He formulated the k as the additive parallel flows through the air (k_0), cell wall ($\rho_0 a_0$) and water ($\rho_0 a_1 \phi$):

$$k = k_0 + \rho_0 \left(a_0 + a_1 \phi \right) \tag{3.3}$$

where ρ_0 is the oven-dry density, ϕ is the moisture content, and k_0, a_0, and a_1 are the constants. ϕ is defined as:

$$\phi = \frac{m - m_{dry}}{m_{dry}} \tag{3.4}$$

where m is the mass of the material and m_{dry} is the oven-dried mass of the material. Following the expression presented by MacLean, TenWolde et al. obtained an equation based on measurements conducted by several investigators on various wood species as follows (TenWolde et al., 1988):

$$k = \left[\rho(0.1941 + 0.4064\phi) + 18.64 \right] \times 10^{-6} \ (\mathrm{kW \cdot m^{-1} \cdot K^{-1}}) \tag{3.5}$$

Figure 3.4 shows the calculated result of the k using Eq. (3.5).

Material temperature is another factor affecting the thermal conductivity of wood. The effect of thermal radiation inside the pores increases as the temperature rises, leading to an increase in the effective thermal conductivity (Gronli, 1996). An experiment on the k of several types of wood reported an increase of about 40%, with an increase in the ambient temperature by 250K (Harada et al., 1998). When exposed to even higher temperatures in a fire environment, a key consideration for organic materials such as wood is

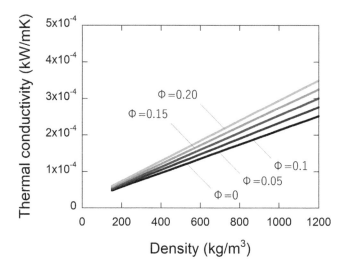

Figure 3.4 Calculated thermal conductivity of wood materials in the direction perpendicular to the grain (Eq. (3.5)).

their thermal decomposition. Thermal decomposition of organic materials releases volatile components into the gas phase, leaving less volatile char on the material surface. This is generally recognized as the formation of a char layer on the virgin layer. The thermo-physical properties of the residual char change significantly from those of virgin wood as it becomes even more porous. Thus, Eq. (3.5) for virgin wood is not applicable to estimating the k of charred materials.

(2) Specific heat of wood materials

The specific heat capacity of a material, $c(\mathrm{kJ \cdot kg^{-1} \cdot K^{-1}})$, is defined as the amount of thermal energy needed to increase a unit temperature of a material per unit mass. Literature data for the c of wood materials as a function of temperature are consistent. This is because measuring the c of solid materials is easier than for the other thermo-physical parameters, such as thermal conductivity, even at higher temperatures (Janssens, 1994). As in the case of the k, the c of wood materials is also affected by their temperature and moisture content. TenWolde et al. derived an equation for the c of wood materials by aggregating extensive survey data as follows (TenWolde et al., 1988):

$$c = \frac{c_0 + \phi \cdot c_W}{1 + \phi} + \Delta c \tag{3.6}$$

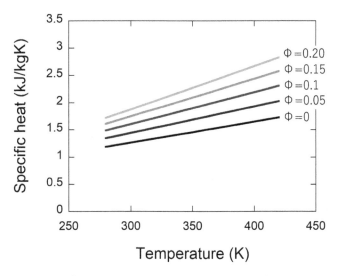

Figure 3.5 Calculated specific heat of wood materials as a function of temperatures (Eq. (3.6)).

where, c_0 is the specific heat of dry wood, c_W is the specific heat of water, 4.186 $kJ \cdot kg^{-1} \cdot K^{-1}$, ϕ is the moisture content ($kg \cdot kg^{-1}$), and Δc is the correction term. c_0 was also obtained by regressing data from several references (TenWolde et al., 1988):

$$c_0 = 0.003867T + 0.1031 \tag{3.7}$$

where T is the temperateure (K). The Δc is also a function of T and ϕ, which takes the form (TenWolde et al., 1988):

$$\Delta c = (0.02355T - 1.326\phi - 6.191)\phi \tag{3.8}$$

Unlike the relationship for the k in Eq. (3.5), the effect of ρ does not appear in this model. The applicability is limited to below fiber saturation point at temperatures between 280 and 420K. Figure 3.5 shows the calculation results of the c using Eq. (3.6) as a function of T.

3.1.3 Steady conduction

Heat conduction in a solid has two different phases, transient and steady. While the temperature changes with time in the transient phase, it no longer changes when shifted to the steady phase. In a strict sense, steady heat conduction rarely occurs in a fire environment, given a large temporal and spatial variability in the temperature field and a state change of the material itself. However, as discussed later, analytical solutions are available only in limited cases for transient heat conduction problems. Thus, for practical purposes, the approximate solutions are often obtained from the steady heat conduction equation.

(1) Plane wall

We assume a plane wall with a thickness of L, as shown in Figure 3.10. The surface temperatures on both sides of the plane wall are kept at T_1 at $x = 0$ and $T_2 (< T_1)$ at $x = L$. In the steady phase, the time derivative and heat generation terms are zero in Eq. (3.2):

$$\frac{\partial}{\partial x}\left(k\frac{\partial T}{\partial x}\right) = 0 \tag{3.9}$$

This indicates that the heat flux within the plane wall, $k \cdot \partial T / \partial x$, is constant regardless of the location and time. According to Fourier's law in Eq. (3.1), the heat flux through the plane wall, $\dot{q}''$, can be expressed as:

$$\dot{q}'' = k\frac{T_1 - T_2}{L} \tag{3.10}$$

or in an alternative form:

$$\dot{q}'' = \frac{T_1 - T_2}{L / k} = \frac{T_1 - T_2}{R} \tag{3.11}$$

where R is the thermal resistance. This name comes from the analogy of Ohm's law of electric conduction, where $\dot{q}''$ and $T_1 - T_2$ correspond to the electrical current and the potential difference (voltage), respectively. For this reason, the equivalent circuit is often used to describe the thermal resistance in a solid, as shown in Figure 3.6.

(2) Composite plane wall

The concept of thermal resistance is useful for considering steady heat conduction in a composite plane wall. As a simple example, we consider a composite plane wall consisting of three materials, A, B, and C, as shown in Figure 3.7. The thicknesses of these walls are L_1, L_2, and L_3, respectively, and the thermal conductivities are k_1, k_2, and k_3, respectively. There is no thermal contact resistance between the comprising walls, and thus there is no temperature gap at the contact surface. In this case, the heat fluxes through each comprising wall are identical:

$$\dot{q}'' = k_1\frac{T_1 - T_2}{L_1} = k_2\frac{T_2 - T_3}{L_2} = k_3\frac{T_3 - T_4}{L_3} \tag{3.12}$$

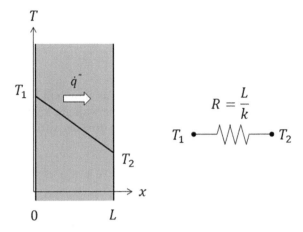

Figure 3.6 Steady heat conduction in a plane wall.

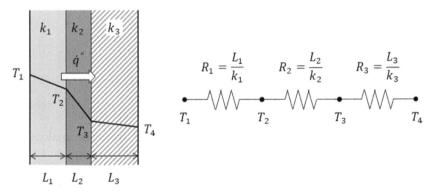

Figure 3.7 Steady heat conduction in a composite plane wall.

By eliminating the temperatures at the contact surface, T_2 and T_3, we obtain:

$$\dot{q}'' = \frac{T_1 - T_4}{\dfrac{L_1}{k_1} + \dfrac{L_2}{k_2} + \dfrac{L_3}{k_3}} \tag{3.13}$$

We now define the denominator of Eq. (3.13) as the total thermal resistance, R_t:

$$R_t = \frac{L_1}{k_1} + \frac{L_2}{k_2} + \frac{L_3}{k_3} \tag{3.14}$$

Using R_t, the heat flux through the composite wall, $\dot{q}''$, can be expressed as:

$$\dot{q}'' = \frac{T_1 - T_4}{R_t} \tag{3.15}$$

As R_t in Eq. (3.14) does not involve the temperature of any comprising wall, $\dot{q}''$ can be calculated only from the temperatures of both ends of the composite plane wall, T_1 and T_4. The above relationship can be generalized to composite walls with an arbitrary number of components.

The above concept can be extended to a steady heat conduction problem in a composite wall between hot and cold air, as shown in Figure 3.8. Considering that the heat fluxes at both ends of the composite wall are equivalent to the heat flux by conduction, $\dot{q}''$, we obtain:

$$\dot{q}'' = h_H \left(T_H - T_1 \right) = k_1 \frac{T_1 - T_2}{L_1} = k_2 \frac{T_2 - T_3}{L_2} = k_3 \frac{T_3 - T_4}{L_3} = h_L \left(T_4 - T_L \right) \tag{3.16}$$

where T_H and T_L are the hot and cold air temperatures, respectively, and h_H and h_L are the heat transfer coefficients on both ends of the composite wall. By eliminating the temperatures at the contact surfaces, T_1, T_2, T_3, and T_4, Eq. (3.16) can be transformed as follows:

$$\dot{q}'' = \frac{T_H - T_L}{\dfrac{1}{h_H} + \dfrac{L_1}{k_1} + \dfrac{L_2}{k_2} + \dfrac{L_3}{k_3} + \dfrac{1}{h_L}} \tag{3.17}$$

Thus, the total thermal resistance, R_t, takes the form:

$$R_t = \frac{1}{h_H} + \frac{L_1}{k_1} + \frac{L_2}{k_2} + \frac{L_3}{k_3} + \frac{1}{h_L} \tag{3.18}$$

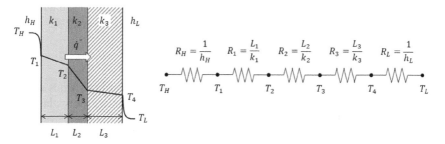

Figure 3.8 Steady heat conduction in a composite plane wall sandwiched by hot and cold air.

In other words, the R_t can also include the effect of heat transfer at the surface boundaries by simply connecting the component thermal resistances in series.

WORKED EXAMPLE 3.1

Assume a flat plate of Douglas fir with a thickness of 1 cm. As a result of exposure to external heating from one side of the plate, the temperatures of the heated and unheated surfaces are $T_1 = 500$ K and $T_2 = 350$ K, respectively. Calculate the internal heat flux, $\dot{q}''$, if the heat conduction in the plate is steady. Use the thermo-physical parameters listed in Table 3.1. The dependency on temperature and moisture content can be ignored.

SUGGESTED SOLUTION

From Table 3.1, the thermal conductivity of Douglas fir is $k = 0.14 \times 10^{-3}$ kW·m^{-1}·K^{-1}. By substituting the values in Eq. (3.10), we obtain the $\dot{q}''$ as follows:

$$\dot{q}'' = k\frac{T_1 - T_2}{L} = 0.14 \times 10^{-3} \times \frac{500 - 350}{0.01} = 2.1 \text{ kW·m}^{-2}$$

WORKED EXAMPLE 3.2

In outdoor fires, fire spread between buildings often occurs when the window panes are broken due to external heating, allowing flames and firebrands to enter the interior. Assume a single-pane window glass on an exterior wall of a building subject to radiative heating from an external fire source, as shown in Figure 3.9. Calculate the maximum possible temperature of the window glass. However, assume that the heat transfer is at a steady state and that convective heat transfer is the dominant mode of heat loss from the surfaces. The width of the window glass is $d = 0.005$ m, the thermal conductivity is $k = 0.001$ kW·m^{-1}·K^{-1}, and the emissivity is $\varepsilon = 1$. The heat transfer coefficients are $h = 0.015$ kW·m^{-1}·K^{-1} on both sides of the window glass. The exterior and interior air temperatures are $T_\infty = T_R = 20$ °C.

SUGGESTED SOLUTION

Assuming that T_1 and T_2 are the surface temperatures on both sides of the window glass ($T_1 > T_2$), the heat flux passing through the window glass, $\dot{q}''$, can be expressed using Eq. (3.16) as follows:

$$\dot{q}'' = \varepsilon\dot{q}_R'' + h\left(T_\infty - T_1\right) = \frac{k}{d}\left(T_1 - T_2\right) = h\left(T_2 - T_R\right)$$

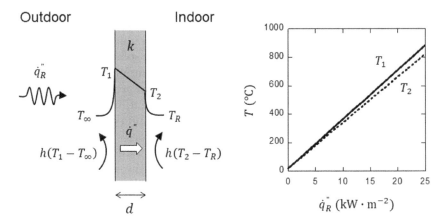

Figure 3.9 Temperature elevation of a window glass due to external radiative heating.

From this equation, we obtain the following expression for the temperature differences:

$$T_\infty - T_1 = \frac{\dot{q}'' - \varepsilon \dot{q}_R''}{h}$$

$$T_1 - T_2 = \frac{\dot{q}'' d}{k}$$

$$T_2 - T_R = \frac{\dot{q}''}{h}$$

By summing these equations, $\dot{q}''$ can be calculated by eliminating the unknown surface temperatures, T_1 and T_2, which takes the form:

$$\dot{q}'' = \frac{T_\infty - T_R + \varepsilon \dot{q}_R'' / h}{2/h + d/k}$$

T_1 and T_2 can be calculated by reverting this equation to the above three equations for the temperature differences. The calculation result is shown in Figure 3.9. The T_1 and T_2 change linearly with the $\dot{q}_R''$. However, a change in the temperature over time is not considered. Despite such a limitation, steady-state solutions are still useful from a practical point of view. This is because they are greater than or equal to the temperature at any time after the heating starts, i.e., steady-state temperatures provide results on a safer side.

3.1.4 Transient conduction

If an object receives continuous external heating, temporal variation of the internal temperature, which is initially large, gradually decreases and approaches a steady state. As discussed above, simple solutions are available to analyze steady conduction problems. However, an important heat conduction problem involved in outdoor fires, the ignition process of solid fuels, is a transient heat conduction process. The solution to transient problems can be obtained by solving the heat conduction equation under given initial and boundary conditions. However, analytical solutions to the heat conduction equation are available only when the thermal conductivity is constant, the boundary conditions are linear, and the shape of the solid fuel is simple. Nevertheless, they still provide valuable insights into the analysis of ignition processes. Engineering correlations widely used to predict the time to ignition of solid fuels subject to external heating are derived by extending such analytical solutions.

In this subsection, we consider 1D heat conduction problems of a semi-infinite solid subject to external heating. As mentioned above, the change in the internal temperature distribution of a solid can be obtained by applying the initial and boundary conditions to the heat conduction equation. We discuss three boundary conditions under which analytical solutions are available, and additionally, a special boundary condition where two objects are in contact.

(1) Specified temperature boundary condition

In the first boundary condition, the temperature of the boundary surface, T_S, is specified. The initial and boundary conditions can be expressed as:

$$\text{Initial condition}: T\big|_{t=0} = T_0 \tag{3.19a}$$

$$\text{Boundary condition}: T\big|_{x=0} = T_S \text{ and } T\big|_{x\to\infty} = T_0 \tag{3.19b}$$

One may find it difficult to assume a situation where these conditions hold in actual fires. These conditions may be valid, for example, when fuels are instantly enveloped by hot gas due to a change in wind direction, or when one attempts to evaluate the temperature elevation on a safer side.

Applying the above initial and boundary conditions to the heat conduction equation (Eq. (3.2)) without heat generation ($\dot{q}''' = 0$), we obtain the following equation for the internal temperature profile, $T(x,t)$ (Carslaw et al., 1959; Incropera et al., 2006):

$$\text{Internal temperature}: \frac{T(x,t) - T_0}{T_S - T_0} = 1 - \text{erf}\left(\frac{x}{2\sqrt{\alpha t}}\right) = \text{erfc}\left(\frac{x}{2\sqrt{\alpha t}}\right) \tag{3.20}$$

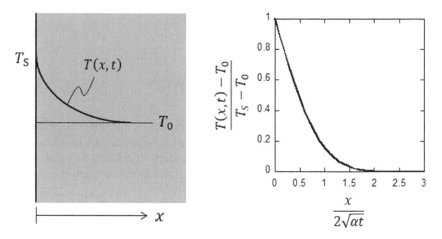

Figure 3.10 Heat conduction in a semi-infinite solid with the specified temperature boundary condition.

where $\operatorname{erf}\left(\dfrac{x}{2\sqrt{\alpha t}}\right)$ is the Gaussian error function defined by:

$$\operatorname{erf}\left(\frac{x}{2\sqrt{\alpha t}}\right) = \frac{2}{\sqrt{\pi}} \int_{0}^{\frac{x}{2\sqrt{\alpha t}}} \exp\left(-\eta^2\right) d\eta \tag{3.21}$$

In Eq. (3.20), $x/2\sqrt{\alpha t}$ is the only thermo-physical parameter that controls the $T(x,t)$. If this parameter is the same, the $T(x,t)$ takes the same profiles as shown in Figure 3.10. As the derivative of the error function is given by:

$$\frac{d}{d\eta}\left\{\operatorname{erf}\left(\eta\right)\right\} = \frac{2}{\sqrt{\eta}} \exp\left(-\eta^2\right) \tag{3.22}$$

the net incident heat flux at the solid surface, $\dot{q}''_{net}$, can be calculated as follows:

$$\dot{q}''_{net} = -k \left.\frac{\partial T}{\partial x}\right|_{x=0} = \sqrt{\frac{k\rho c}{\pi t}}\left(T_S - T_0\right) \tag{3.23}$$

(2) Convection boundary condition

In the second boundary condition, a semi-infinite solid is exposed to a volume of gas at a constant temperature of T_∞. If the dominant mode of heat transfer between the gas and semi-infinite solid is convective heat transfer, the initial and boundary conditions take forms:

$$\text{Initial condition}: T\big|_{t=0} = T_0 \tag{3.24a}$$

$$\text{Boundary condition}: -k\frac{\partial T}{\partial x}\bigg|_{x=0} = h(T_\infty - T_S) \tag{3.24b}$$

If radiative heating/cooling occurs concurrently, the heat transfer coefficient, h, changes nonlinearly. However, the analytical solution can be obtained only when h is constant. Thus, such a boundary condition is valid only when the temperature difference, $T_\infty - T_S$, is relatively small when the radiative effect is negligible.

By applying the above initial and boundary conditions, Eq. (3.2) can be solved to yield $T(x,t)$ and T_S as follows, respectively (Carslaw et al., 1959; Incropera et al., 2006):

$$\text{Internal temperature}: \frac{T(x,t)-T_0}{T_\infty - T_0} = \text{erfc}\left(\frac{x}{2\sqrt{\alpha t}}\right)$$
$$-\exp\left(\frac{h_S x}{k} + \frac{h_S^2 \alpha t}{k^2}\right)\text{erfc}\left(\frac{x}{2\sqrt{\alpha t}} + \frac{h_S\sqrt{\alpha t}}{k}\right) \tag{3.25a}$$

$$\text{Surface temperature}: \frac{T_S - T_0}{T_\infty - T_0} = 1 - \exp\left(\frac{h_S\sqrt{\alpha t}}{k}\right)^2 \text{erfc}\left(\frac{h_S\sqrt{\alpha t}}{k}\right) \tag{3.25b}$$

Figure 3.11 shows the time evolution of T_S when $T_\infty - T_0$ is changed. However, in this calculation, the heat transfer coefficient was $h = 0.015$ kW $\cdot$ K^{-1}, the

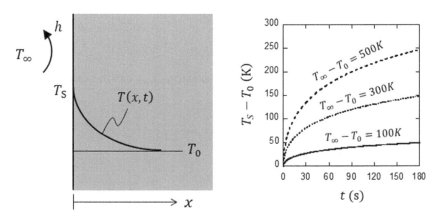

Figure 3.11 Heat conduction in a semi-infinite solid with the convection boundary condition.

thermal conductivity was $k = 0.11 \times 10^{-3}$ kW $\cdot$ m$^{-1} \cdot$ K^{-1}, and the thermal diffusivity was $\alpha = 0.17 \times 10^{-6}$ m$^2 \cdot$ s^{-1}.

(3) Specified heat flux boundary condition

In the third boundary condition, incident heat flux to the surface, $\varepsilon \dot{q}_R''$, is specified. The initial and boundary conditions take forms:

$$\text{Initial condition: } T\big|_{t=0} = T_0 \tag{3.26a}$$

$$\text{Boundary condition: } -k\frac{\partial T}{\partial x}\bigg|_{x=0} = \varepsilon \dot{q}_R'' \tag{3.26b}$$

This boundary condition can be introduced when the radiative heat transfer is dominant relative to convective heat transfer. When $\varepsilon \dot{q}_R'' = 0$, it becomes an adiabatic boundary condition, which is:

$$\text{Boundary condition (adiabatic)} - k\frac{\partial T}{\partial x}\bigg|_{x=0} = 0 \tag{3.27}$$

Similar to the other boundary conditions, we can derive $T(x,t)$ and T_S by applying the initial and boundary conditions to Eq. (3.2), which yields (Carslaw et al., 1959; Incropera et al., 2006):

$$\text{Internal temperature: } T(x,t) - T_0 = \frac{2\varepsilon \dot{q}_R'' \sqrt{\alpha t / \pi}}{k}$$

$$\times \exp\left[-\left(\frac{x}{2\sqrt{\alpha t}}\right)^2\right] - \frac{\varepsilon \dot{q}_R'' x}{k}\operatorname{erfc}\left(\frac{x}{2\sqrt{\alpha t}}\right) \tag{3.28a}$$

$$\text{Surface temperature: } T_S - T_0 = 2\varepsilon \dot{q}_R'' \sqrt{\frac{t}{\pi k \rho c}} \tag{3.28b}$$

Figure 3.12 shows the time evolution of the T_S as a function of $\varepsilon \dot{q}_R''$. However, in the calculation of T_S, we assumed that the thermal inertia of the solid material is $k\rho c = 0.07$ kJ$^2 \cdot$ s$^{-1} \cdot$ m$^{-4} \cdot$ K^{-2}, which is the value for Douglas fir (see Table 3.1).

(4) Interface boundary condition

The three boundary conditions discussed above are common in that semi-infinite solids are all exposed to hot gases. We now consider two semi-infinite solids, 1 and 2, at different temperatures, $T_{1,0}$ and $T_{2,0}$, that are in contact with each other. The two solids exchange heat by conduction through the

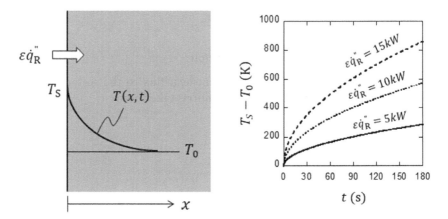

Figure 3.12 Heat conduction in a semi-infinite solid with the specified heat flux boundary condition.

interface. We call this the interface boundary condition. A situation similar to the interface boundary condition in outdoor fires can be found where a hot firebrand is deposited on a combustible on the ground. The analytical solution for the specified temperature boundary condition described in (1) of this subsection can be used to solve this problem.

We assume coordinates in the depth direction of each semi-infinite solid, x_1 and x_2, respectively, and the interface temperature, $T_I(t)$, as shown in Figure 3.13. By substituting $T_I(t)$ into T_S in Eq. (3.20), the internal temperatures of the two solids can be obtained as functions of the elapsed time after contact, t, which are respectively expressed as:

$$T_1(x_1,t) = T_I(t) + \left(T_{1,0} - T_I(t)\right)\mathrm{erf}\left(\frac{x_1}{2\sqrt{\alpha_1 t}}\right) \tag{3.29a}$$

$$T_2(x_2,t) = T_I(t) + \left(T_{2,0} - T_I(t)\right)\mathrm{erf}\left(\frac{x_2}{2\sqrt{\alpha_2 t}}\right) \tag{3.29b}$$

where α_1 and α_2 are the thermal diffusivities. When the two solids are in perfect contact, the temperatures and heat fluxes at their interface should coincide (Figure 3.13). Thus, the boundary conditions should take the forms:

$$\left(T_1\right)_{x=0} = \left(T_2\right)_{x=0} = T_I(t) \tag{3.30a}$$

$$\mathrm{abs}\left(-k_1 \left.\frac{\partial T_1}{\partial x_1}\right|_{x_1=0}\right) = \mathrm{abs}\left(-k_2 \left.\frac{\partial T_2}{\partial x_2}\right|_{x_2=0}\right) \tag{3.30b}$$

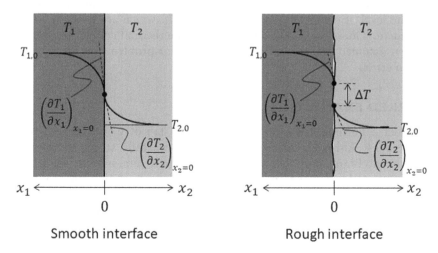

Smooth interface   Rough interface

Figure 3.13 Thermal contact resistance between two semi-infinite solids in contact.

where k_1 and k_2 are the thermal conductivities. The first boundary condition in Eq. (3.30a) is already incorporated into the temperature profiles of the two solids in Eq. (3.29). Thus, by substituting Eq. (3.29) into the second boundary condition in Eq. (3.30b), we obtain the interface temperature as follows (Carslaw et al., 1959; Incropera et al., 2006):

$$T_I = \frac{\sqrt{k_1\rho_1c_1}\,T_{1,0} + \sqrt{k_2\rho_2c_2}\,T_{2,0}}{\sqrt{k_1\rho_1c_1} + \sqrt{k_2\rho_2c_2}} \tag{3.31}$$

It is noteworthy that the boundary conditions in Eq. (3.30) hold only if the two solids are in perfect contact. However, actual solids generally have rough surfaces, and the contact between two solids is only partial in most cases. A partial contact creates thermal resistance at the interface, causing a temperature gap, ΔT, between the two solids. In such a case, the boundary conditions in Eq. (3.30) can be modified by introducing the concept of thermal contact resistance, R_C:

$$\Delta T = T_1\big|_{x_1=0} - T_2\big|_{x_2=0} \tag{3.32a}$$

$$\mathrm{abs}\left(-k_1\frac{\partial T_1}{\partial x_1}\bigg|_{x_1=0}\right) - \mathrm{abs}\left(-k_2\frac{\partial T_2}{\partial x_2}\bigg|_{x_2=0}\right) = \frac{\Delta T}{R_C} \tag{3.32b}$$

Eq. (3.32) is a more realistic boundary condition for the heat conduction problem between two solids in contact. However, quantification of R_C is not that straightforward as it significantly depends on the surface roughness and the level of contact between the two solids, which are generally case-dependent. Thus, Eq. (3.30), which ignores the thermal contact resistance, is often assumed in practical analyses of similar problems while allowing certain errors.

WORKED EXAMPLE 3.3

Consider a burning firebrand deposited on a slab of wood, as shown in Figure 3.14. Although actual firebrands are generally small, we assume that the firebrand and the wood slab are both semi-infinite solids for simplicity. Calculate the interface temperature given that the thermal contact resistance, R_C, is negligibly small. The thermal inertias of the firebrand and the recipient wood slab are $k_C \rho_C c_C = 0.010$ kJ$^2 \cdot$s$^{-1} \cdot$m$^{-4} \cdot$K^{-2} and $k_W \rho_W c_W = 0.113$ kJ$^2 \cdot$s$^{-1} \cdot$m$^{-4} \cdot$K^{-2}, respectively. Representative temperatures of the firebrand and the recipient wood slab are $T_C = 800°$C and $T_W = 20°$C, respectively.

SUGGESTED SOLUTION

Assuming no heat generation at the interface, T_I can be obtained as 198.6°C by substituting the given conditions into Eq. (3.31). However, this would underestimate T_I given that the oxidation reaction in the char layer is the root cause of the hazardous nature of firebrands. Thus, let the heat release rate per unit time and area of the char layer be $\dot{q}_C''$, the boundary condition alternative to Eq. (3.32b) can be given by:

$$-k_1 \frac{\partial T_1}{\partial x_1}\bigg|_{x_1=0} - k_2 \frac{\partial T_2}{\partial x_2}\bigg|_{x_2=0} = \dot{q}_C''$$

T_I can be obtained by substituting Eq. (3.29) into this boundary condition, which yields:

$$T_I = \frac{\sqrt{k_C \rho_C c_C}\, T_{C,0} + \sqrt{k_W \rho_W c_W}\, T_{W,0}}{\sqrt{k_C \rho_C c_C} + \sqrt{k_W \rho_W c_W}} + \frac{\dot{q}_C'' \sqrt{\pi t}}{\sqrt{k_C \rho_C c_C} + \sqrt{k_W \rho_W c_W}}$$

This form of equation follows Eq. (3.31), but with an additional term associated with $\dot{q}_C''$. Figure 3.14 shows the time evolution of T_I with different $\dot{q}_C''$. The larger $\dot{q}_C''$, the higher T_I. However, it should be noted that we assumed a smooth contact between the firebrand and the recipient wood slab in this calculation. Actual increases in the interface temperature are expected to be smaller than these results.

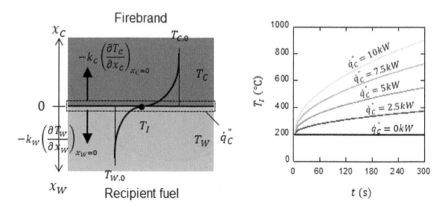

Figure 3.14 Temperature at the interface of a firebrand deposited on a wood slab.

3.2 CONVECTIVE HEAT TRANSFER

Convective heat transfer is a mode of heat transfer in which fluid heated by a hot object surface moves to a cold object surface. As the fluid at the object surface is stationary, heat is transferred from the object surface to the fluid by heat conduction. Subsequently, the heat retained by the fluid is transported by its convective motion. In other words, convective heat transfer is a mode of heat transfer that combines heat conduction and convective flow of the fluid. Thus, unlike the thermal conductivity of a solid, the heat transfer coefficient is not determined solely by the physical properties of the fluid. It also depends on the object geometry and the convective flow condition.

3.2.1 Heat transfer coefficient

Assume that low-temperature fluid flows around a high-temperature object and that heat is transferred from the object surface to the fluid, as shown in Figure 3.15. If the temperature of the object surface is T_S and the bulk temperature of the fluid (the uniform fluid temperature away from the solid surface) is T_∞, the heat flux is proportional to the temperature difference, $T_S - T_\infty$, according to Newton's law of cooling:

$$\dot{q}'' = h\left(T_S - T_\infty\right) \tag{3.33}$$

where h is the heat transfer coefficient. As the flow condition changes over the object surface, h is not material-specific, but varies with location. Thus, the rate of heat transfer from the entire surface of the object to the fluid can be expressed as:

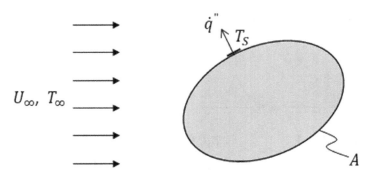

Figure 3.15 Fluid flow and heat transfer around an object.

$$\dot{Q} = \int_A h(T_S - T_\infty) dA = (T_S - T_\infty)\int_A h dA \qquad (3.34)$$

where A is the surface area of the object. Now, let us introduce the mean heat transfer coefficient, which can be defined as:

$$\bar{h} = \frac{\int_A h dA}{A} \qquad (3.35)$$

Using the mean heat transfer coefficient, $\bar{h}$, the mean heat flux, $\bar{q}''$, can be expressed as:

$$\bar{q}'' = \frac{\dot{Q}}{A} = \bar{h}(T_S - T_\infty) \qquad (3.36)$$

In contrast to $\bar{h}$, h in Eq. (3.33) is defined locally on the surface. Thus, it is called the local heat transfer coefficient. Recall that the fluid molecules are constrained at the solid surface, and the fluid velocity is zero. Thus, Fourier's law holds on the solid surface, and the heat flux can be given by:

$$\dot{q}'' = -k\left(\frac{\partial T}{\partial y}\right)_{y=0} \qquad (3.37)$$

where k is the thermal conductivity of the fluid. By comparing this relationship with Eq. (3.33), h can be expressed as:

$$h = -\frac{k(\partial T / \partial y)_{y=0}}{T_S - T_\infty} \qquad (3.38)$$

Namely, h is determined by the temperature gradient of the fluid at the solid surface, which is associated with the fluid motion. In the above discussion, we assumed that the temperature of the object, T_S, is higher than that of the gas, T_∞. However, the relationship in Eq. (3.33) still holds for the reversed case.

3.2.2 Forced convection

Convective heat transfer is closely related to the behavior of fluid motion. The convection mode significantly differs between forced convection, in which the fluid is mechanically driven, and natural or free convection, in which the fluid is driven by buoyancy. In some cases, the fluid motion is combinedly affected by the forced and buoyancy flows, which is called mixed convection or combined convection. With an increased flow velocity and object size, the flow transition occurs from viscosity-dominated laminar flow to inertia-dominated turbulent flow. In turbulent flows, the random movement of fluid (turbulent mixing) becomes the dominant mechanism of momentum and heat transfer. The intensity of momentum mixing and the rate of heat transfer in turbulent flows are much greater than in laminar flows, where molecular diffusion is dominant.

(1) Boundary layer approximation

Assume that a flat plate with a constant temperature, T_S, is placed horizontally in a flow with uniform velocity, U_∞, and uniform temperature, T_∞, as shown in Figure 3.16. The flow velocity at the object surface is zero, but gradually approaches the main flow velocity with separation from the object. The velocity changes sharply in a narrow range called the velocity boundary layer. The fluid temperature similarly shows a rapid change within a narrow range called the thermal boundary layer. Changes in the velocity and temperature are not strictly limited to within the boundary layers, but theoretically extend infinitely far. The boundary layer is generally defined as the point where the velocity or temperature is 99% of the mainstream value.

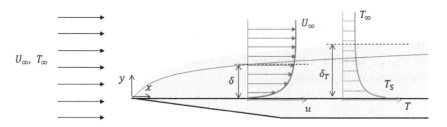

Figure 3.16 Thermal and velocity boundary layers formed along a horizontal flat plate.

Namely, the thicknesses of the velocity and thermal boundary layers, δ and δ_T, are respectively expressed as:

$$\left(\frac{u}{U_\infty}\right)_{y=\delta} = 0.99 \quad \text{and} \quad \left(\frac{T_S - T}{T_S - T_\infty}\right)_{y=\delta_T} = 0.99 \tag{3.39}$$

If the thermo-physical properties of the fluid are constant and there is no internal heat generation, the conservation equations for mass, momentum (x and y), and energy are, respectively, expressed as follows:

$$\frac{\partial u}{\partial x} + \frac{\partial v}{\partial y} = 0 \tag{3.40a}$$

$$u\frac{\partial u}{\partial x} + v\frac{\partial u}{\partial y} = -\frac{1}{\rho}\frac{\partial p}{\partial x} + v\left(\frac{\partial^2 u}{\partial x^2} + \frac{\partial^2 u}{\partial y^2}\right) \tag{3.40b}$$

$$u\frac{\partial v}{\partial x} + v\frac{\partial v}{\partial y} = -\frac{1}{\rho}\frac{\partial p}{\partial y} + v\left(\frac{\partial^2 v}{\partial x^2} + \frac{\partial^2 v}{\partial y^2}\right) \tag{3.40c}$$

$$u\frac{\partial T}{\partial x} + v\frac{\partial T}{\partial y} = \alpha\left(\frac{\partial^2 T}{\partial x^2} + \frac{\partial^2 T}{\partial y^2}\right) \tag{3.40d}$$

The velocity and thermal boundary layers are thin if the flow along the plate surface is sufficiently fast. Thus, the following boundary layer approximations hold:

$$|u| \gg |v| \tag{3.41a}$$

$$\left|\frac{\partial u}{\partial y}\right| \gg \left|\frac{\partial u}{\partial x}\right|, \left|\frac{\partial v}{\partial x}\right|, \left|\frac{\partial v}{\partial y}\right| \tag{3.41b}$$

$$\left|\frac{\partial T}{\partial y}\right| \gg \left|\frac{\partial T}{\partial x}\right| \tag{3.41c}$$

The velocity component along the object surface, u, is much larger than the velocity component perpendicular to it, v. The changes in u and T are significant in the direction perpendicular to the object surface and small along the object. In such a boundary layer flow, the following boundary layer equations can be obtained:

$$\frac{\partial u}{\partial x} + \frac{\partial v}{\partial y} = 0 \tag{3.42a}$$

$$u\frac{\partial u}{\partial x} + v\frac{\partial u}{\partial y} = -\frac{1}{\rho}\frac{\partial p}{\partial x} + v\frac{\partial^2 u}{\partial y^2} \qquad (3.42b)$$

$$u\frac{\partial T}{\partial x} + v\frac{\partial T}{\partial y} = \alpha\frac{\partial^2 T}{\partial y^2} \qquad (3.42c)$$

The conservation equation for mass, Eq. (3.40a), is unchanged by the boundary layer approximation. Among the two momentum equations, only the one in the x direction is retained as Eq. (3.42b).

(2) Similarity law and dimensionless parameters

For the analysis of flow behavior in the boundary layer, the flow velocity components, u and v, coordinate components, x and y, pressure, p, and temperature, T, in the boundary layer equation can be non-dimensionalized as follows:

$$u^* = \frac{u}{U_\infty}, \ v^* = \frac{v}{U_\infty}, \ x^* = \frac{x}{L}, \ y^* = \frac{y}{L}, \ p^* = \frac{p}{\rho U_\infty^2}, \ T^* = \frac{T - T_S}{T_\infty - T_S} \qquad (3.43)$$

where L is the representative length of the object. By substituting these into Eq. (3.42), the boundary layer equations can be transformed into the following dimensionless forms:

$$\frac{\partial u^*}{\partial x^*} + \frac{\partial v^*}{\partial y^*} = 0 \qquad (3.44a)$$

$$u^*\frac{\partial u^*}{\partial x^*} + v^*\frac{\partial u^*}{\partial y^*} = -\frac{\partial p^*}{\partial x^*} + \frac{1}{Re^*}\frac{\partial^2 u^*}{\partial y^{*2}} \qquad (3.44b)$$

$$u^*\frac{\partial T^*}{\partial x^*} + v^*\frac{\partial T^*}{\partial y^*} = \frac{1}{Re^* Pr^*}\frac{\partial^2 T^*}{\partial y^{*2}} \qquad (3.44c)$$

In Eq. (3.44), Re^* is the Reynolds number representing the scale of the inertia force, $\mu U_\infty / L$, relative to the viscous shear force, ρU_∞^2. Pr^* is the Prandtl number representing the scale of the kinematic viscosity coefficient, $v = \mu / \rho$, relative to the thermal diffusivity, $\alpha = k / \rho c$. They are, respectively, defined as follows:

$$Re^* = \frac{\rho U_\infty^2}{\mu U_\infty / L} = \frac{U_\infty L}{v} \qquad (3.45a)$$

$$Pr^* = \frac{v}{\alpha} \qquad (3.45b)$$

Pr^* is closely related to the development of velocity and thermal boundary layers, and affects the ratio of their thicknesses, δ and δ_T ($\dfrac{\delta}{\delta_T} \sim Pr^{*\frac{1}{3}}$ in laminar boundary layers). For reference, the thermo-physical properties of air at atmospheric pressure are listed in Table 3.2. It is noteworthy that the value of the Prandtl number is insensitive to the temperature, i.e., $Pr^* \cong 0.7$.

The dimensionless equation, Eq. (3.44), demonstrates that there is a similarity in terms of the mass and heat transfer in the boundary layer flow. The flow behavior in the boundary layer is governed by the thermo-physical properties of the flow (ρ and μ), the mainstream condition (U_∞ and T), and the object dimension (L). These parameters are integrated into the two dimensionless numbers, Re^* and Pr^*, for characterizing the flow. In other words, if the values of Re^* and Pr^* are identical, the velocity and temperature field will be similar despite differences in the thermo-physical properties, the mainstream condition, and the object dimension.

We introduce another dimensionless number, the Nusselt number, Nu^*, involved in the convective heat transfer of an object. This dimensionless number expresses the scale of heat transfer by convective heat transfer relative to heat conduction, defined as follows:

$$Nu^* = \frac{h(T_S - T_\infty)}{\dfrac{k}{L}(T_S - T_\infty)} = \frac{hL}{k} \tag{3.46}$$

By introducing Eq. (3.46) into Eq. (3.38), we obtain:

$$h = -\frac{k(\partial T / \partial y)_{y=0}}{T_S - T_\infty} = \frac{\dfrac{k}{L}(T_S - T_\infty)}{T_S - T_\infty}\left(\frac{\partial T^*}{\partial y^*}\right)_{y^*=0} \to Nu^* = \left(\frac{\partial T^*}{\partial y^*}\right)_{y^*=0} \tag{3.47}$$

According to Eq. (3.44), the dimensionless temperature, T^*, can be expressed as a function of several dimensionless numbers. If the effect of $\partial p^* / \partial x^*$ can be ignored, Eq. (3.44) can be expressed as (Incropera et al., 2007):

$$T^* = \text{function}(x^*, y^*, Re^*, Pr^*) \tag{3.48}$$

By substituting Eq. (3.48) into Eq. (3.47), the relationship can be viewed as:

$$Nu^* = \text{function}(x^*, Re^*, Pr^*) \tag{3.49}$$

Furthermore, for the mean heat transfer coefficient:

$$\overline{Nu}^* = \int_0^1 Nu^* dx^* = \text{function}(Re^*, Pr^*) \tag{3.50}$$

Table 3.2 Thermophysical properties of air at atmospheric pressure (JSTP, 2000).

Temperature T(K)	Density ρ(kg·m⁻³)	Specific heat c (kJ·kg⁻¹·K⁻¹)	Viscosity coefficient μ (μPa·s⁻¹)	Kinematic viscosity coefficient ν (m²·s⁻¹)	Thermal conductivity k (kW·m⁻¹·K⁻¹)	Thermal diffusivity k (m²·s⁻¹)	Prandtl number Pr^*
260	1.358	1.006	–	–	–	–	–
273.15	1.292	1.006	17.24	1.334×10^{-5}	2.421×10^{-5}	1.861×10^{-5}	0.7170
280	1.261	1.007	17.59	1.395×10^{-5}	2.473×10^{-5}	1.949×10^{-5}	0.7158
300	1.176	1.007	18.57	1.579×10^{-5}	2.623×10^{-5}	2.215×10^{-5}	0.7128
400	0.8818	1.015	23.10	2.619×10^{-5}	3.328×10^{-5}	3.719×10^{-5}	0.7042
500	0.7053	1.030	27.13	3.847×10^{-5}	3.971×10^{-5}	5.464×10^{-5}	0.7041
600	0.5878	1.052	30.82	5.244×10^{-5}	4.573×10^{-5}	7.397×10^{-5}	0.7089
800	0.4408	1.099	37.47	8.499×10^{-5}	5.699×10^{-5}	11.76×10^{-5}	0.7227
1000	0.3527	1.142	43.43	12.31×10^{-5}	6.763×10^{-5}	16.80×10^{-5}	0.7331
1200	0.2939	1.175	48.91	16.64×10^{-5}	7.792×10^{-5}	22.56×10^{-5}	0.7377

Thus, the mean Nusselt number, $\overline{Nu}^*$, from which the mean convective heat transfer coefficient, $\overline{h}$, is calculated, can be expressed as a function of the two dimensionless numbers, Re^* and Pr^*.

(3) Mean Nusselt number

As mentioned above, the heat transfer coefficient, h, is not a material-specific property of the object, but is closely related to the fluid motion. Therefore, we basically need to solve the governing equations of the fluid motion for obtaining the h. However, due to the high cost of computation for solving the governing equations, it is not practical to keep re-calculating h every time the flow condition changes. Therefore, according to the concept presented in Eq. (3.50), the relationships between $\overline{Nu}^*$ and the two dimensionless numbers, Re^* and Pr^*, have been derived for some typical flow conditions either theoretically or experimentally. Unfortunately, the range of applicability of such relationships is limited. Examples of the relationships that could be used to analyze the behavior of fire spread in outdoor fires are listed below.

The $\overline{Nu}^*$ for the turbulent forced convection from a flat plate placed parallel to the horizontal flow, as shown in Figure 3.16, can be expressed as follows:

$$\text{Horizontal plate}: \overline{Nu}^* = 0.037 Re^{*4/5} Pr^{*1/3} \quad \left(5 \times 10^5 < Re^* < 10^8\right) \quad (3.51)$$

The $\overline{Nu}^*$ for the forced convection from a cylinder placed perpendicular to the flow was obtained experimentally, which takes the form (Churchill et al., 1977):

$$\text{Circular cylinder}: \overline{Nu}^* = 0.3 + \frac{0.62 Re^{*1/2} Pr^{*1/3}}{\left\{1 + \left(0.4 / Pr^*\right)^{2/3}\right\}^{1/4}} \quad (3.52)$$
$$\times \left\{1 + \left(\frac{Re^*}{282000}\right)^{5/8}\right\}^{4/5} \quad \left(Re^* Pr^* > 0.2\right)$$

In the above equations, the thermo-physical properties of the fluid required for the calculation are assumed to be constant. They are not necessarily constant in actual boundary layers as they vary with temperature (see Table 3.2). However, if the change in thermo-physical properties is taken into account, the problem becomes strongly nonlinear and causes difficulty in obtaining a solution. For this reason, the following film temperature, the average of the mainstream and the object surface temperatures, is generally used as the representative temperature when determining the temperature-dependent thermo-physical properties:

$$T_f = \frac{T_S + T_\infty}{2} \tag{3.53}$$

WORKED EXAMPLE 3.4

Assume that a square plate with a side length of 3 m is placed horizontally on the ground. The surface temperature, T_S, is uniform at 300 K. Calculate the mean convective heat flux, $\overline{q}''$, when a hot airflow with a temperature of T_∞ = 500 K passes over the plate at a velocity of 5 m/s. Use the values in Table 3.2 for the thermo-physical properties of the air.

SUGGESTED SOLUTION

We first identify the thermo-physical properties of the air at the film temperature, T_f = 400 K. From Table 3.2, the thermal conductivity is $k = 3.328 \times 10^{-5}$ kW·m^{-1}·K^{-1}, the kinematic viscosity is $v = 2.619 \times 10^{-5}$ m^2·s^{-1}, and the thermal diffusivity is $\alpha = 3.719 \times 10^{-5}$ m^2·s^{-1}. With these values, the Reynolds number of the airflow passing over the plate, Re^*, can be calculated using Eq. (3.45a):

$$Re^* = \frac{U_\infty L}{v} = \frac{5 \times 3}{2.619 \times 10^{-5}} = 5.73 \times 10^5$$

As $Re^* > 5 \times 10^5$, the airflow is turbulent. Thus, the mean Nusselt number, $\overline{Nu}^*$, can be calculated using Eq. (3.51). Another variable remained in Eq. (3.51), the Prandtl number, Pr^*, can be calculated using Eq. (3.45b):

$$Pr^* = \frac{v}{\alpha} = \frac{2.619 \times 10^{-5}}{3.719 \times 10^{-5}} = 0.704$$

By substituting Re^* and Pr^* into Eq. (3.51), the $\overline{Nu}^*$ can be obtained as:

$$\overline{Nu}^* = 0.037 Re^{*4/5} Pr^{*1/3} = 1.33 \times 10^3$$

Further, from the definition of the Nusselt number in Eq. (3.46), the mean heat transfer coefficient, $\overline{h}$, can be calculated as:

$$\overline{h} = \overline{Nu}^* \frac{k}{L} = 1.33 \times 10^3 \times \frac{3.328 \times 10^{-5}}{3} = 1.48 \times 10^{-2} \text{ kW·K}^{-1}$$

Thus, the convective heat flux, $\overline{q}''$, is:

$$\overline{q}'' = \overline{h}(T_S - T_\infty) = 1.48 \times 10^{-2} \times (-200) = -2.96 \text{ kW}$$

As $\bar{q}''$ takes a negative value, the convective heat transfer occurs in the direction from the hot airflow to the plate.

3.2.3 Natural convection

The density of most fluids decreases as the temperature increases. Therefore, when a mass of fluid is heated, a change in the density causes buoyancy-induced convection. Such a flow is called natural convection. In general, natural convection has a lower velocity than forced convection, resulting in a lower heat transfer rate.

(1) Similarity law and dimensionless parameters

Consider a 2D steady natural convection along a hot vertical plate, as shown in Figure 3.17. The thermo-physical properties of the fluid are assumed to be constant, and there is no internal heat generation. As in the case of forced convection, we use the conservation equations for mass, momentum, and energy with the boundary layer approximation. A similar form of equations can be obtained as follows:

$$\frac{\partial u}{\partial x} + \frac{\partial v}{\partial y} = 0 \tag{3.54a}$$

$$u\frac{\partial u}{\partial x} + v\frac{\partial u}{\partial y} = \beta g\left(T - T_\infty\right) + v\frac{\partial^2 u}{\partial y^2} \tag{3.54b}$$

$$u\frac{\partial T}{\partial x} + v\frac{\partial T}{\partial y} = \alpha\frac{\partial^2 T}{\partial y^2} \tag{3.54c}$$

where β is the coefficient of volumetric expansion, which is given by:

$$\beta = -\frac{1}{\rho}\left(\frac{\partial \rho}{\partial T}\right)_p \cong \frac{1}{T_\infty} \tag{3.55}$$

In forced convection, the velocity field is not affected by the temperature field, as the temperature, T, is not included in the conservation equation for momentum (see Eq. (3.42)). On the other hand, in natural convection, T is included in the conservation equation for momentum (Eq. (3.54)). This indicates that the energy conservation equation must be coupled to solve the conservation equation for momentum.

As in the case of forced convection, the flow velocity components, u and v, coordinate components, x and y, pressure, p, and temperature, T, in the boundary layer equations are non-dimensionalized as follows:

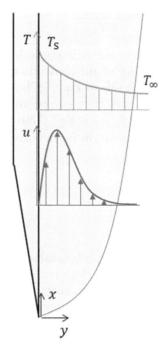

Figure 3.17 Natural convection along a vertical flat plate.

$$u^* = \frac{u}{U_\infty}, \quad v^* = \frac{v}{U_\infty}, \quad x^* = \frac{x}{L}, \quad y^* = \frac{y}{L}, \quad T^* = \frac{T - T_\infty}{T_S - T_\infty} \qquad (3.56)$$

where L is the representative length of the object. By introducing Eq. (3.56) into Eq. (3.54), we obtain the governing equations in a dimensionless form as:

$$\frac{\partial u^*}{\partial x^*} + \frac{\partial v^*}{\partial y^*} = 0 \qquad (3.57a)$$

$$u^* \frac{\partial u^*}{\partial x^*} + v^* \frac{\partial u^*}{\partial y^*} = \frac{Gr^*}{Re^{*2}} T^* + \frac{1}{Re^*} \frac{\partial^2 u^*}{\partial y^{*2}} \qquad (3.57b)$$

$$u^* \frac{\partial T^*}{\partial x^*} + v^* \frac{\partial T^*}{\partial y^*} = \frac{1}{Re^* Pr^*} \frac{\partial^2 T^*}{\partial y^{*2}} \qquad (3.57c)$$

where Gr^* is the Grashof number representing the scale of the buoyancy force relative to the viscous forces, which is defined as:

$$Gr^* = \frac{\beta g (T_S - T_\infty) L^3}{v^2} \cong \frac{g (T_S - T_\infty) L^3}{v^2 T_\infty} \tag{3.58}$$

Gr^* and Re^* respectively represent the strength of natural and forced convections. The predominancy of the convection mode can be judged by the first term of the right-hand side of Eq. (3.57b), which is the ratio of the two. In other words, natural convection is dominant if $Gr^* \gg Re^{*2}$, whereas forced convection is dominant if $Gr^* \ll Re^{*2}$. Thus, from Eq. (3.57), we can deduce the following expression for the mean Nusselt number, $\overline{Nu}$, for natural convection, $Gr^* \gg Re^{*2}$:

$$\overline{Nu}^* = \text{function}(Gr^*, Pr^*) \tag{3.59}$$

(2) Mean Nusselt number

As in the case of forced convection, relationships between the $\overline{Nu}^*$ and the two dimensionless numbers, Gr^* and Pr^*, have been derived for some typical natural convection cases. Among them, correlations that could be used to analyze the behavior of fire spread in outdoor fires are listed below.

In 2D natural convection along a hot vertical plate, the flow is initially laminar at the bottom tip of the plate. However, the boundary layer becomes unstable, causing a transition to a turbulent flow at a higher elevation. In the regime where this transition occurs, the product of Gr^* and Pr^* is approximately $Gr^* Pr^* \cong 10^9$. Analytical solutions are generally not available for turbulent flows, as it is difficult to accurately approximate the velocity and temperature distributions in the turbulent boundary layer. However, the following equation was derived based on extensive experimental data that range from laminar to turbulent regimes (Churchil et al., 1975):

$$\text{Vertical plate}: \overline{Nu}^* = \left[0.825 + \frac{0.387 (Gr^* Pr^*)^{1/6}}{\left\{ 1 + (0.492 / Pr^*)^{9/16} \right\}^{8/27}} \right]^2 \tag{3.60}$$
$$(10^4 < Gr^* Pr^* < 10^{13})$$

An expression for natural convection around a horizontal circular cylinder was also derived experimentally (Churchil et al., 1975):

$$\text{Circular cylinder}: \overline{Nu}^* = \left[0.60 + \frac{0.387 (Gr^* Pr^*)^{1/6}}{\left\{ 1 + (0.559 / Pr^*)^{9/16} \right\}^{8/27}} \right]^2 (Gr^* Pr^* < 10^{12})$$
$$\tag{3.61}$$

Another relationship was derived for natural convection at the upper surface of a hot horizontal plate, which is given below:

$$\text{Upper surface of hot plate}: \overline{Nu}^{*} = \begin{cases} 0.54\left(Gr^{*}Pr^{*}\right)^{1/4}\left(10^{4} < Gr^{*}Pr^{*} < 10^{7}\right) \\ 0.15\left(Gr^{*}Pr^{*}\right)^{1/3}\left(10^{7} < Gr^{*}Pr^{*} < 10^{11}\right) \end{cases} \quad (3.62)$$

WORKED EXAMPLE 3.5

Assume that a square plate with a side length of 2 m is placed vertically, as shown in Figure 3.17. The surface temperature is uniform at $T_S = 700\,\text{K}$. Calculate the mean convective heat flux, $\overline{q}''$, when the ambient air at $T_\infty = 300\,\text{K}$ passes over the plate. Use Table 3.2 for the thermo-physical properties of the air.

SUGGESTED SOLUTION

We first specify the thermo-physical properties of the air at the film temperature, $T_f = 500\,\text{K}$. From Table 3.2, the thermal conductivity is $k = 3.971 \times 10^{-5}\,\text{kW} \cdot \text{m}^{-1} \cdot \text{K}^{-1}$, the kinematic viscosity is $v = 3.847 \times 10^{-5}\,\text{m}^2 \cdot \text{s}^{-1}$, and the thermal diffusivity is $\alpha = 5.464 \times 10^{-5}\,\text{m}^2 \cdot \text{s}^{-1}$. The Grashof number, Gr^*, and the Prandtl number, Pr^*, of the flow along the flat plate can be calculated using Eqs. (3.58) and (3.45b), respectively, as follows:

$$Gr^{*} \cong \frac{g\left(T_S - T_\infty\right)L^3}{v^2 T_\infty} = \frac{9.8 \times 400 \times 2^3}{\left(3.847 \times 10^{-5}\right)^2 \times 300} = 7.06 \times 10^{10}$$

$$Pr^{*} = \frac{v}{\alpha} = \frac{3.847 \times 10^{-5}}{5.464 \times 10^{-5}} = 0.704$$

As the product of the two dimensionless numbers is above the transition point, $Gr^*Pr^* = 4.97 \times 10^{10} > 10^9$, the flow is turbulent. The mean Nusselt number, $\overline{Nu}^*$, can be calculated using Eq. (3.62) as follows:

$$\overline{Nu}^{*} = 0.15\left(Gr^{*}Pr^{*}\right)^{1/3} = 5.51 \times 10^2$$

The mean heat transfer coefficient, $\overline{h}$, can be obtained using the definition of the Nusselt number, Eq. (3.46) as follows:

$$\overline{h} = \overline{Nu}^{*}\frac{k}{L} = 5.51 \times 10^2 \times \frac{3.971 \times 10^{-5}}{2} = 1.09 \times 10^{-2}\,\text{kW} \cdot \text{m}^{-2} \cdot \text{K}^{-1}$$

The mean heat flux can be calculated using Newton's law of cooling, Eq. (3.33):

$$\overline{q}'' = \overline{h}(T_S - T_\infty) = 1.09 \times 10^{-2} \times 400 = 4.38 \text{ kW} \cdot \text{m}^{-2}$$

As $\overline{q}''$ takes a positive value, the convective heat transfer occurs in the direction from the plate to the air.

3.3 RADIATIVE HEAT TRANSFER

A fraction of the internal energy of an object is converted into electromagnetic waves, such as visible light and infrared rays, and emitted from its surface. Electromagnetic wave with a wavelength that falls into a certain range is called thermal radiation. Electromagnetic wave emitted from an object propagate through space and reach other objects, where they are converted into a form of internal energy. As the emissive power is greater when an object is hotter, an electromagnetic wave transfers heat from a hot object to a cold object. Such a mode of heat transfer is called radiative heat transfer.

Radiation is the transfer of energy in the form of electromagnetic waves. Unlike convection or heat conduction, it does not require a medium to transport heat from a heat source to a recipient. As the emissive power of a radiant heat source is proportional to the fourth power of its absolute temperature, radiative heat transfer gains its importance in a fire environment where hot materials and substances are involved. Thus, radiative heat transfer often becomes a dominant factor of fire spread between combustibles in outdoor fires.

3.3.1 Thermal radiation

Every substance consists of molecules and atoms, which are in motion in accordance with their absolute temperature. The motion of molecules and atoms is accompanied by the emission of electromagnetic waves of various wavelengths. In Figure 5.18, electromagnetic waves are classified according to their wavelength, λ. The classification bounds are not definite, but rather approximate. Visible light ranges from 0.38 to 0.78 µm. The colors of visible light are violet, blue, green, yellow, orange, and red, in ascending order of wavelength, λ. It is important to note that electromagnetic waves are not colored. They are merely perceived as different colors when they are incident on our eyes at different wavelengths. If λ is shorter than violet light, electromagnetic waves become ultraviolet radiation, X-rays, and gamma-rays. On the other hand, if λ is longer than that of red light, electromagnetic waves become infrared radiation and radio waves. Radiation is the general term for these electromagnetic waves. Among these, electromagnetic waves in the range of wavelengths that can be detected as heat or light are specifically called thermal radiation. As shown in Figure 3.18, thermal radiation covers the range of wavelengths from visible light to about 100 µm.

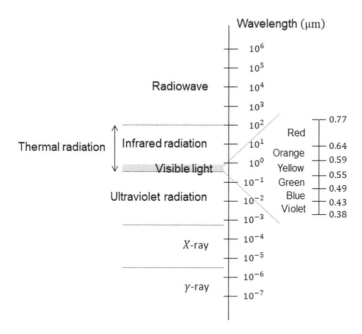

Figure 3.18 Classification of electromagnetic waves in terms of wavelength.

In a fire, thermal energy (internal energy) attributed to the molecular motion inside a fire source is converted into radiation (electromagnetic waves). When the emitted radiation propagates through space and reaches an adjacent object, it is again converted into thermal energy in the form of molecular motion. Thus, a process of radiative heat transfer is completed. As radiative heat transfer does not require a medium such as a solid or air, heat can be transferred via electromagnetic waves even in a vacuum. In this respect, the heat transfer mechanism differs from conduction or convection as both require a medium. Heat transfer is often classified into heat conduction, convective heat transfer, and radiative heat transfer. However, from the viewpoint of the heat transfer mechanism, there are two broad classifications: heat conduction and radiative heat transfer. Convective heat transfer is a variant of heat conduction in which the rate of heat transfer depends on the temperature gradient within the medium.

Consider the case where an object receives radiation, as shown in Figure 3.19. When radiation is incident on an object, it is partially reflected, partially absorbed, and the rest is transmitted. The absorption of the radiation causes an increase in the temperature. However, the radiation does not cause any increase in the temperature if it is completely reflected or transmitted. The ratios of reflected, absorbed, and transmitted energy to

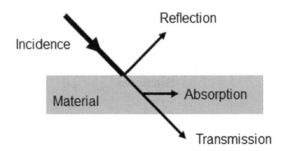

Figure 3.19 Reflection, absorption, and transmission of radiation incident on an object.

the incident radiation energy are called reflectivity, ρ, absorptivity, α, and transmissivity, τ, respectively. From the energy conservation, their sum must equal unity:

$$\rho + \alpha + \tau = 1 \qquad (3.63)$$

Suppose the object is opaque, $\tau = 0$. Thus, there is a relationship between ρ and α, which is:

$$\rho + \alpha = 1 \qquad (3.64)$$

According to Eq. (3.64), α of an opaque object can be determined when the amount of incident and reflected radiant energies can be measured. In evaluating the hazard of outdoor fires, almost all objects involved in fire spread can be treated as opaque objects. In such a case, ρ and α depend only on the state of the object surface. In other words, the radiative heat transfer between objects can be calculated by regarding them as assemblies of geometric surfaces at different temperatures. This can significantly reduce the computational load compared to the case where the objects are not opaque.

3.3.2 Radiation intensity and emissive power

Consider an infinitesimal element, dA, on the surface of an object emitting energy by thermal radiation, $d\dot{Q}$ (kW), as shown in Figure 3.20. In general, the surface of objects emits thermal radiation in all directions. However, its intensity varies by direction. For further discussion, consider another surface element, dA', on a hemisphere of unit radius with dA at its center. The location of dA' can be identified with the polar and azimuthal angles, ϕ and θ, respectively. Given that the projected area of dA in the direction of emission is $dA \cos\phi$, the radiative energy per unit area is $d\dot{Q} / dA \cos\phi$. On the surface of a hemisphere of unit radius, the surface element is

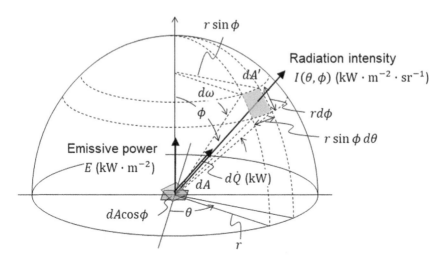

Figure 3.20 A hemisphere of unit radius and radiation intensity.

$dA' / r^2 = dA' = d\omega$. This indicates that dA' is equivalent to the solid angle, $d\omega$ (sr: steradian). Thus, the radiation intensity emitted per unit area, unit time, and unit solid angle, I (kW·m⁻²·sr⁻¹), can be expressed as:

$$I = \frac{d\dot{Q}/dA\cos\phi}{dA'} = \frac{d\dot{Q}/dA\cos\phi}{d\omega} \tag{3.65}$$

On the contrary, if I is given, the radiative energy emitted within the $d\omega$ can be expressed as:

$$d\dot{Q} = I \cdot dA\cos\phi \cdot d\omega \tag{3.66}$$

Referring to Figure 3.20, $d\omega$ can be expressed using ϕ and θ as:

$$d\omega = \frac{dA'}{r^2} = \frac{r\sin\phi d\theta \cdot rd\phi}{r^2} = \frac{1 \cdot \sin\phi d\theta \cdot 1 \cdot d\phi}{1^2} = \sin\phi d\theta d\phi \tag{3.67}$$

By substituting Eq. (3.67) into Eq. (3.66), we obtain the following expression for $d\dot{Q}$:

$$d\dot{Q} = I \cdot dA\cos\phi \cdot d\omega = I \cdot dA\sin\phi\cos\phi \cdot d\theta d\phi \tag{3.68}$$

Thermal radiation emitted from an object comprises components of various wavelengths. The radiation intensity, I, varies depending on its wavelength, λ. However, λ is not the unique parameter that controls I, but is also dependent on the surface roughness of the object. Thus, I can be expressed in the form of $I_\lambda(\lambda, \theta, \phi)$, which is a function of wavelength, λ, polar angle, ϕ, and azimuthal angle, θ. On this basis, the radiative energy emitted from the dA, which is $\dot{Q}$ (kW), can be obtained by integrating Eq. (3.68) over the entire surface of the unit hemisphere:

$$\dot{Q} = dA \int_0^\infty \int_0^{2\pi} \int_0^{\pi/2} I_\lambda(\lambda, \theta, \phi)\sin\phi\cos\phi\, d\theta d\phi d\lambda \tag{3.69}$$

Thermal radiation energy emitted from the unit surface area of the object, $\dot{Q}/dA$, is termed as the emissive power, $E(\text{kW}\cdot\text{m}^{-2})$, which is:

$$E = \frac{\dot{Q}}{dA} = \int_0^\infty \int_0^{2\pi} \int_0^{\pi/2} I_\lambda(\lambda, \theta, \phi)\sin\phi\cos\phi\, d\theta d\phi d\lambda \tag{3.70}$$

Note that a surface where I is uniform regardless of its emitting direction is called a diffusively emitting surface or a Lambert surface. On the other hand, a surface where I depends on the emitting direction is called a directionally emitting surface. In the case of a diffusively emitting surface, the E in Eq. (3.70) can be simplified as:

$$E = \int_0^\infty I_\lambda(\lambda, \theta, \phi)d\lambda \int_0^{2\pi}\int_0^{\pi/2} \sin\phi\cos\phi\, d\theta d\phi = \pi I \tag{3.71}$$

An infinitesimal element, dA, of a diffusively emitting surface emits thermal radiation hemispherically. Thus, I is independent of the emission angle, $I(\theta, \phi) = I$. The emissive power per unit solid angle, $dE/d\omega$, can therefore be expressed by using Eq. (3.66) as:

$$\frac{dE}{d\omega} = \frac{d\dot{Q}/dA}{d\omega} = I\cdot\cos\phi \tag{3.72}$$

This equation indicates that the emissive power, dE, from an infinitesimal element, dA, that passes through a solid angle, $d\omega$, decreases with ϕ, and becomes zero at $\phi = 90°$. This is called Lambert's cosine law. The reason why $dE/d\omega$ decreases with the value of $\cos\phi$, even if the radiation intensity, I, is hemispherically uniform, is that the projected area, $dA\cos\phi$, decreases with ϕ.

3.3.3 Blackbody radiation

Electromagnetic waves perceived with our eyes are those within the visible light spectrum, as shown in Figure 3.18. A surface that perfectly absorbs light is called 'black' in the visible light spectrum, while a surface that perfectly reflects light is called 'white'. In radiative heat transfer, a blackbody or black surface is defined as a virtual object or surface that absorbs radiation (electromagnetic waves) of all incident wavelengths ($\alpha = 1$).

The emissive power of electromagnetic waves emitted from a blackbody is not uniform over all the wavelength bands, but follows a certain distribution. According to Planck's law, the spectral emissive power emitted from a blackbody with an absolute temperature of T (K) within a small wavelength ranging from λ to $\lambda + d\lambda$ is expressed by the following equation:

$$E_{b,\lambda} = \frac{C_1}{\lambda^5 \left[\exp\left(\dfrac{C_2}{\lambda T}\right) - 1 \right]} (W \cdot m^{-2} \cdot \mu m^{-1}) \tag{3.73a}$$

$$C_1 = 2\pi h c_0^2 = 3.742 \times 10^8 \ W \cdot \mu m^4 \cdot m^{-2} \tag{3.73b}$$

$$C_2 = \frac{h c_0}{k} = 1.439 \times 10^4 \ \mu m \cdot K \tag{3.73c}$$

where C_1 and C_2 are the first and second radiation constants, respectively, $h = 6.6256 \times 10^{-34} \ J \cdot s$ is Planck constant, $k = 1.3805 \times 10^{-23} \ J \cdot K^{-1}$ is Boltzmann constant, and c_0 is the speed of light in a vacuum. Figure 3.21 shows the relationship between the spectral emissive power in Eq. (3.71),

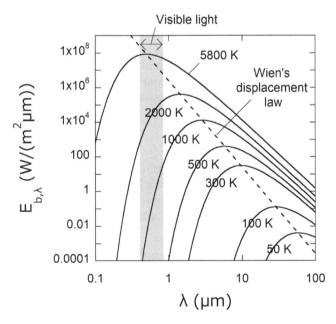

Figure 3.21 Spectral emissive power of radiation emitted from a blackbody at different temperatures (Planck distribution).

$E_{b,\lambda}$, and the wavelength, λ, for different absolute temperatures, T. At any T, $E_{b,\lambda}$ once increases and then decreases as λ increases. Focusing on a specific wavelength, $E_{b,\lambda}$ increases monotonically as T increases. Moreover, the emission of electromagnetic waves with shorter λ increases as T increases.

When T is given, $E_{b,\lambda}$ takes a unimodal distribution (Figure 3.21). At the unique peak of the distribution where $E_{b,\lambda}$ takes the maximum value, the wavelength, λ_{max}, should satisfy the following condition:

$$\frac{\partial E_{b,\lambda}}{\partial \lambda} = 0 \qquad (3.74)$$

By solving this, a relationship between λ_{max} and T can be obtained as:

$$\lambda_{max} T = 2897\,\mu\text{m} \cdot \text{K} \qquad (3.75)$$

This is called Wien's displacement law, shown as a dashed line in Figure 3.21. According to this law, the higher T, the smaller λ_{max}. For example, the sun can be regarded as a blackbody with an absolute temperature of approximately 5800K. The strong radiation spectrum extends over the entire visible bands and is perceived as white light (Figure 3.21). In contrast, the spectrum of radiation emitted from an object at $T = 1000$ K is mainly in the infrared band. The radiation is stronger at longer wavelengths in the visible band, which is perceived as red light. For $T < 800$ K, the radiation spectrum is mostly in the infrared band, which cannot be perceived as light by our eyes and appears as black.

$E_{b,\lambda}$ shown in Figure 3.21 and Eq. (3.71) is the amount of energy emitted from a blackbody per unit wavelength and time. The total emissive power, which is the total energy of radiation emitted from a blackbody at T, can be obtained by integrating the spectral emissive power over all wavelengths:

$$E_{b,\lambda} = \int_0^\infty E_{b,\lambda} d\lambda = \sigma T^4 \qquad (3.76)$$

where $\sigma = 5.667 \times 10^{-11}\,\text{kW} \cdot \text{m}^{-2} \cdot \text{K}^{-4}$ is the Stefan-Boltzmann constant. Eq. (3.76) is called the Stefan-Boltzmann's law. This indicates that the radiation energy emitted per unit area and time (heat flux) from a blackbody at a temperature T is proportional to the fourth power of T.

3.3.4 Radiation properties of real surfaces

Blackbody surface is the surface that gives the maximum emissive power under a given absolute temperature. The spectral emissive power, $E_{b,\lambda}$, of radiation emitted from a blackbody surface follows Planck's law. However, the wavelength distribution of the spectral emissive power, E_λ, of radiation emitted from a real surface does not necessarily follow Planck's law, as

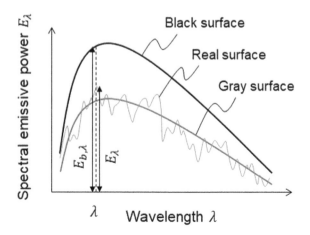

Figure 3.22 Spectral emissive powers of blackbody, graybody, and real surfaces at a certain absolute temperature T.

schematically illustrated in Figure 3.22. The spectral emissivity, ε_λ, is defined as the ratio of the spectral emissive power of a real surface, $E_\lambda(\lambda, T)$, to that of a black surface at the same temperature, $E_{b,\lambda}(\lambda, T)$, which takes the form:

$$\varepsilon_\lambda = \frac{E_\lambda(\lambda, T)}{E_{b,\lambda}(\lambda, T)} \tag{3.77}$$

As is evident from the definition, ε_λ takes a value of between 0 and 1.

To this point, we have independently discussed the emission and absorption of radiant energy. However, they are closely related to each other. Consider a closed space maintained at a constant temperature, T, inside which a small object is contained. Let α_λ and ε_λ be the emissivity and absorptivity of the small object, respectively. In this situation, the radiant energy received by the small object at a wavelength, λ, which is $\alpha_\lambda E_{b\lambda}(\lambda, T)$, should be equivalent to the emitted radiant energy, $E_\lambda(\lambda, T) = \varepsilon_\lambda E_{b,\lambda}(\lambda, T)$. Thus, we obtain:

$$\alpha_\lambda = \varepsilon_\lambda \tag{3.78}$$

This is Kirchhoff's law, which states the equivalency of the absorptivity and emissivity at the same temperature. The implications of Kirchhoff's law could be more comprehensible if we consider that thermal radiation is associated with the thermal motion of molecules. A molecular system that emits thermal radiation at a specific wavelength is more likely to be excited at a similar wavelength. In turn, the molecular system is more likely to absorb radiation of a similar wavelength.

The Stefan-Boltzmann's law in Eq. (3.76) was derived by integrating the blackbody emissive power, $E_{b,\lambda}$, over all wavelengths, λ. However, in real surfaces, the spectral emissivity, ε_λ, varies depending on λ. Therefore, the total emissive power, E, cannot be obtained without identifying the value of ε_λ for all wavelengths. Let us assume that the emissivity of a real surface is approximately constant irrespective of the wavelength, as shown in Figure 3.22. Such an object whose emissivity is independent of wavelength is called a graybody or gray surface. By introducing this assumption, ε_λ can be excluded from the integral calculation of E, as follows:

$$E = \int_0^\infty E_\lambda d\lambda = \int_0^\infty \varepsilon_\lambda E_{b,\lambda} d\lambda = \varepsilon \int_0^\infty E_{b,\lambda} d\lambda = \varepsilon \sigma T^4 \tag{3.79}$$

where ε is the total emissivity integrated over the entire spectrum. For gray surfaces, the following relationship can be derived from Kirchhoff's law and the definitions of absorptivity and emissivity:

$$\alpha = \alpha_\lambda = \varepsilon_\lambda = \varepsilon \tag{3.80}$$

where α is the total absorptivity of the object surface. The thermal radiation emitted from a gray surface is similar to that from a blackbody surface. Although almost no real surfaces are perfectly gray, most objects have similar properties within a certain band of wavelengths. However, as discussed in the following subsections, the gray surface assumption facilitates the calculation of radiative heat transfer. Thus, the gray surface assumption has been introduced for practical purposes rather than to follow rigorous theoretical background.

3.3.5 Radiation by gases

Hot gases, such as flame and smoke, emit and absorb thermal radiation in a similar manner as solids. The calculation of radiative heat transfer for gases is more complicated than for solids for several reasons: thermal radiation passes through gases and cannot be considered as a surface-to-surface radiation exchange; thermal radiation of gases exhibits selective emission and absorption for specific wavelengths; gases are mixtures of many chemical species with different radiation properties. The emissivity and absorptivity of solids and liquids exhibit a continuous spectrum, depending on the wavelength. On the other hand, those of gases exhibit a selective absorptivity. This is because the molecular interaction in gases is weaker than that in solids and liquids. Thus, thermal radiation is emitted and absorbed only in a limited range of wavelengths associated with the vibration and rotation of molecules. Figure 3.23 shows different spectra related to solar radiation (Thekaekara, 1965). Among these spectra, the top smooth spectrum represents the theoretical energy distribution for a blackbody at 6000 K.

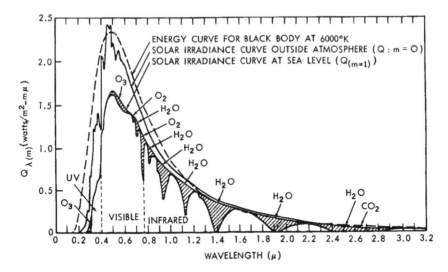

Figure 3.23 Attenuation by Earth's atmosphere of incident solar spectral energy flux (Thekaekara, 1965; Siegel et al., 2002).

The bottom spectrum, with a large number of sharp dips, represents the spectral radiant flux observed on the ground. Such a difference is because the radiative energy from the sun is absorbed predominantly by carbon dioxide (CO_2) and water vapor (H_2O) in the atmosphere. The atmosphere is mainly composed of oxygen (O_2; approximately 21% by volume) and nitrogen (N_2; 79% by volume). These symmetric diatomic molecules can neither emit nor absorb energy by thermal radiation. However, molecules composed of three or more atoms, such as carbon dioxide and water vapor, asymmetric diatomic molecules such as carbon monoxide (CO), and solid particles such as soot generated by combustion, all emit and absorb energy by thermal radiation.

Let us assume that there is a layer of gas with a geometric thickness of L receiving spectral radiation of a wavelength, λ, and radiation intensity, $I_{\lambda,0}$, as shown in Figure 3.24. As the spectral radiation passes through the gas layer, the radiation intensity is attenuated due to absorption. Given that the attenuation through an infinitesimally thin layer, dI_λ, is proportional to the incident radiation intensity, I_λ, and the thickness of the infinitesimally thin layer, dx, we obtain:

$$dI_\lambda = -\kappa_\lambda I_\lambda dx \qquad (3.81)$$

where κ_λ is the spectral absorption coefficient. By integrating this from $x = 0$ to L, the radiation intensity after passing through the gas layer, $I_{\lambda,L}$, can be obtained as:

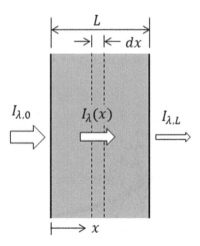

Figure 3.24 Attenuation of radiation intensity due to absorption by gas layer.

$$\frac{I_{\lambda,L}}{I_{\lambda,0}} = \exp(-\kappa_\lambda L) \tag{3.82}$$

where $\kappa_\lambda L$ is termed as the optical thickness. The relationship described in Eq. (3.82) is called Beer's law, which expresses the exponential decay of radiation intensity as a function of the optical thickness. Eq. (3.82) indicates that if the κ_λ is large, $\kappa_\lambda L$ can be thick even if L is small. Consequently, the radiation intensity can be significantly attenuated. The rate of attenuation calculated by Eq. (3.82) can be regarded as the transmissivity of this gas layer, τ_λ. Thus, the absorptivity, α_λ, can be expressed as:

$$\alpha_\lambda = 1 - \tau_\lambda = 1 - \exp(-\kappa_\lambda L) \tag{3.83}$$

If Kirchhoff's law is applicable, the emissivity can be obtained as:

$$\varepsilon_\lambda = \alpha_\lambda = 1 - \exp(-\kappa_\lambda L) \tag{3.84}$$

Furthermore, for a gray gas whose absorptivity is independent of wavelength, Eq. (3.84) can be extended to take the following form:

$$\varepsilon = \alpha = 1 - \exp(-\kappa L) \tag{3.85}$$

Table 3.3 summarizes the effective absorption coefficient, κ, measured for flames when various fuels are burnt. It should be noted that measured values of κ generally involve variations depending on the measurement environment.

Table 3.3 Effective absorption coefficients for flames of various fuels (Atallah et al., 1971; Yuen et al., 1977; de Ris, 1979).

	Fuel	Effective absorption coefficient $\kappa(m^{-1})$	Reference
Gas	Methane	6.45	Yuen et al., 1977
	Ethane	6.39	
	Propane	13.32	
Liquid	Methanol	4.6	Atallah et al., 1971
	Ethanol	0.37	
	Gasoline	2.0	
	Kerosene	2.6	
Solid	Wood	0.8	Yuen et al., 1977
	Polypropylene	1.2	
	PMMA	1.3	de Ris, 1979
	Polypropylene	1.8	
	Polystyrene	5.3	

The above 1D relationship is based on the consideration of unidirectional radiation. Thus, the emissivity, ε, and the absorptivity, α, are called the directional emissivity and directional absorptivity, respectively. However, radiation from real gases is not necessarily as idealistic as assumed. In general, we need to account for the passage of radiation from multiple directions with different path lengths, given that the geometry of the mass of gas is complex. Thus, to obtain the emissivity, ε (or absorptivity), the absorption of the radiative energy in all directions needs to be integrated over the entire volume of gas. If the emissivity obtained by integration, ε, is equivalent to the directional emissivity of the mass of gas with the thickness $L = \bar{L}$, then $\bar{L}$ is called its mean optical length. The $\bar{L}$ is a value determined solely by the geometry of the mass of gas, independent of the gas properties. Following is an engineering equation for the $\bar{L}$ applicable to a wide range of conditions:

$$\bar{L} \cong \frac{3.6V}{A} \tag{3.86}$$

Interested readers are referred to a summary by Tien et al. for the values of $\bar{L}$ for specific geometries (Tien et al., 2002).

WORKED EXAMPLE 3.6

Assume that methane is burnt using a square burner with a side length of $W = 0.5$ m. When the mean flame height is $L_{fl} = 1.08$ m, calculate the emissivity, ε.

SUGGESTED SOLUTION

For the calculation of ε, the mean optical length of the flame, $\bar{L}$, is required. If we consider the shape of the flame as a square cylinder with a base side length of $W = 0.5$ m and a height of $L_{fl} = 1.08$ m, Eq. (3.86) can be used to calculate $\bar{L}$ as follows:

$$\bar{L} \cong \frac{3.6V}{A} = \frac{3.6 \times (0.5 \times 0.5 \times 1.08)}{2 \times 0.5 \times 0.5 + 4 \times 0.5 \times 1.08} = 0.365 \text{ m}$$

Substituting the obtained $\bar{L}$ and the effective absorption coefficient of methane $\kappa = 6.45$ m^{-1} from Table 3.3 into Eq. (3.85), ε can be calculated as follows:

$$\varepsilon = 1 - \exp(-\kappa\bar{L}) = 1 - \exp(-6.45 \times 0.365) = 0.905$$

3.3.6 Radiative heat exchange between surfaces

Based on the above discussion, we now consider the calculation of radiative heat exchange between discrete solid objects. It involves not only the radiation characteristics of the objects (temperature and emissivity), but also their geometric configurations (shape, orientation, and distance between each other). When considering radiative energy exchange between objects, complex calculations are unavoidable if the object surfaces are neither black nor gray, or if any radiation-absorbing gases exist between the object surfaces. Thus, in the following subsections, we assume that all object surfaces are either black or gray, and that the intervening gases in the space are not involved in thermal radiation.

When surfaces i and j of different radiation characteristics are facing each other, radiative heat exchange occurs between them. The emissive power of each surface is determined solely by its absolute temperature, T, and emissivity, ε. However, the proportion of the emissive power transferred to the other surface is determined by their geometric relationships, such as shape, orientation, and distance. Consider a case where gray surfaces, i and j, are facing each other, as shown in Figure 3.25. If the temperature of each surface is constant at T_i and T_j, their radiation intensities, I_i and I_j, can be expressed as:

$$I_i = \frac{E_i}{\pi} = \frac{\varepsilon_i \sigma T_i^4}{\pi} \quad \text{and} \quad I_j = \frac{E_j}{\pi} = \frac{\varepsilon_j \sigma T_j^4}{\pi} \tag{3.87}$$

where E_i and E_j are the emissive power (kW·m^{-2}), and ε_i and ε_j are the emissivity (=absorptivity). Next, consider infinitesimal elements, dA_i and dA_j, on the surfaces i and j, respectively. The length of the straight line connecting the two elements is r, and the angles between the straight line

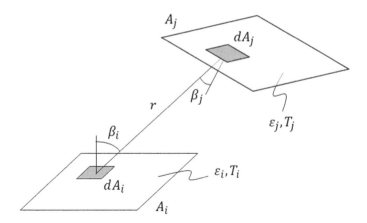

Figure 3.25 Exchange of radiative energy between two surfaces.

and the normal of each surface are β_i and β_j, respectively. If the radiative energy emitted from dA_i that is incident on and absorbed by dA_j is $d\dot{Q}_{i \to j}$, then $d\dot{Q}_{i \to j}$ can be expressed using Lambert's cosine law as:

$$d\dot{Q}_{i \to j} = \varepsilon_j \left(\frac{\varepsilon_i \sigma T_i^4}{\pi} \right) dA_i \cos\beta_i \left(\frac{\cos\beta_j dA_j}{r^2} \right) = \varepsilon_j \varepsilon_i \sigma T_i^4 \left(\frac{\cos\beta_i \cos\beta_j}{\pi r^2} \right) dA_i dA_j \quad (3.88)$$

Integrating this over the entire surface of i and j, we obtain the total radiative energy transferred from i to j as follows:

$$\dot{Q}_{i \to j} = \varepsilon_j \varepsilon_i \sigma T_i^4 \int_{A_i} \int_{A_j} \left(\frac{\cos\beta_i \cos\beta_j}{\pi r^2} \right) dA_j dA_i \quad (3.89)$$

Conversely, the total radiative energy transferred from j to i is given by:

$$\dot{Q}_{j \to i} = \varepsilon_i \varepsilon_j \sigma T_j^4 \int_{A_j} \int_{A_i} \left(\frac{\cos\beta_i \cos\beta_j}{\pi r^2} \right) dA_i dA_j \quad (3.90)$$

From Eqs. (3.89) and (3.90), the net radiative energy transferred from i to j can be expressed as:

$$\dot{Q}_{i \to j, net} = \varepsilon_j \varepsilon_i \sigma \left(T_i^4 - T_j^4 \right) \int_{A_i} \int_{A_j} \left(\frac{\cos\beta_i \cos\beta_j}{\pi r^2} \right) dA_j dA_i \quad (3.91)$$

It is noteworthy that the radiative energy emitted from the gray surface i and transferred to the other gray surface j is reflected by a fraction of $1 - \varepsilon_j$, and a part of it is further returned to the surface i. A part of this returned radiative energy is similarly reflected by a fraction of $1 - \varepsilon_i$, thus resulting in an infinite exchange of radiative energy between the gray surfaces i and j. However, in most cases, such an effect is negligible, except when surfaces i and j constitute two surfaces of a closed space. Calculation using Eq. (3.91) provides sufficient accuracy for engineering purposes, despite ignoring the infinite exchange effect. The above equations can also be used for radiative heat exchange between blackbody surfaces by assuming $\varepsilon_i = \varepsilon_j = 1$, without losing any generality.

3.3.7 View factor

In calculating the radiative energy transfer, $\dot{Q}_{i \to j}$, using Eq. (3.89), we need to divide the surfaces i and j into infinitesimal elements and integrate each element-to-element relationship over the entire surfaces. Numerical computation is generally unavoidable when employing such a calculation procedure to solve actual radiative heat transfer problems. However, when the two surfaces have uniform radiation characteristics, the concept of view factor can be introduced to alleviate the computational complexity. In the geometrical configuration of two surfaces shown in Figure 3.25, the view factor is defined as the fraction of the radiative energy incident on surface j out of the radiative energy emitted from the surface $i (= \varepsilon_i \sigma T_i^4 A_i)$. Denoting the view factor as F_{ij}, Eq. (3.89) can be expressed as follows:

$$\dot{Q}_{i \to j} = \varepsilon_j \left(\varepsilon_i \sigma T_i^4 A_i \right) F_{ij} \tag{3.92}$$

By comparing Eq. (3.92) with Eq. (3.89), F_{ij} can be expressed as:

$$F_{ij} = \frac{1}{A_i} \int_{A_i} \int_{A_j} \left(\frac{\cos\beta_i \cos\beta_j}{\pi r^2} \right) dA_j dA_i \tag{3.93}$$

Eq. (3.92) only substitutes the integral calculation in Eq. (3.89) with the view factor, F_{ij}. The substitution itself does not facilitate the numerical computation. However, F_{ij} is determined only by the geometric configuration of the two surfaces i and j, the two angles, β_i and β_j, and the distance, r. Relatively simple analytical expressions are available for calculating the view factor between two surfaces of some typical geometric configurations.

View factors have two important characteristics: reciprocity and summation laws.

The reciprocity law is a relationship between the two view factors, F_{ij} (the view factor of the surface j as viewed from the surface i) and F_{ji} (the view factor of the surface i as viewed from the surface j). Following the derivation procedure of F_{ij} in Eq. (3.93), F_{ji} can be expressed as:

$$F_{ji} = \frac{1}{A_j} \int_{A_j} \int_{A_i} \left(\frac{\cos\beta_i \cos\beta_j}{\pi r^2} \right) dA_i dA_j \tag{3.94}$$

As the integrals in Eqs. (3.93) and (3.94) are common:

$$\int_{A_j} \int_{A_i} \left(\frac{\cos\beta_i \cos\beta_j}{\pi r^2} \right) dA_i dA_j = A_i F_{ij} = A_j F_{ji} \tag{3.95}$$

Using this relationship, $\dot{Q}_{i \to j}$ in Eq. (3.92) can be calculated in two different ways:

$$\dot{Q}_{i \to j} = \left(\varepsilon_j \varepsilon_i \sigma T_i^4 \right) A_i F_{ij} = \left(\varepsilon_j \varepsilon_i \sigma T_i^4 \right) A_j F_{ji} \tag{3.96}$$

In other words, we can calculate $\dot{Q}_{i \to j}$ using either of the two view factors, F_{ij} or F_{ji}, whichever is easier to calculate.

The summation law states that the sum of the view factors of all the surfaces for a given surface never exceeds unity. Consider a closed space composed of N surfaces. The radiation emitted from a surface i must be transferred to one of the surfaces between the 1st and the N th. Thus, the sum of the fractions transferred to each surface in the closed space must be unity:

$$\sum_{j=1}^{N} F_{ij} = 1 \ \left(j = 1, 2, \cdots, N \right) \tag{3.97}$$

Fortunately, Eq. (3.97) holds even if the surfaces do not form a closed space. This is obvious if we consider a fraction of radiative energy transferred to a virtual surface that does not actually exist. It is also noteworthy that Eq. (3.97) does not exclude the possibility that the radiation surface i and the heat receiving surface j are identical. This is because the radiation emitted from the surface i may be transferred to the surface i itself when the surface i is concave.

WORKED EXAMPLE 3.7

A hot blackbody with a uniform absolute temperature, T, is emitting radiation. The radiative heat flux measured by a sensor placed at a distance from the heat source is $\dot{q}'' = 10 \ \mathrm{kW \cdot m^{-2}}$. Calculate T if the view factor of the heat source as viewed from the sensor is 0.2. The emissivity of the sensor is unity.

SUGGESTED SOLUTION

If the surfaces i and j in Figure 3.25 represent the heat source and the sensor, respectively, the rate of radiative energy transfer can be calculated using

Eq. (3.92). In this case, we need to know the view factor of the surface j as viewed from the surface i, F_{ij}. However, the given view factor is its reciprocal view factor, F_{ji}. So, we use the reciprocity law in Eq. (3.96) to calculate the $\dot{Q}_{i \to j}$. As the incident heat flux to the sensor, $\dot{q}''$, can be obtained by dividing $\dot{Q}_{i \to j}$ by the A_j :

$$\dot{q}'' = \left(\sigma T_i^4 \right) F_{ji}$$

By substituting $\dot{q}'' = 10 \ \mathrm{kW \cdot m^{-2}}$, $F_{ji} = 0.2$, and $\sigma = 5.667 \times 10^{-11} \ \mathrm{kW \cdot m^{-2} \cdot K^{-4}}$ into this equation, we obtain the absolute temperature of the heat source as $T = T_i = 969.2$ K.

3.3.8 Analytical solutions for view factors

Eqs. (3.93) and (3.94) define the view factors between two surfaces i and j. In general, the calculation of view factors using these equations is not that straightforward. However, analytical solutions that can be readily solved by hand calculation or spreadsheet software are available for some typical geometrical configurations. In this subsection, we introduce analytical solutions for the following geometrical configurations:

- A rectangular radiating surface A_j parallel to a target dA_i.
- A rectangular radiating surface A_j perpendicular to a target dA_i.
- A circular radiating surface A_j right in front of a target dA_i.
- A cylindrical radiating body A_j with the axis perpendicular to a target dA_i.

These analytical solutions are often used to evaluate the hazard of fire spread in outdoor fires. The temperature rise and ignition of an object exposed to external heating are estimated by calculating the incident heat flux to a point on the heat-receiving object.

(1) dA$_i$ parallel to a rectangle A$_j$

Consider that a target dA_i is in a parallel position as to a rectangular radiating surface A_j. The lengths of the two sides of A_j are a and b, and the distance between the two surfaces is d. The foot of the perpendicular line from dA_i to A_j coincides with one of its vertices. If we introduce ratios, $X = a / d$ and $Y = b / d$, we can calculate the view factor of A_j as viewed from dA_i as follows (Siegel et al., 2002):

$$F_{ij} = \frac{1}{2\pi} \left(\frac{X}{\sqrt{1+X^2}} \tan^{-1} \frac{Y}{\sqrt{1+X^2}} + \frac{Y}{\sqrt{1+Y^2}} \tan^{-1} \frac{X}{\sqrt{1+Y^2}} \right) \qquad (3.98)$$

The upper left panel of Figure 3.26 shows the calculation results of F_{ij} under different combinations of X and Y. F_{ij} increases as the ratios, X and Y, increase. However, the maximum value of F_{ij} is 0.25. This is obvious because only one quadrant of the front side of the target dA_i is covered by the radiating surface A_j, while the remaining three quadrants are uncovered.

(2) dA_i perpendicular to a rectangle A_j

Consider that a target dA_i is in a perpendicular position as to a radiating surface A_j. The lengths of the two sides of A_j are a and b, and the distance between the two surfaces is d. The foot of the perpendicular line from dA_i to A_j coincides with one of its vertices. If we introduce ratios, $X = a / d$ and $Y = b / d$, we can calculate the view factor of A_j as viewed from dA_i as follows (Siegel et al., 2002):

$$F_{ij} = \frac{1}{2\pi}\left(\tan^{-1} Y - \frac{1}{\sqrt{1 + X^2}} \tan^{-1} \frac{Y}{\sqrt{1 + X^2}} \right) \quad (3.99)$$

The upper right panel of Figure 3.26 shows the calculation results of F_{ij} under different combinations of X and Y. Similar to the first case, F_{ij} increases as the ratios, X and Y, increase. However, compared to the first case, the increase in F_{ij} is more gradual. This is because the normal vectors of the two surfaces are orthogonal, and thus the incidence angle of the radiation emitted from A_j that is incident on the dA_i is larger than the first case.

(3) dA_i right opposite a circle A_j

Consider that a target dA_i is in the right opposite position to a circular radiating surface A_j. The radius of the radiating surface is r, and the distance between the two surfaces is d. We can calculate the view factor of A_j as viewed from dA_i with the following equation (Siegel et al., 2002):

$$F_{ij} = \frac{r^2}{d^2 + r^2} \quad (3.100)$$

The bottom left panel of Figure 3.26 shows the calculation results of F_{ij} under different combinations of d and r. Unlike the cases mentioned above for the rectangular radiating surfaces, dA_i is in the right front of A_j. Thus, the maximum value of F_{ij} is unity when d is zero. Although F_{ij} decreases as d increases, it is more gradual for a larger r.

(4) dA_i perpendicular to a circular cylinder A_j

Consider a target dA_i and a cylindrical radiator A_j with a radius of r and a height of h. The normal vector of dA_i is perpendicular to the axis of the cylinder. The distance between the target and the axis of the cylinder is d. The

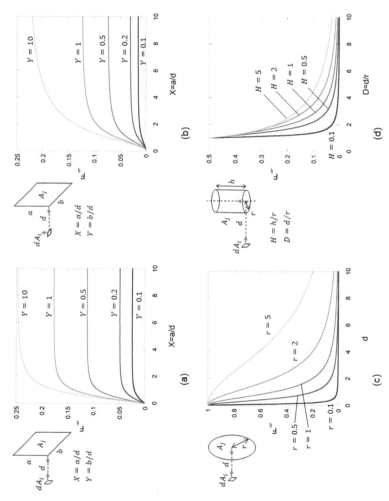

Figure 3.26 Analytical solutions for the view factor of a radiating surface or a body as viewed from an infinitesimal element. (a) dA_i parallel to a rectangle A_j (b) dA_i perpendicular to a rectangle A_j (c) dA_i right opposite to a circle A_j (d) dA_i perpendicular to a circular cylinder A_j.

foot of the perpendicular line from dA_i to the A_j is at the center of the cylinder base. By introducing the ratios, $H = h / r$, $D = d / r$, $X = (1 + D)^2 + H^2$, and $Y = (1 - D)^2 + H^2$, we can calculate the view factor of A_j as viewed from dA_i as follows (Siegel et al., 2002):

$$F_{ij} = \frac{1}{\pi D} \tan^{-1} \frac{H}{\sqrt{D^2 - 1}} + \frac{H}{\pi} \left[\frac{X - 2D}{D\sqrt{XY}} \tan^{-1} \sqrt{\frac{X(D-1)}{Y(D+1)}} \right.$$
$$\left. - \frac{1}{D} \tan^{-1} \sqrt{\frac{D-1}{D+1}} \right] (D \geq 1) \tag{3.101}$$

The bottom right panel of Figure 3.26 shows the calculation results of F_{ij} under different combinations of H and D. If $D < 1$, F_{ij} is unavailable because dA_i is included in the radiator A_j. Thus, the solutions are available only for $D \geq 1$. When $D = 1$, at which dA_i and A_j are at the closest position, the F_{ij} takes its maximum value of 0.5. This is because only one-half of the front side of the target dA_i is covered by the radiating cylinder A_j, while the remaining half is uncovered.

WORKED EXAMPLE 3.8

Consider two geometrical configurations, cases (a) and (b), where an infinitesimal element, dA, is positioned in a parallel position to a rectangle, as shown in Figure 3.27. In case (a), the foot of the perpendicular line from dA to the rectangle is included in the rectangle, whereas in case (b), it is not included in the rectangle. Calculate the view factors of the rectangle as viewed from dA in each case.

SUGGESTED SOLUTION

In both Eqs. (3.98) and (3.99), the foot of the perpendicular line from dA_i to A_j coincides with one of the four vertices of the rectangle A_j. Although this may seem to narrow the scope of application, the two equations can be applied to various situations related to outdoor fires. More specifically, we divide A_j into several rectangles and calculate the view factor for each subdivided rectangle. According to the summation law in Eq. (3.97), the sum of all these view factors becomes the view factor of the entire A_j before division.

For case (a), divide the rectangle into four small rectangles A_1, A_2, A_3, and A_4 separated by two straight lines that pass through the foot of the perpendicular line and parallel to each side of the rectangle. If the view factors of each subdivided rectangle viewed from the element dA are F_1, F_2, F_3, and F_4,

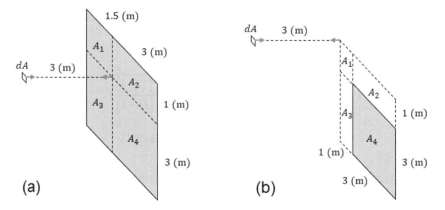

Figure 3.27 Two sample geometrical configurations for view factor calculation.

respectively, the view factor, F, can be obtained based on the summation law as follows:

$$F = F_1 + F_2 + F_3 + F_4$$

As the infinitesimal element dA and each subdivided rectangle are parallel, the view factors for each subdivided rectangle can be calculated using Eq. (3.98), which are $F_1 = 0.043$, $F_2 = 0.064$, $F_3 = 0.139$, and $F_4 = 0.090$. Thus, the view factor for the entire rectangle is $F = 0.336$.

For case (b), we similarly draw straight lines parallel to each side of the rectangle and subdivide the rectangle into four, A_1, A_2, A_3, and A_4. However, as A_2, A_3, and A_4 are outside the radiating surface, the calculation requires a little modification. We combine multiple subdivided rectangles so that one of the vertices always coincides with the foot of the perpendicular line. According to the summation law, the view factor for the entire radiating surface, F, can be obtained as:

$$F = F_{1+2+3+4} - F_{1+2} - F_{1+3} + F_1$$

Note that the subscript identifies the subdivided rectangles. For example, F_{1+2} represents the view factor of the combined rectangles A_1 and A_2. Each term of the right-hand side of this equation can be calculated using Eq. (3.98), which gives, $F_{1+2+3+4} = 0.172$, $F_{1+2} = F_{1+3} = 0.071$, and $F_1 = 0.031$. Thus, the view factor of the radiating surface can be obtained as $F = 0.061$.

WORKED EXAMPLE 3.9

Let us assume the same fire source as the one in Worked example 3.6. Methane is burnt using a square burner with a side length of $W = 0.5$ m.

The mean flame height is $L_{fl} = 1.08$ m, and the emissivity is $\varepsilon = 0.905$. Calculate the radiative heat flux received on the ground at 3 m from the center of the flame base, $\dot{q}''$. However, the normal of the heat receiving surface is perpendicular to the axis of the flame. The temperature of the flame is $T_{fl} = T_{\infty} + \Delta T = 300 + 880 = 1180$ K.

SUGGESTED SOLUTION

We calculate the view factor, F, by approximating the methane flame as a circular cylinder. The equivalent radius of the base of the cylinder, r, is determined so that the base area is equivalent to that of the square burner:

$$r = \sqrt{\frac{0.5 \times 0.5}{\pi}} = 0.282 \text{ m}$$

The variables required for Eq. (3.101) are $H = 1.08 / 0.282 = 3.82$, $D = 3 / 0.282 = 10.6$, $X = (1 + 10.6)^2 + 3.82^2 = 149$, and $Y = (1 - 10.6)^2 + 3.82^2 = 107$. Substituting these into Eq. (3.101), we obtain $F = 0.0212$. Thus, the radiative heat flux is:

$$\dot{q}'' = \left(\varepsilon \sigma T_{fl}^4\right) F = 0.905 \times \left(5.667 \times 10^{-11}\right) \times 1180^4 \times 0.0212 = 2.11 \text{ kW} \cdot \text{m}^{-2}$$

Note that the flame height used in the above calculations, L_{fl}, is the mean flame height that includes a part of the intermittent flame regime, whereas the flame temperature, T_{fl}, is the value in the continuous flame regime (refer to Chapter 5). So the obtained result involves a certain inconsistency.

3.3.9 Point source models for radiative heat transfer

One of the examples where radiative heat transfer calculations are required in outdoor fires is the calculation of the radiative heat flux, $\dot{q}''$, from an independent flame to an object away from the flame. One approach to this problem is to consider the flame as a gray homogeneous radiator, as discussed in Worked examples 3.6 and 3.9. They are often referred to as the 'solid flame model', which has been widely used in practice (Mudan, 1984). In the solid flame model, $\dot{q}''$ is generally calculated using the analytical solutions of view factors, such as those discussed in the previous subsection. However, the calculation procedure becomes complex when the geometrical configuration of the flame and the heat-receiving surface is irregular (Tien et al., 2002). Such a situation often appears when the flame is tilted due to crosswind in outdoor fires. As an alternative to the solid flame model, we can consider the point source model, where the radiant energy of the flame is represented by one or more points, as schematically described in Figure 3.28.

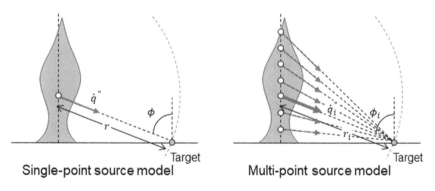

Single-point source model   Multi-point source model

Figure 3.28 Schematic representation of point source models.

(1) Single-point source model

Although a flame emits radiation from its entire body, the point source model represents its radiation characteristics by an internal point. Assuming that the radiation from the representative point is omnidirectional, the radiative heat flux transferred to a target, $\dot{q}''$, which is located at a distance, r, from the representative point, can be considered equivalent to the heat flux passing through the surface of a sphere with an identical radius. Thus, $\dot{q}''$ can be calculated by dividing the radiative energy, $\chi_R \dot{Q}$, by the surface area of the sphere, $4\pi r^2$, as follows (Modak, 1977):

$$\dot{q}'' = \frac{\chi_R \dot{Q}}{4\pi r^2} \cos\phi \qquad (3.102)$$

where χ_R is the radiative fraction (refer to Chapter 5 for χ_R), $\dot{Q}$ is the heat release rate (HRR), and ϕ is the incidence angle at the target. $\dot{q}''$ decreases as r and the surface area of the sphere increase.

As discussed above, the emissive power of hot objects can be evaluated by Stefan-Boltzmann's law in Eq. (3.76). However, the emissive power is often evaluated in the form of a radiative fraction of the HRR, $\chi_R \dot{Q}$, in the point source model. Given that the flame temperature required for Stefan-Boltzmann's law is determined by the HRR, there would be a rationale for directly using the HRR to evaluate $\dot{q}''$. As r is the only geometrical condition required in Eq. (3.102), the calculation procedure is simplified compared to the solid flame model. However, it should be noted that the point source model is basically applicable only when r relative to the flame dimensions is large (Modak, 1977), as it approximates the 3D flame geometry with a single representative point.

(2) Multi-point source model

The multi-point source model has been proposed to extend the single-point source model. As the name indicates, the multi-point source model represents the source of radiant energy by multiple points within the flame (Markstein, 1977; Hankinson et al., 2012; Zhou et al., 2017). In the multi-point source model, the overall effect of multiple representative points is evaluated by summing the heat fluxes from each representative point to the target. When each representative point is identified by i, the incident heat flux to the target, $\dot{q}''$, can be expressed by extending equation (3.102) as follows:

$$\dot{q}'' = \sum_i \left(\frac{w_i \cdot \chi_R \dot{Q}}{4\pi r_i^2} \cos \phi_i \right) \tag{3.103}$$

where w_i is the weight coefficient for the i th representative point, a sum of which satisfies, $\sum w_i = 1$. The multi-point source model definitely requires more computational procedures than the single-point source model because the number of representative points increases. However, it has an advantage that the weight coefficient, w_i, can be adjusted according to the radiation characteristics of the flame. This enables flexible modeling of flame radiation.

Figure 3.29 shows the normalized weight coefficient, $w_{N,i}$, of different liquid pool fire flames measured in a laboratory experiment (Zhou et al., 2017). The relationship between the w_i and $w_{N,i}$ can be given by:

$$\frac{w_{N,i}}{w_i} = N \tag{3.104}$$

where N is the total number of representative points. Thus, the w_i is in proportion to the $w_{N,i}$. The distributions of $w_{N,i}$ exhibit the difference in radiation characteristics of flames when burning fuels with different chemical structures and sooting propensities. For all the fuels, the $w_{N,i}$ decreased toward the flame tip. Although the shape of the distribution varied among fuels, they were similar for those with similar chemical structures. Based on the experimental observation, Zhou et al. suggested that the difference and similarity in the distribution were mainly attributed to those of the flame geometry (Zhou et al., 2017).

WORKED EXAMPLE 3.10

Propane is burnt using a diffusion burner with a side length of 0.5 m in an open quiescent environment. The HRR is $\dot{Q} = 230\,\text{kW}$, and the mean flame height is $L_{fl} = 1.51$ m. If the radiative fraction of the HRR is $\chi_R = 0.3$,

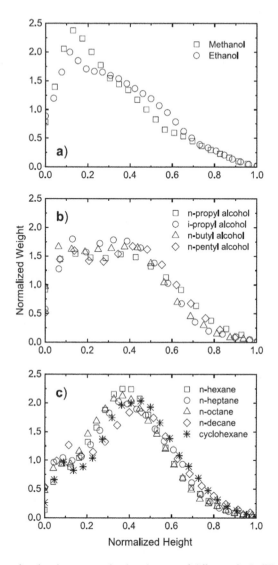

Figure 3.29 Normalized radiation weight distribution of different fuels (Zhou et al., 2017).

calculate the heat flux, $\dot{q}''$, incident on the ground 2 m away from the center of the burner.

SUGGESTED SOLUTION

We use the single-point source model in this calculation. If we let the center of the flame be the representative point of radiation, its height is 0.755 m.

Thus, we can calculate the distance, r, from the representative point to the target as follows:

$$r = \sqrt{0.755^2 + 2^2} = 2.14\,\text{m}$$

The cosine of the incidence angle, $\cos\phi$, can be obtained as:

$$\cos\phi = \frac{0.755}{2.14} = 0.353$$

By substituting these values into Eq. (3.102), we obtain $\dot{q}''$ as follows:

$$\dot{q}'' = \frac{0.3 \times 230}{4 \times 3.14 \times 2.14^2} \times 0.353 = 0.423\,\text{kW} \cdot \text{m}^{-2}$$

Chapter 4

Fire sources

Most combustibles in outdoor fires are solid. Due to pyrolysis under external heating, a solid combustible releases flammable gas into the gas phase. The flammable gas mixes and reacts with oxygen in the air, forming a flame above the solid combustible. In this book, a solid combustible and a flame formed above it are collectively called a fire source. Thermal energy generated during the combustion of pyrolyzed flammable gas is used in several ways. A fraction of it is used to heat the solid combustible to sustain its pyrolysis. It also raises the gas temperature near the fire source, forming a buoyancy-induced updraft. The rest is transferred to adjacent objects by radiation.

The characteristics of fire sources are a key factor in determining the fire spread behavior of outdoor fires. However, the fuel conditions vary significantly between wildlands, urban areas, and their interfaces. Vegetative fuels are predominant in wildlands. Their structure is generally stratified into canopy, surface, and ground fuel layers in the vertical direction, each exhibiting a different form of combustion. They are distributed either continuously or discretely in the horizontal direction. On the other hand, the primary fuels in urban areas are buildings and combustibles stored inside them, including some non-vegetative fuels. As a fire occurs in an enclosed space in urban areas, the burning behavior of combustibles is distinct from that of fires in the open environment but dominantly controlled by the rate of fresh air supply through openings.

This chapter first explains the fundamentals of solid fuel combustion as a common basis for evaluating the fire source characteristics in outdoor fires. We then describe fire sources that comprise wildland and urban fires in separate sections.

4.1 FORM OF COMBUSTION

Combustion is an oxidation reaction that occurs in a mixture of fuel and oxygen. The heat generation in the oxidation reaction raises the temperature

DOI: 10.1201/9781003096689-4

of the system, resulting in strong radiation emission perceived by our eyes as light. Rusting of iron is also caused by the oxidation reaction. However, the reaction rate is much slower, involving negligible temperature rise and light emission. Such a reaction is oxidation, but not combustion. The presence of fuel and oxygen is essential for combustion but is not sufficient alone. The onset of the combustion of a mixture of fuel and oxygen additionally requires some energy supply from an ignition source. Hence, fuel, oxygen, and an ignition source are the three elements of combustion. If any one of the elements is absent, the combustion can neither occur nor continue.

Among the three elements of combustion, oxygen is virtually inexhaustible in outdoor fires because the atmosphere is its supply source. In contrast, a wide variety of fuels exist in nature. Chemical elements other than inert ones can compound with oxygen. Chemical elements that do not generate heat when compounded with oxygen (nitrogen, chlorine, fluorine, and gold) and those that generate little heat (silver and mercury) cannot become fuels. Conversely, anything other than these can become fuel. Chemical compounds with more than two chemical elements can also become fuels, except those that have already been fully compounded with oxygen and have no capacity for further oxygen incorporation. Most organic compounds, such as natural vegetative materials and fossil fuel originated materials, are combustibles. There are various ignition sources in outdoor fires, either anthropogenic or non-anthropogenic. However, once combustion starts, the heat generation by the combustion itself becomes the ignition source, sustaining the combustion until the fuel supply ends.

The form of combustion changes depending on the state of the fuel, i.e., gaseous, liquid, or solid. In outdoor fires, primary fuels are solid vegetative materials, whether in wildlands, urban areas, or their interfaces. It may sound paradoxical, but there are only a few solid fuels whose combustion occurs in the solid state. Rather, flammable gaseous components released into the gas phase by evaporation or pyrolysis are generally burnt to form flames. We explain the combustion of gaseous fuels first, followed by the combustion of liquid and solid fuels.

4.1.1 Combustion of gaseous fuels

Gaseous fuels are not common fuels in outdoor fires. However, liquid and solid fuels generally burn in the gas phase after being gasified. Thus, the combustion characteristics of liquid and solid fuels have much in common with those of gaseous fuels. The combustion of gaseous fuels can be divided into two major types: premixed combustion and diffusion (non-premixed) combustion.

Combustion in which a flammable mixture (pre-mixture) of gaseous fuel and oxygen (or fresh air) is supplied to the reaction zone is called premixed combustion. In premixed combustion, the combustion that starts at the ignition source propagates by chain-igniting the surrounding flammable mixture that is not yet burning. Premixed combustion is not a common form of combustion in outdoor fires. Rather, it generally applies to mechanically controlled combustion, such as combustion in gas stoves or gasoline engines, from which we benefit daily. Premixed flames caused by the ignition of a stationary pre-mixture may even cause a sudden thermal expansion of flammable gas, resulting in a large pressure increase to further expand the surrounding gas (i.e., explosion). Premixed combustion is characterized by the formation of a reaction zone called the flame interface in the premixture, which propagates by its own heat release. Flame propagation cannot occur when the mixing ratio is too high or low. The range of mixing ratios over which the flame can propagate depends on the temperature, pressure, and direction in which the flame propagates.

In contrast to premixed combustion, non-premixed combustion is a form of combustion in which the fuel and oxygen, which exist separately, are transported to the reaction zone by diffusion. Non-premixed combustion is the most common form of combustion not only in outdoor fires but also in fires in general. A flame interface is also formed in non-premixed combustion, but unlike in premixed combustion, it does not propagate. As non-premixed combustion requires some time and distance to mix flammable gas and oxygen, the flame can be shortened or lengthened depending on the mixing condition.

In premixed combustion, the combustion is homogeneous because the mixing ratio of fuel and oxygen is uniform at all reaction points. In non-premixed combustion, the combustion is heterogeneous because the flammable gas mixture burns only when its status meets the required mixing ratio, temperature, and pressure conditions.

4.1.2 Combustion of liquid and solid fuels

With some exceptions, such as surface combustion of solid fuels, combustion of liquid and solid fuels in outdoor fires is generally non-premixed combustion. Once gasified, the combustion process of liquid and solid fuels is essentially the same as gaseous fuels. The gasification process of liquid and solid fuels makes the overall combustion process more complicated.

Figure 4.1 schematically illustrates the general combustion process of a liquid and solid combustible in a fire. The combustible heated by the flame gasifies either by evaporation or pyrolysis, releasing the flammable components into the gas phase. A flammable gas mixture is formed when they are mixed with the surrounding fresh air. When the mixture ratio of the flammable mixture is within its flammable range and an ignition source is present, ignition occurs in the simultaneous presence of all three elements of combustion.

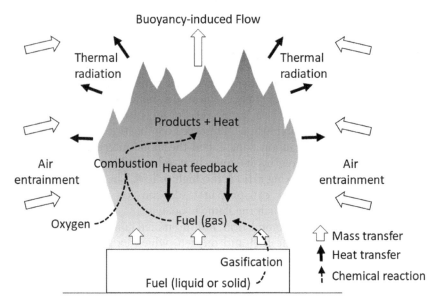

Figure 4.1 General burning process of liquid and non-charring solid fuels in fire (modified from Drysdale, 2011).

The flame heats and gasifies the fuel bed, thus forming a sustained cycle of the combustion process.

The gasification process of fuel is dependent on its type. Liquid fuels evaporate when their temperature exceeds the evaporation temperature due to heating. Solid fuels with relatively low melting points, such as wax and paraffinic hydrocarbons (hydrocarbons with a high number of carbons), melt when their temperature exceeds the melting temperature. The subsequent process is similar to that of liquid fuels. Other solid fuels, including vegetative fuels, thermally decompose (pyrolyze) upon heating. They similarly release gaseous pyrolysate into the gas phase. The difference between evaporation and pyrolysis is that the former is a physical process that involves a phase change. In contrast, the latter is a chemical process that involves an irreversible change in chemical composition (Rein, 2013). The gases produced by pyrolysis are a mixture of flammable gases, such as hydrogen (H_2), carbon monoxide (CO), and hydrocarbon (HC), and non-flammable gases, such as carbon dioxide (CO_2) and water vapor (H_2O). Unlike liquid fuels or solid fuels with a low melting point, a layer of carbon residue remains on the surface when burnt. The formation of a char layer complicates the overall combustion process because it generates heat due to glowing combustion on the surface while reducing the incident heat flux to the virgin layer due to the thermal insulation effect.

(1) Mass loss rate (mass burning rate)

When a liquid or solid fuel subject to heating gasifies, the weight of the original fuel decreases. The rate of decrease in weight is called the mass burning rate or mass loss rate (MLR), an important measure representing the fire condition. An Arrhenius-type reaction rate equation is often used to express the rate per unit area, $\dot{m}''_B$ (Quintiere, 2006; Drysdale, 2011; Turns, 2011):

$$\dot{m}''_B = A \exp\left(-\frac{E}{RT}\right) \tag{4.1}$$

where A and E are the fuel-dependent frequency factor and activation energy, respectively, R is the universal gas constant, and T is the temperature. Although the Arrhenius-type reaction rate equation expresses the essential features of phase transition and chemical reaction, it requires a separate calculation for solving T. Alternatively, $\dot{m}''_B$ is often calculated based on experimental observations that it is proportional to the incident heat flux to the fuel (Quintiere, 2006; Drysdale, 2011):

$$\dot{m}''_B = \frac{q''_{fl} + q''_{ex} - q''_{loss}}{L_G} \tag{4.2}$$

where q''_{fl} is the incident heat flux from the flame, q''_{ex} is the incident heat flux from the other heat sources, q''_{loss} is the heat loss due to radiation from the fuel surface, and L_G is the gasification energy required for the phase transition (the latent heat of gasification). If the numerator of Eq. (4.2) is greater than zero, i.e., the net incident heat flux to the fuel is greater than zero, flammable gas supply due to gasification is sustained until the combustible is fully consumed. Table 4.1 summarizes the q''_{loss} and L_G of various materials. If the fuel has a low melting point, the surface temperature remains approximately constant at the evaporation temperature when subject to heating. Thus, the change in q''_{loss} during the evaporation process is expected to be minor. However, for charring solid fuels, q''_{loss} changes with time as the thickness of the char layer increases. For this reason, the values shown in Table 4.1 should not be taken as definitive. L_G is generally smaller for liquids than for solids. However, only distilled water has an exceptionally large L_G. The high energy requirement for vaporization, high chemical stability, and high availability are the reasons for the wide use of water in fire suppression.

(2) Combustion of charring materials

Many solid fuels, including vegetative fuels, produce carbon residue and form a char layer on their surface upon pyrolysis. As shown in Figure 4.2,

Table 4.1 Surface re-radiation and heat of gasification of various materials (abstracted from Tewarson, 2002).

Materials	Surface re-radiation $q''_{loss}(kW \cdot m^{-2})$	Heat of gasification, $L_G(kJ \cdot g^{-1})$
Distilled water	0.63	2.58
Hexane	0.50	0.50
Heptane	0.63	0.55
Octane	0.98	0.60
Filter paper	10	3.6
Corrugated paper	10	2.2
Wood (Douglas fir)	10	1.8
Plywood/FR	10	1.0
Polypropylene	15	2.0
Polyethylene	15	1.8–2.3
Polymethylmethacrylate	11	1.6
Polycarbonate	11	2.1
Flexible polyurethane foam	16–19	1.2–2.7
Rigid polyurethane foam	14–22	1.2–5.3

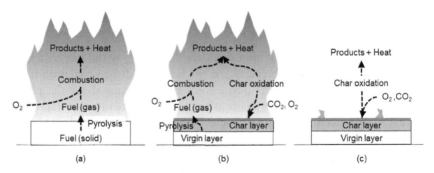

Figure 4.2 Burning phase transition of charring material. (a) Pyrolysis dominant phase (b) Transitional phase (c) Char oxidation dominant phase.

the combustion process of solid fuels forming a char layer can be divided into three successive phases.

In the first phase, the char layer is still thin. There is virtually no difference in the combustion behavior from non-charring materials. The temperature rise due to heating from the flame causes pyrolysis of the fuel. The yielded flammable gas mixes and reacts with oxygen in the gas phase to sustain the flame.

In the second phase, the effect of char layer formation starts to be exhibited. Char oxidation at the surface becomes a heat source to the virgin layer of the fuel, in addition to the heating from the flame. However, the pyrolysis rate of the fuel is dominated by the heating from the flame as char oxidation is inhibited due to the blockage of oxygen supply by the flame. At the same time, the char layer reduces heat transfer from the flame to the virgin layer. This is because the thermal conductivity of char is lower than for virgin materials due to its high porosity.

A transition to the third phase occurs when the combustion in the gas phase ceases. This is caused when the concentration of the flammable gas mixture falls out of the flammable range due to a reduction in the flammable gas supply. In the third phase, the oxidation reaction at the char surface becomes the dominant form of combustion. The intensity and duration of surface combustion depend on the fuel temperature and partial pressures of oxidants (e.g., O_2 and CO_2).

Char oxidation (surface combustion) is a common form of combustion in fuels containing few volatile components and composed almost entirely of carbon. Carbon residues yielded in pyrolysis are not the only examples; the burn of charcoal and coke also takes a form of surface combustion. In surface combustion, oxidants diffuse into internal pores of solid material to burn glowingly. As carbon has a very high melting point of over 3770 K, decomposition, melting, or evaporation do not occur before combustion. Accordingly, the fuel is not released into the gas phase, but rather the diffusion of oxidants into the solid dominates the combustion behavior. However, if the oxygen supply by diffusion becomes insufficient, the intermediate products (such as CO) produced by incomplete combustion may burn in the gas phase above the surface. The surface reaction rate, $\dot{m}_C''$, is often expressed in the form of MLR per unit area, which has the same dimension as the MLR by pyrolysis, $\dot{m}_B''$. If sufficient oxygen is present in the vicinity of the char surface, $\dot{m}_C''$ takes the form (Mizutani, 1989):

$$\dot{m}_C'' = p(O_2) \cdot A \exp\left(-\frac{E}{RT}\right) \tag{4.3}$$

If oxygen is scarcely present, oxidation by CO_2 takes place:

$$\dot{m}_C'' = p(CO_2) \cdot A \exp\left(-\frac{E}{RT}\right) \tag{4.4}$$

where p is the partial pressure near the char surface. In Eqs. (4.3) and (4.4), the pre-exponential factor, A, is determined by the type and structure of the carbon residue. Given that the higher the ambient temperature and the more intense the pyrolysis reaction, the more porous the char layer becomes, it should be noted that the value of A is affected by the fire environment in which the char layer is formed.

In Figure 4.2, the transition from pyrolysis-dominant combustion to char-oxidation-dominant combustion occurs when the flammable gas supply becomes too low to sustain non-premixed combustion in the gas phase.

However, fuels may burn without experiencing pyrolysis-dominant combustion, such as the case of smoldering combustion. Smoldering combustion is described as the slow, low-temperature, flameless burning of porous fuels (Rein, 2013). It occurs when live vegetative fuel or peat buried underground burns with a relatively weak heat release. It occurs when the mixture ratio of flammable gas in the gas phase is low, either due to a lack of flammable components in the solid fuel or the flushing of flammable gas out of the reaction zone by the airflow. Another condition is when incomplete combustion occurs upon a limited supply of oxygen relative to pyrolyzed flammable gas. However, if the conditions inhibiting the combustion of flammable gas are removed, smoldering combustion may shift to flaming combustion.

4.2 HEAT RELEASE IN COMBUSTION

In assessing the hazard of outdoor fires, it is often important to identify the heat release rate (HRR) at a given time rather than the total heat release (THR) of the fire. The HRR, $\dot{Q}_B$, is representative of the combustion intensity of the fire source that governs the behavior of the updraft formed above the fire source. $\dot{Q}_B$ generally takes the form:

$$\dot{Q}_B = \Delta H \left(\dot{m}_B'' A_B \right) \tag{4.5}$$

where ΔH is the heat of combustion, $\dot{m}_B''$ is the MLR per unit mass, and A_B is the burning area. Eq. (4.5) can also be used to calculate $\dot{Q}_B$ for surface combustion by substituting $\dot{m}_B''$ with the surface oxidation rate, $\dot{m}_C''$.

Given that combustion is an exothermic reaction that consumes oxygen, $\dot{Q}_B$ in Eq. (4.5) can alternatively be expressed as:

$$\dot{Q}_B = \Delta H_O \dot{m}_O \tag{4.6}$$

where ΔH_O is the heat of combustion for unit consumption of oxygen, and $\dot{m}_O$ is the consumption rate of oxygen. Further, given that oxygen required for the combustion in outdoor fires is exclusively supplied from fresh air and that the mass fraction of oxygen in fresh air is constant, Eq. (4.6) can also be expressed as:

$$\dot{Q}_B = \Delta H_A \dot{m}_A \tag{4.7}$$

where ΔH_A is the heat of combustion for unit consumption of fresh air, and $\dot{m}_A$ is the consumption rate of fresh air.

4.2.1 Heat of combustion

An elementary reaction is a chemical reaction in which one or more chemical species react directly to form products in a single reaction step and with a single transition state. Chemical reactions in combustion are composed of a number of elementary reactions. These chemical reactions eventually produce stable chemical compounds through which the excess energy is released as heat. For example, the chemical reaction of carbon (C) with oxygen (O_2) can be expressed as:

$$C + O_2 = CO_2 + 33.90 \text{ kJ} \cdot \text{g}^{-1} \tag{4.8}$$

The value on the right-hand side represents the heat release per gram of the combustion component reacted. The combustion reaction is exothermic, and the amount of heat generation is called the calorific value. It is the amount of heat extracted from the system after the complete combustion of the fuel until the system returns to the reference temperature. It generally refers to the value when combustion is started under atmospheric pressure (1 atm) and temperature (298.15 K). Due to a change in the number of molecules in the system during combustion, the calorific values are slightly different between combustions under a constant volume and pressure. In combustion under a constant volume, the calorific value, ΔH_V, is expressed as the change in the internal energy:

$$\Delta H_V = u_2 - u_1 \tag{4.9}$$

where u_1 and u_2 are the internal energy of the system before and after combustion, respectively. In combustion under constant pressure, the calorific value, ΔH_P, is expressed as the difference in enthalpies:

$$\Delta H_P = h_2 - h_1 \tag{4.10}$$

where h_1 and h_2 are the enthalpy of the system before and after combustion, respectively. Combustion in fires, especially outdoor fires, can generally be regarded as a chemical reaction under constant pressure. The amount of heat generation from the combustion of actual fuels can be measured using a fixed-volume container called the oxygen bomb calorimeter. Thus, the calorific value measured by the oxygen bomb calorimeter is equivalent to ΔH_V. ΔH_P can be obtained by adjusting the measured value of ΔH_V for the work done by the volume change. Theoretically, the two calorific values, ΔH_V and ΔH_P, should be distinguished. However, the difference is generally minor and is unlikely to be an issue from the practical point of view of fire hazard evaluation.

Most fuels that burn in fires contain hydrogen (H) as an element. When these fuels burn, water (H_2O) is produced as the combustion product. If H_2O is included in the products, the calorific value of the fuel can be treated in two different ways. One is to measure the calorific value after restoring the product system to the reference temperature. In this process, the latent heat of vaporization is recovered during the phase transition of the H_2O

from the gaseous state to the liquid state, thus resulting in a higher amount of heat generation. This is called the higher calorific value. Another is to measure the calorific value without restoring the product system to the reference temperature. In this process, the latent heat of vaporization is not recovered, thus resulting in a lower amount of heat generation. This is called the lower calorific value. In outdoor fires, HRR is calculated using lower calorific values because H_2O, included in the combustion products as water vapor, is transported in the gaseous state. Thus, lower calorific values are commonly called the heat of combustion.

Table 4.2 summarizes the heat of combustion of common fuels, in which ΔH, ΔH_O, and ΔH_A are the heat of combustion for the unit mass consumption of fuel, oxygen, and fresh air, respectively (Tewarson, 2002). Values of ΔH_A are calculated by multiplying the values of ΔH_O by 0.233, which is the mass fraction of oxygen in fresh air. Regarding the value of ΔH, there is a large variation between fuels. For example, the values of ΔH of normal alkanes, which are commonly used as fuels for heating appliances and internal combustion engines, are much higher than those of woods, which are common fuels in outdoor fires. The actual heat of combustion is smaller than the ideal heat of combustion for all fuels. This is due to the incompleteness of the combustion reaction. The greater difference in solid fuels relative to gaseous and liquid fuels is due to the complexity of the combustion reaction, which can be divided into the combustion of pyrolyzed gas in the gas phase and char oxidation on the fuel surface (Quintiere, 2006). However, interestingly, the values of ΔH_O (and ΔH_A) are generally constant irrespective of the fuel type. Thornton showed that this relationship applies to organic liquids and gases in 1917 (Quintiere, 2006). Huggett demonstrated that the relationship also applies to organic solids where the average, with a few exceptions, is $\Delta H_O \cong 13.1 \text{ kJ} \cdot \text{g}^{-1}$ (Huggett, 1980). This gives several important implications for the analysis of fire behavior. First, if we know the stoichiometric value of oxygen consumption in the combustion reaction of a fuel, we can estimate its heat of combustion. This is a technically valuable finding from a practical point of view that led to the development of important measurement techniques in fire experiments, such as the oxygen consumption method for HRR measurement. In addition, even if the fuel consists of more than one type of material, the HRR can be estimated with good accuracy if the amount of oxygen or fresh air consumed in combustion is available (Tanaka, 1993). This is particularly useful in analyzing the behavior of fires, in which the composition of fuels involved in combustion is generally unidentifiable.

Although Tewarson did not specify which part of the wood he used in the measurement of the heat of combustion in Table 4.2, it is presumable that they were lumber cuts from the trunk section for construction usage. In contrast, in a series of experiments by Etlinger et al. in which entire trees of six different species were burnt, the foliage was reportedly the primary

Table 4.2 Net heats of complete combustion per unit mass of fuel and oxygen consumed (abstracted from Tewarson, 2002).

Fuel	Formula	Heat of combustion ΔH (kJ·g⁻¹)		Heat of combustion per unit mass of oxygen ΔH_O (kJ·g⁻¹)	Heat of combustion per unit mass of air ΔH_A (kJ·g⁻¹)*
		Ideal	Actual		
Normal alkanes					
Methane	CH_4	50.1	49.6	12.5	2.91
Ethane	C_2H_6	47.1	45.7	12.7	2.96
Propane	C_3H_8	46.0	43.7	12.9	3.01
Heptane	C_7H_{16}	44.6	41.2	12.7	2.96
Kerosene	$C_{14}H_{30}$	44.1	40.3	12.7	2.96
Alcohols					
Methyl alcohol	CH_4O	20.0	19.1	13.4	3.12
Ethyl alcohol	C_2H_6O	27.7	25.6	13.2	3.08
Isopropyl alcohol	C_3H_8O	31.8	29.0	13.3	3.10
Polymeric materials					
Polymethylmethacrylate	$(CH_{1.6}O_{0.4})_n$	25.2	24.2	13.1	3.05
Red oak	$(CH_{1.7}O_{0.72}N_{0.001})_n$	17.1	12.4	13.2	3.08
Douglas fir	$(CH_{1.7}O_{0.74}N_{0.002})_n$	16.4	13.0	12.4	2.89
Flexible polyurethane foam (GM21)	$(CH_{1.8}O_{0.30}N_{0.05})_n$	26.2	17.8	12.1	2.82
Flexible polyurethane foam (GM29)	$(CH_{1.1}O_{0.23}N_{0.10})_n$	26.0	16.4	12.6	2.94

Table 4.3 Net heats of complete combustion per unit mass of wildland fuels at constant pressure (abstracted from de Dios Rivera et al., 2012).

| Fuel | Heat of combustion ΔH (kJ · g⁻¹) | |
	Dry base	Dry, ash-free base
Herbaceous	16.674	18.354
Hardwood	17.524	18.104
Shrub	18.465	19.145
Softwood	19.425	20.204

combustible with the average heat of combustion of $\Delta H = 17.3$ kJ·g⁻¹ (Etlinger et al., 2004). The similar measurement results imply that the heat of combustion of wood does not vary significantly by species or part. de Dios Rivera et al. summarized values of the net heat of combustion of wildland fuels at constant pressure measured by various investigators, as shown in Table 4.3 (de Dios Rivera et al., 2012). The heats of combustion in Table 4.3 are those of fuels without water and water-ash contents. The latter values are larger than the former values. Increased ash contents reduce flammability due to lower concentrations of combustible organic matter and inhibit the formation of volatile compounds by pyrolysis (de Dios Rivera et al., 2012).

4.2.2 Estimation of the heat of combustion using chemical formulae

The heats of combustion of materials summarized in Table 4.2 were measured by the oxygen bomb calorimeter. However, there are various other fuels that are expected in fires, but the measured heats of combustion are not necessarily available. Such heat of combustion of materials can be estimated from their chemical formula if allowing for a certain margin of error. Combustible materials involved in a fire are mostly organic compounds from elements such as carbon (C), hydrogen (H), and oxygen (O). Let us take the combustion reaction of methane (CH_4) as an example. The overall reaction equation for the complete combustion of CH_4 can be expressed as follows:

$$CH_4 + 2O_2 \rightarrow CO_2 + 2H_2O \tag{4.11}$$

This equation shows that 1 mole of CH_4 reacts with 2 moles of O_2 to produce 1 mole of CO_2 and 2 moles of H_2O. The mixture ratio based on the mass is expressed as the oxygen-to-fuel ratio, r. In particular, the oxygen-to-fuel ratio required for complete combustion is called the stoichiometric oxygen-to-fuel ratio, r_{st}. r_{st} can be regarded as the minimum oxygen required for the complete combustion of a unit mass of fuel. As the atomic weights are conserved in the overall reaction equation, the mass ratio of CH_4 and O_2 required for complete combustion should be $CH_4 : O_2 = (12 \times 1 + 1 \times 4) : 2(16 \times 2) = 1 : 4$ given that the atomic weights are 12 for C, 1 for H, and 16 for O. In other words, the stoichiometric oxygen-to-fuel ratio for CH_4, which is r_{st,CH_4}, can be given by:

$$r_{st,CH_4} = \frac{\text{mass of oxygen}}{\text{mass of fuel}} = 4 \qquad (4.12)$$

Recall that Eqs. (4.5) and (4.6) calculate the HRR, $\dot{Q}_B$, focusing on the consumption rates of fuel and oxygen, respectively. As they express the same combustion reaction, the $\dot{Q}_B$ obtained by both equations should yield an identical result:

$$\dot{Q}_B = \Delta H \left(\dot{m}_B'' A_B \right) = \Delta H_O \dot{m}_O \qquad (4.13)$$

Referring to the definition, r_{st} can also be given as the ratio of the mass consumption rate of oxygen, $\dot{m}_O$, relative to that of fuel (mass burning rate), $\dot{m}_B'' A_B$, which takes the form:

$$r_{st} = \frac{\dot{m}_O}{\dot{m}_B'' A_B} \qquad (4.14)$$

Substituting this into Eq. (4.13), the relationship between the heat of combustion of fuel, ΔH, and the heat of combustion per unit mass of oxygen consumed, ΔH_O, can be expressed as:

$$\Delta H = r_{st} \cdot \Delta H_O \qquad (4.15)$$

Substituting $r_{st,CH_4} = 4$ and $\Delta H_O \cong 13.1 \ kJ \cdot g^{-1}$ into Eq. (4.15), we can estimate the heat of combustion of CH_4 as follows:

$$\Delta H_{CH_4} \cong 4 \times 13.1 = 52.4 \ kJ \cdot g^{-1} \qquad (4.16)$$

The measured value of CH_4 shown in Table 4.2 is $50.1 \ kJ \cdot g^{-1}$, the error of which is 4.4% as to the estimated value. The error could be attributed to the difficulty in reproducing complete combustion even inside an oxygen bomb calorimeter. The combustion reaction is a complex chain reaction comprising many elementary processes involving various molecules and free radicals, even for a single substance such as CH_4 (Turns, 2011).

The above estimation procedure can be used when the chemical formula of the fuel is known, and its complete combustion can be expressed as an

Table 4.4 Ultimate analysis of dry and ash-free wood (weight %) (Ragland et al., 1991).

Wood type	C	H	O	N	S
Hardwoods (average of 11 hardwoods)	50.2	6.2	43.2	0.1	-
Softwoods (average of 9 softwoods)	52.7	6.3	40.8	0.2	0.0
Oak bark	52.6	5.7	41.5	0.1	0.1
Pine bark	54.9	5.8	39.0	0.2	0.1

overall reaction equation. Such requirements can be met only by pure gaseous or liquid fuels. In general, combustibles that burn in outdoor fires are natural or synthetic organic solid compounds with no clear chemical formula available. However, even if the chemical formula is unknown, mass fractions of the major constituent elements such as carbon (C), hydrogen (H), oxygen (O), and nitrogen (N) are available by the ultimate analysis, as shown in Table 4.4. The elemental composition of organic compounds can be used to estimate the r_{st}, which can be further converted to the ΔH by substituting it into Eq. (4.15). Some combustibles contain sulfur (S) or chlorine (Cl) as their constituent, but their effect can be ignored as their mass fraction is generally small. In such a case, the overall chemical reaction equation for the complete combustion of organic compounds can be expressed in a general form as follows:

$$C_a H_b O_c N_d + \left(a + \frac{b}{4} - \frac{c}{2}\right) O_2 \rightarrow a CO_2 + \frac{b}{2} H_2 O + \frac{d}{2} N_2 \qquad (4.17)$$

Based on the conservational relationship of atomic weight in this equation, the mass ratio of the fuel $C_a H_b O_c N_d$ versus oxygen (O_2) in complete combustion is $C_a H_b O_c N_d : O_2 = (12a + b + 16c + 14d) : 32(a + b/4 - c/2)$. Thus, r_{st} can be calculated from the following equation:

$$r_{st} = \frac{\text{mass of oxygen}}{\text{mass of fuel}} = \frac{32(a + b/4 - c/2)}{12a + b + 16c + 14d} \qquad (4.18)$$

Given that the mass fraction of oxygen in fresh air is constant (0.233 for dry air), the air-to-fuel ratio can be used as an alternative to the oxygen-to-fuel ratio when the oxidant for combustion is from fresh air. In such a case, $r_{st,A}$ can be calculated by dividing r_{st} by the mass fraction of oxygen as follows:

$$r_{st,A} = \frac{\text{mass of fresh air}}{\text{mass of fuel}} = \frac{r_{st}}{0.233} \qquad (4.19)$$

WORKED EXAMPLE 4.1

Calculate the heat of combustion, ΔH, for hardwoods and softwoods using the elemental compositions shown in Table 4.4.

SUGGESTED SOLUTION

Substituting the molar fractions of carbon (C), hydrogen (H), oxygen (O), and nitrogen (N) shown in Table 4.4 into Eq. (4.18), r_{st} is obtained as follows:

$$\text{Hardwoods}: r_{st} = \frac{32\left(\dfrac{50.2}{12} + \dfrac{6.2}{4} - \dfrac{43.2}{2 \times 16}\right)}{12 \times \dfrac{50.2}{12} + 6.2 + 16 \times \dfrac{43.2}{16} + 14 \times 0.1} = 1.41 \text{ kg}$$

$$\text{Softwoods}: r_{st} = \frac{32\left(\dfrac{52.7}{12} + \dfrac{6.3}{4} - \dfrac{40.8}{2 \times 16}\right)}{12 \times \dfrac{52.7}{12} + 6.3 + 16 \times \dfrac{40.8}{16} + 14 \times 0.2} = 1.50 \text{ kg}$$

The results indicate that the mass of oxygen required for the complete combustion of a unit mass of wood is 1.41 kg for hardwoods and 1.50 kg for softwoods. By multiplying the obtained r_{st} by $\Delta H_O \cong 13.1 \text{ kJ} \cdot \text{g}^{-1}$ following Eq. (4.15), we obtain:

$$\text{Hardwoods}: \Delta H = 18.4 \text{ kJ} \cdot \text{g}^{-1}$$

$$\text{Softwoods}: \Delta H = 19.7 \text{ kJ} \cdot \text{g}^{-1}$$

As shown in Table 4.4, the molar fraction of carbon is higher, and that of oxygen is lower in softwoods than in hardwoods. This is the reason why the ΔH of softwoods is higher than that of hardwoods.

4.2.3 Estimation of the heat of combustion during incomplete combustion using equivalent ratio

The overall reaction described in Eq. (4.17) assumes the complete combustion of fuels with oxygen. However, complete combustion rarely occurs in a fire environment due to various possible causes, such as insufficient oxygen supply or a partial drop in temperature. Incomplete combustion often becomes problematic from the viewpoint of toxicity and corrosiveness of combustion products. If a fire occurs in a building where the supply of fresh air is limited, incomplete combustion can easily occur, causing harmful effects on occupants and firefighters. One may consider that it is unlikely to occur in outdoor fires as there is an abundant supply of fresh air. However, sooty smoke observed in outdoor fires is nothing but a sign of its occurrence. The far-reaching effect of air pollution in outdoor fires is becoming a subject of increasing concern. On the contrary, incomplete combustion accompanies a reduction in the effective heat of combustion of fuels, ΔH_{eff}, which virtually leads to a reduction in the HRR. Thus, if we assume complete combustion for the evaluation of the HRR, we can avoid an underestimation of the hazard of fire spread in outdoor fires. However, a correct understanding of an actual fire environment is always necessary to maintain the rationality of evaluation.

The effect of incomplete combustion on the net heat of combustion of a flammable gas mixture can be discussed in terms of the equivalence ratio, ϕ. ϕ is defined as the ratio of the stoichiometric oxygen-to-fuel ratio, r_{st}, to the actual oxygen-to-fuel ratio, r, which is given by:

$$\phi = \frac{r_{st}}{r} = \frac{(O/F)_{st}}{O/F} \tag{4.20}$$

When $\phi = 1$, complete combustion under the stoichiometric mixing ratio occurs. $\phi < 1$ represents a fuel-lean state, where there is less fuel (F) compared to oxygen (O). $\phi > 1$ represents a fuel-rich state, where there is more fuel (F) compared to oxygen (O). The change in the ΔH_{eff} associated with the incompleteness of combustion can be expressed in the form of combustion efficiency, ζ. ζ is the ratio of the heat of incomplete combustion to that of complete combustion given by:

$$\zeta = \frac{\Delta H_{eff}}{\Delta H} \tag{4.21}$$

Tewarson et al. conducted combustion experiments using six different polymeric materials. They obtained a relationship between the ζ and ϕ as shown in Figure 4.3. When $\phi \leq 1$, ζ is approximately unity, but when $\phi > 1$, ζ decreases monotonically as ϕ increases. Such a relationship is common for all six types of polymeric materials studied. A regression equation of the data points is obtained as follows (Tewarson et al., 1993):

$$\zeta = \frac{\Delta H_{eff}}{\Delta H} = 1 - \frac{0.97}{\exp(2.5\phi^{-1.2})} \tag{4.22}$$

Figure 4.3 also shows that combustion in the gas phase ceases when $\phi > 4$. However, ζ is still greater than zero. This is due to the surface oxidation of the solid fuel that continued after the cessation of the gas phase combustion.

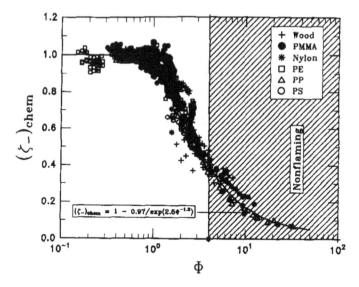

Figure 4.3 Ratio of the combustion efficiency to the equivalence ratio (Tewarson et al., 1993).

WORKED EXAMPLE 4.2

Calculate the ΔH_{eff} of red oak when $\phi = 2$. Use the value of ΔH (ideal) as shown in Table 4.2.

SUGGESTED SOLUTION

According to Eq. (4.22), the combustion efficiency is $\zeta = 0.673$ at $\phi = 2$. Multiplying this by $\Delta H = 17.1$ kJ·g^{-1}, we obtain $\Delta H_{eff} = 11.5$ kJ·g^{-1}.

4.3 FIRE SOURCE IN WILDLAND FIRES

Wildland fires involve the simultaneous combustion of combustibles of different combustibility. If burning combustibles are located close to each other, they would burn more intensely in an integrated manner. The augmentation effect on the burning intensity due to the integrated combustion can be quantified based on the geometry and configuration of burning combustibles, in addition to their combustibility. However, it is not straightforward to generalize such an effect, given the large variability in the underlying conditions. Thus, for quantitative evaluation of outdoor fire behavior, a collection of diverse plants and trees is often described as a homogeneous distribution of combustibles with unique combustibility.

4.3.1 Type and structure of fuels in wildland fires

Fuels in wildland fires are mainly live and dead vegetation in forests. Attributes such as species composition, morphology, and function of vegetative fuels in forests vary significantly between regions. However, the characteristics that have a particularly large impact on fire behavior are the spatial distribution and physical structure of the fuels. Thus, it is possible to classify the characteristics of forest fuels from this perspective (Sullivan et al., 2012). Although there are differences among investigators in the way they describe the structure of wildland fuels, they share a common approach of categorizing the fuels into a stratum of layers in the vertical direction (Pyne et al., 1996; Nelson, 2001; Gould et al., 2008; Sullivan et al., 2012; Penney et al., 2020). In this section, we divide the fuel structure into three hierarchical layers: ground fuel, surface fuel, and canopy fuel layers, as schematically shown in Figure 4.4.

The ground fuel layer is the lowest layer of the stratum comprising materials such as duff, root, and rotten buried logs. In some cases, underground peat is also included in this layer. Duff is the decomposed organic compounds comprising needles, leaves, twigs, branches, bark, and other castoff vegetation materials on the mineral soil that have begun to decompose (Pyne et al., 1996). The ground fuel layer has higher water retention and humidity than the other layers. It is often densely compacted and has less exposure to the air. For this reason, the form of combustion that often

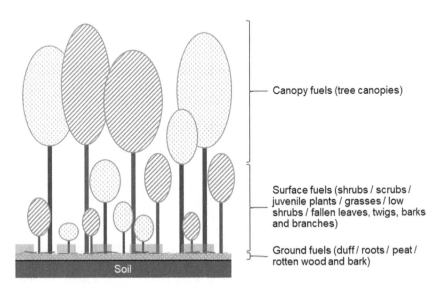

Figure 4.4 Fuel components are categorized according to ground, surface, and crown fuels in the wildland.

occurs in the ground fuel layer is smoldering combustion that spreads slowly within the layer.

The canopy fuel layer is the top layer of the stratum comprising leaves, twigs, and branches as the primary fuels. Burn of canopy fuels often leads to extensive fire damage. When flames reach the canopy layer, rapid fire growth may occur, resulting in the formation of elongated flames and the generation of numerous firebrands that are further dispersed downwind. Among the fuels belonging to the canopy layer, the primary fuel involved in wildland fires is the canopy of coniferous trees. Although the canopy of deciduous trees can also be regarded as combustibles in the canopy layer, the possibility of fire spread is relatively low as they generally grow at larger spacings between each other and have a higher moisture content than coniferous trees (Nelson, 2001). However, there are exceptions. Fires can occur in deciduous forests during dry and windy seasons when their moisture content drops. Fires can even become vigorous in deciduous forests with a high composition of flammable components, such as *Eucalyptus* trees (Nelson, 2001). It is noteworthy that when the canopy fuel layer is on fire, the trunk portion of trees burns only partially and rarely burns out. This is because the large volume relative to the exposed surface area of a trunk prevents thermal penetration of external heating into the body and reduction of its moisture content.

The surface fuel layer is the layer of fuel between the ground fuel and canopy fuel layers, which comprises even more diverse types of fuel. The lowest layer of the surface fuel layer comprises horizontally layered fallen leaves, twigs, barks, and branches that are have been on the ground for a relatively short time. These are called dead fuels as they are decoupled from plant growth. These dead fuels account for a large portion of the total combustible volume in wildlands. They are often primarily responsible for the fire spread behavior, including the flame spread rate and flame thickness (Sullivan et al., 2012). In contrast, live fuels comprise grasses, shrubs, scrubs, and juvenile understory plants. Compared to dead fuels, live fuel components include a mixture of orientations from horizontal to vertical (Penney et al., 2020). The burn of surface fuels can bridge a fire from the bottom to the upper canopy layer and thus accelerating its momentum.

Fuel complexes vary from place to place. They are rarely continuous and uniform over a wide area. It may be a herbaceous meadow dominated by grasses, or a dense forest dominated by coniferous trees. The habitat type also varies considerably by region and continent (Brown et al., 1986; Cruz et al., 2003; Reinhardt et al., 2006; Mitsopoulos et al., 2007; Küçük et al., 2008; Keane et al., 2012; Sullivan et al., 2012). The composition of wildland fuels can be classified according to their burn characteristics, an example of which is shown in Table 4.5 (Keane et al., 2012). In Table 4.5, wildland fuels are classified into two categories: the canopy fuels located above 2 m

Table 4.5 An example description of the canopy and surface fuel characteristics (Keane et al., 2012).

General fuel type	Fuel component	Size	Fuel characteristics
Canopy fuels			
Canopy	Aerial fuels > 2 m above ground surface	All dead and live biomass less than 3 mm in diameter	• Canopy fuel load (CFL) (kg/m²) • Canopy bulk density (CBD) (kg/m³)
	All material > 2 m above ground	All canopy biomass	• Canopy cover (CC) (%)
Surface fuels			
Downed dead woody	1-hour woody (twigs) 10-hour woody (branches) 100-hour woody (large branches) 1000-hour woody (logs)	< 1 cm diameter 1–2.5 cm diameter 2.5–7.0 cm diameter > 7.0 cm diameter	• Fuel load (kg/m²) • Particle density/ bulk density (kg/m³) • Mineral content (%)
Shrubs	All burnable shrubby biomass	< 5 cm	
Herbs	All live and dead grass, forb, and fern biomass	All sizes	
Duff	Partially decomposed biomass whose origins cannot be determined	All sizes	
Litter	Freshly fallen, non-woody material that includes leaves, cones, and pollen	All sizes	

from the ground, and the surface fuels between the ground and the canopy fuels. The ground fuels are included in the surface fuels in this example.

For the canopy fuels, the fuel characteristics are summarized in terms of the canopy fuel load (CFL), canopy bulk density (CBD), and canopy cover (CC). CFL is defined as the mass per unit area of fuel, an indicator related to the THR and burn duration. In Table 4.5, particles larger than 3 mm are not included in this classification, following the idea that they are rarely involved in the combustion in wildland fires (Keane et al., 2012). However, the maximum dimensions of fuels contributing to combustion are controversial (Stocks et al., 2004), as they depend on the burning intensity and duration of a fire. Thus, the value of 3 mm is not a definitive threshold, and there are cases where other thresholds have been adopted. CBD is defined as the mass per unit volume of fuel. As the CBD is typically distributed in the vertical direction, the maximum value of CBDs obtained at a certain interval is adopted as the representative value. CC is a measure of canopy density, defined as the ratio of the projected area of the forest canopy to the area of the ground. CC indicates the likelihood of fire spreading between trees in canopy fires. It is also an important indicator for understanding changes in the humidity of surface fuels as it affects the transfer of solar radiation to the ground. An indicator of the canopy fuels not included in Table 4.5 is the canopy base height (CBH). CBH is the vertical distance between the ground and live canopy fuel layers. It is an important indicator to evaluate the possibility of a fire in the surface fuel layer spreading to the canopy fuel layer (Van Wagner, 1977).

In contrast, indicators for the surface fuels are grouped by the fuel type, such as downed dead woody, shrubs, herbaceous, duff, and litter. Among these groups, downed dead woody is often divided into four diameter classes, i.e., 1-hour, 10-hours, 100-hours, and 1000-hours, based on its drying time (Fosberg, 1970). In Table 4.5, these four categories correspond to particles with diameters of less than 1 cm, 1–2.5 cm, 2.5–7.0 cm, and greater than 7.0 cm, respectively. Indicators adopted to describe the characteristics of each fuel group include the fuel load (mass per unit area), and particle density (mass per unit volume for downed dead woody components) or bulk density (mass per unit volume for shrubs, herbs, duff, and litter). The difference between surface and canopy fuels is that the mineral content can be considered as an additional indicator representing the condition of the former. Mineral content is the ratio of inorganic biomass in the downed fuel components; the higher the ratio, the greater the retardation effect of combustion (Rothermel, 1972).

Figure 4.5 summarizes the results of a survey of wildland fuel distributions in six selected sites in the northern Rocky Mountain ecosystems (Keane et al., 2012) using the indicators listed in Table 4.5. Note that litter and duff are combined in Figure 4.5 due to difficulty distinguishing them in actual wildland sites (Keane et al., 2012). The fuel complexes in the six surveyed sites are significantly different. Site A is the forest composite of ponderosa pine, Douglas fir, and western larch with dominant fuel of partially decomposed light thinning slash. Site B is the lodgepole pine forest with dominant fuel of low-live shrubs and scattered downed woody particles. Site C is the forest composite of ponderosa pine and Douglas fir with dominant fuel of grass and widely scattered thinning slash. Site D is grassland dominantly covered with sagebrush. Site E is the forest composite of pinyon pine and juniper with dominant patchy, light herbaceous fuels. Site F is the ponderosa pine savanna with dominant fuel of grass and scattered woody particles (Keane et al., 2012). As is evident in Figure 4.5, the composition of wildland fuels varies by location in terms of both quality and quantity. However, the proportion of litter and duff to the total fuel load is commonly large in all survey sites, suggesting their importance in evaluating the hazard of wildland fires. It is also important to note that the amount of combustibles is not the sole determinant of the fire behavior but also depends on the fuel composition.

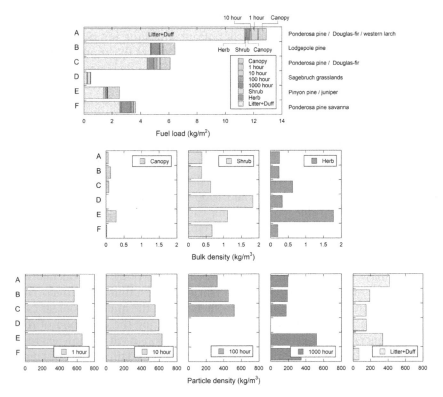

Figure 4.5 Examples of fuel characteristics (fuel load, bulk density, and particle density) in selected ecosystems in western North America (modified from Keane et al., 2012).

4.3.2 Transient burning process of individual fuel components

There are two main approaches to modeling the propagation of flames through a combustible space. One is to consider the target combustible space comprising a set of independent combustibles. The growth and decay of fires of individual fuels are predicted, and the results are aggregated over the entire target combustible space to draw a complete picture of the fire. This approach can be termed as discrete fuel modeling. The other ignores the unevenness in the spatial distribution of combustible characteristics due to the variability in the shape and combustibility of individual combustibles. The flame propagation is modeled as the movement of the leading edge of the burning region in a continuously distributed homogeneous fuel bed. This approach can be termed as continuous fuel modeling. Historically, the latter approach has often been adopted to analyze wildland fires. However, this subsection describes the characteristics of fire sources from the viewpoint of discrete fuel modeling.

As summarized in the previous subsection, various fuels are involved in combustion in wildland fires. The burning behavior of combustibles differs from one to another, as it depends not only on the shape and combustibility at an individual level but also is thermally affected by the surrounding burning combustibles. Thus, the burning behavior of a group of combustibles differs from that of a single combustible. Nevertheless, understanding the behavior of an individual combustible that burns independently is a premise for obtaining an overall picture of wildland fires. The hazard of a burning combustible can be typically represented by the geometry and temperature rise of the flame or fire plume formed above it. The MLR or HRR is the dominant factor in determining such burning behavior of combustibles.

To analyze the combustibility of vegetative fuels, investigators have often focused on the combustibility of extracted fuel components from canopy and surface fuels. There are relatively few examples of the measurement of the MLR or HRR by burning a whole plant in an upright position (Etlinger et al., 2004; Evans et al., 2004; Mell et al., 2009; Li et al., 2017; Zhou et al., 2018; Morandini et al., 2019). Figure 4.6 shows the mass loss history of whole Douglas fir trees involved in fire (Mell et al., 2009). The specimens were approximately 2 m high. At the beginning of the experiment, the average moisture content was 49% for the three specimens in the upper panel and 14% for the six specimens in the lower panel. The MLR increased almost linearly after ignition in both cases of moisture content. They began to decrease at a similar rate as they increased immediately after reaching their peak. When the average moisture content was 14%, the MLR was almost zero after the burnout of needles. This indicates that the heat generation by the combustion of needles was insufficient to cause a continuous burn of branches and trunks that were relatively large in diameter.

There are two main reasons for the absence of the quasi-steady phase in the burning of trees, which would be observed in enclosure fires. First, the main component of a tree involved in combustion is needles, which have a large surface-area-to-volume ratio. This implies that the Biot number of needles is low; material with a low Biot number burns up intensely as it quickly reaches its pyrolysis temperature when subject to external heating (refer Chapter 7). Second, fuels are exposed to the open air in outdoor fires, providing virtually unlimited fresh air for combustion. Thus, flammable gases released into the gas phase by pyrolysis react with oxygen in the air without time delay. At the same time, a thermal environment that could intensify the combustion of fuels and retain the generated heat is unlikely to be formed without an enclosure.

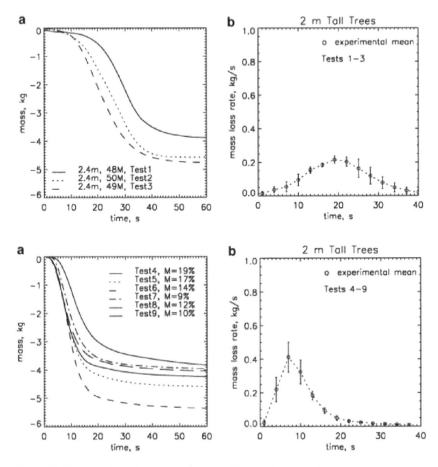

Figure 4.6 Results of the burn experiment of 2 m tall trees with an average moisture content of 49% in the top and 14% in the bottom. Mass loss history curves are on the left (a), and average MLRs are on the right, with error bars indicating standard deviations (b) (Mell et al., 2009).

The fire duration of the Douglas fir trees was approximately 40 s and 20 s when the average moisture content was 49% and 14%, respectively. The difference indicates a significant moisture content effect on fire duration. This is because the combustion of vegetative fuels undergoes two stages of endothermic reactions: the evaporation of water and the pyrolysis of organic compounds. Thus, fuels with higher moisture content take longer to complete the evaporation before shifting to the pyrolysis phase. It is noteworthy that the simultaneous progress of water evaporation and pyrolysis could occur, which diffuses the flammable pyrolyzed gas and suppress the exothermic combustion reaction (Albini, 1980). This is an additional suppression effect of combustion that may cause a difference in the fire duration when the fuel moisture content is high.

Given that burn of vegetative fuels in the open air does not undergo a quasi-steady combustion mode, the peak HRR, $\dot{Q}_{B,peak}$, is the key parameter in evaluating the hazard of a fire source. Babrauskas derived the following regression equation for $\dot{Q}_{B,peak}$ based on various experimental results for Douglas fir (Babrauskas, 2002a):

$$\frac{\dot{Q}_{B,peak}}{W} = \exp(5.84 - 0.017\phi) \tag{4.23}$$

where W is the mass of the tree, and ϕ is the moisture of needles as a percentage (%).

Fire duration is another important factor in evaluating the hazard of a fire source. However, the MLR of an individual plant is generally unsteady, as shown in Figure 4.6. A practical approach to evaluating the fire duration is to assume a burn scenario. We approximate the time variation of MLR by an isosceles triangle with its top angle at the peak HRR point. In such a case, the THR can be approximated by the area of the triangle:

$$\Delta H W_F \cong \frac{1}{2} \dot{Q}_{B,peak} t_{fire} \tag{4.24}$$

where ΔH is the heat of combustion, $W_F (\leq W)$ is the weight of burnable fuel, and t_{fire} is the fire duration. Eq. (4.24) can be transformed to derive an expression for t_{fire} as follows:

$$t_{fire} \cong \frac{2\Delta H W_F}{\dot{Q}_{B,peak}} \tag{4.25}$$

If the HRR can be determined according to the above relationships, the flame geometry can be estimated using the equations given in Chapters 5 and 6. The flame geometry can be used to evaluate the heat transfer from the flame to its surrounding fuels. This allows us to determine the time to ignition of the heat receiving fuels in response to their thermal properties.

WORKED EXAMPLE 4.3

Calculate the t_{fire} of a Douglas fir tree with a total weight $W = 10\,\text{kg}$, a burnable weight $W_F = 3\,\text{kg}$, and moisture content $\phi = 20\%$.

SUGGESTED SOLUTION

The $\dot{Q}_{B,peak}$ of fire involving a Douglas fir tree can be calculated using Eq. (4.23):

$$\dot{Q}_{B,peak} = 10 \times \exp(5.84 - 0.017 \times 20) = 2446.9 \text{ kW}$$

t_{fire} can be obtained by substituting $\dot{Q}_{B,peak}$ into Eq. (4.25). According to Table 4.2, the ideal heat of combustion of Douglas fir is $\Delta H = 16.4 \text{ kJ} \cdot \text{g}^{-1}$. However, as the use of the actual value is recommended (Babrauskas, 2002a), we use $\Delta H = 13.0 \text{ kJ} \cdot \text{g}^{-1}$ instead. By substituting these values in Eq. (4.25), we obtain t_{fire} as follows:

$$t_{fire} \cong \frac{2 \times (13.0 \times 1000) \times 3}{2446.9} = 31.9 \text{ s}$$

4.3.3 Combustion of homogeneous porous fuel

A continuously distributed homogeneous fuel is a hypothetical form of fuel distribution that does not exist in actual wildlands in a strict sense. Actual wildland fuels are a group of fuels of different morphology and combustibility. Overall fire behavior of such a group of fuels can be evaluated by aggregating that of individual fuels. However, if they are densely assembled and the unevenness of fuel characteristics is ignored, they can be assumed as a continuously distributed homogeneous fuel. Such an assumption enables a flexible treatment in the theoretical analysis of fire spread. In fact, this has been a successful representation of actual surface and ground fuel distributions in wildlands. It would even be possible to apply this approximation to some crown fuels, depending on the spacing between individual plants.

A fire that starts in a continuously distributed homogeneous fuel will spread concentrically if the effects of topography and wind are negligible. As the fuel burns out from the part ignited earlier, a line-shaped or belt-shaped burning zone is formed at the fire front. Such a fire source can be referred to as a line fire source. Figure 4.7 shows the result of MLR and HRR measurements during a fire spread through a designed continuous fuel bed of pine needles (2 m in the spreading direction and 1 m in its orthogonal direction) under no slope and no wind conditions. This figure compares the results in two fuel load conditions, 0.6 and 1.2 $\text{kg} \cdot \text{m}^{-2}$. A higher fuel load resulted in a higher HRR, $\dot{Q}_B$, and a shorter fire duration, t_{fire}. However, in starting the experiment, the shorter side of the rectangular fuel arrangement was ignited at once to simulate a line fire source. Therefore, the measured $\dot{m}_B$ and $\dot{Q}_B$ shown in Figure 4.7 do not represent the burning behavior of the entire fuel bed when fully covered by flame, but that of the line fire

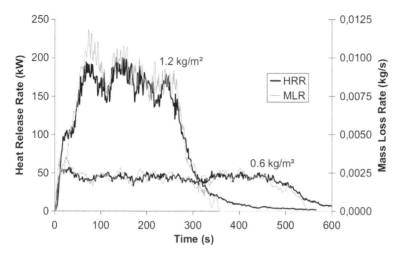

Figure 4.7 MLR and HRR during a fire spread through a fuel bed of pine needles under no slope and no wind conditions (Morandini et al., 2013).

source starting from one end and approaching the other end. On the other hand, the measured t_{fire} does not represent the burning behavior of the line fire source, but that of the entire fuel bed depending on the size of the rectangular fuel arrangement. Thus, the measurement results of the continuously distributed homogeneous fuel bed in Figure 4.7 need to be interpreted differently from those of independent fuels in Figure 4.6.

The HRR per unit length, $\dot{Q}_B'$, is often used to describe the burning behavior of a line fire source, which takes the form:

$$\dot{Q}_B' = \Delta H \cdot \dot{m}_B' = \Delta H \cdot \dot{m}_B'' d_B \qquad (4.26)$$

where $\dot{m}_B'$ and $\dot{m}_B''$ are the MLR per unit length and area, respectively, and d_B is the depth of the line fire source in the propagating direction. $\dot{m}_B'$ or $\dot{Q}_B'$ can also be used to evaluate the flame length of a line fire source by substituting them into the correlations given in Chapters 5 or 6. However, it is not straightforward to directly measure the values of $\dot{m}_B'$ or $\dot{Q}_B'$ from a moving line fire source. As shown in Figure 4.7, it is generally the values integrated over the specimen area, which are $\dot{m}_B$ and $\dot{Q}_B$, that are directly measurable in burn experiments using line fire sources. Thus, Byram termed $\dot{Q}_B'$ as the fireline intensity, and associated it with the rate of flame propagation or fire spread, v_{fl}, as below (Byram, 1959):

$$\dot{Q}_B' = \Delta H \cdot w_a v_{fl} \qquad (4.27)$$

where ΔH is the heat of combustion of the fuel, and w_a is the fuel consumed in the active flaming front. An advantage of the expression in Eq. (4.27) is that the parameter, v_{fl}, can be identified from the visual observation of fire spread. In turn, Eq. (4.27) can be viewed as a relationship that formulates v_{fl} as a function of $\dot{Q}'_B$. We will discuss the rate of fire spread, v_{fl}, in a continuously distributed homogeneous fuel in Chapter 7. A similar but another form of equation alternative to Eq. (4.26) expresses the $\dot{Q}'_B$ in terms of the residence time, t_r, as follows:

$$\dot{Q}'_B = \dot{Q}''_B \cdot t_r v_{fl} = \Delta H \cdot \dot{m}''_B \left(t_r v_{fl} \right) \tag{4.28}$$

t_r is the time during which the burn of a specific portion of a fuel bed is undertaken. This is a basic form of equation incorporated into one of the well-known modeling systems of fire spread, in which $\dot{Q}''_B$ is termed as the reaction intensity (Rothermel, 1972; Albini, 1976). $t_r v_{fl}$ in Eq. (4.28) corresponds to the depth of the line fire source, d_B, in Eq. (4.26).

4.4 FIRE SOURCE IN URBAN FIRES

Buildings are valuable assets that need to be protected from the hazard of outdoor fires. At the same time, they can become fire sources heating their surrounding buildings once ignited. In wildland fires, combustibles burn in an open environment. In contrast, fires generally occur in an enclosed space in urban fires. The fire behavior in an enclosed space significantly differs from that in an open environment. A building generally consists of multiple rooms (compartments). When a fire starts inside a building, fire spread occurs progressively from one compartment to another. However, it is not straightforward to generalize the fire behavior of an entire building, given the variety of connections between compartments and the complexity of the mutual effects when multiple compartments burn simultaneously. This section provides an overview of the behavior of compartment fire as the basic component comprising a fire in a building.

4.4.1 Fuels in compartment fires

Combustibles that contribute to fire behavior in a building can be divided into structural members, interior furnishings, and stored items. Each of these combustible affects fire behavior in a distinctive way.

Structural members consist of frames (columns and beams) and face members (walls, floors, and windows). The proportion of structural members among all combustibles in terms of THR is large, especially in wooden buildings. The involvement of structural members in a fire is problematic as it would considerably extend fire duration and lead to a collapse of the entire building due to the thermal degradation of load-bearing members. To avoid this, they are often covered with fire-resistant materials protecting them from heat exposure.

The proportion of interior furnishings among all combustibles in terms of THR is generally small. However, the fire growth rate at the initial fire stage strongly depends on the condition of the interior furnishings. Its contribution to fire behavior depends on the part used in the compartment: ceilings, walls, and floors. Hot gas produced by the fuel combustion in a compartment accumulates under the ceiling exposing the interior furnishings to strong heating. Thus the flammability of interior furnishings in the upper part of the compartment is especially important in controlling the fire growth rate following ignition. The part used in the compartment becomes less sensitive once the fire is fully developed as all the interior surfaces are covered by flame.

Combustibles that occupants bring into the building after construction, such as household goods, clothes, furniture, and electrical appliances, are referred to as stored items. Compared to structural members and interior furnishings, materials used for stored items are diverse. They have a high proportion of materials that originate from fossil fuels and have high calorific values. To adequately assess the amount of stored items brought into the compartment, it is generally converted to an equivalent value in terms of the calorific value of a reference fuel such as wood.

The fire load density, the lower calorific value of combustibles per unit floor area, can be used to evaluate the fire duration. The greater the fire load, the longer the fire duration, which leads to greater damage to structural members and risk of fire spread to adjacent buildings. However, the fire load density significantly varies between buildings. This variation is attributed to the use of the building and the preference of the occupants. This is especially true for stored items, as they are brought into the building freely by the occupants. In contrast, the variation is relatively small for structural members and interior furnishings as they are not frequently changed after construction. Various investigators have surveyed the fire load density of buildings (Culver, 1978; Chalk et al., 1980; Thomas PH, 1986; Aburano et al., 1999; Zalok et al., 2013; Khorasani et al., 2014). Table 4.6 shows one of the results for residential buildings. Differences in the values may be due to inherent national differences, but may also result from different sampling and evaluation techniques (Thomas PH, 1986).

4.4.2 Development process of a compartment fire

In a compartment fire, flammable gases produced from the pyrolysis of combustible materials react with fresh air (oxygen) that flows into the compartment through openings. The heat released by this reaction raises the gas temperature inside the compartment. Thermal radiation from the compartment gas and the flame ejected from an opening become major causes of

Table 4.6 Variable fire load densities in dwellings (abstracted from Thomas, 1986).

Single value	Average $(MJ \cdot m^{-2})$	S.D. $(MJ \cdot m^{-2})$
Swedish data		
3 rooms	720	104
2 rooms	780	128
European data		
6 rooms	500	180
5 rooms	540	125
3 rooms	670	133
2 rooms	780	129
1 room	720	104
Swiss risk evaluation		
Flat	330	–
USA data		
Living room	350	104
Family room	250	58
Bedroom	390	104
Dining room	330	92
Kitchen	290	71
All rooms	320	88
USA data		
Residence	750*	–
Max for linen closet	4440*	–
Range of maximum values for single occupied room	730–1270*	–

* Total fire load including permanent fire load.

fire spread to surrounding buildings and objects. The behavior of a fire in a compartment significantly differs from that of a fire in an open environment. Figure 4.8 schematically shows the time evolution of compartment gas temperature after ignition. The progressive changes in the state inside the compartment from ignition to extinction are often classified into growth (or initial), fully developed, and decay phases.

The growth phase is the phase that follows an ignition in the compartment. In this phase, combustibles in the compartment burn locally, causing a spatially uneven temperature distribution inside the compartment; locally high temperatures near the fire source and relatively low temperatures elsewhere. The fire plume formed above the fire source ascends due to buoyancy and accumulates under the ceiling, forming an upper hot gas layer. The continuous supply of hot gas to the upper layer increases its thickness and temperature with time, which in turn increases the heating intensity on the interior furnishings and stored items. This leads to a flashover which involves the simultaneous onset of ignition of combustibles engulfing the entire compartment by flame within a short period of time.

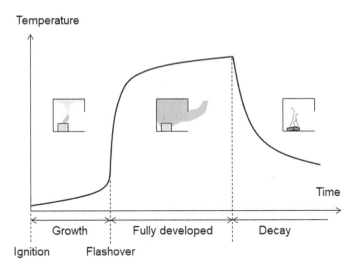

Figure 4.8 Schematic representation of the fire development process in a compartment.

Compartment fire behavior in the fully-developed phase is controlled by the supply rate of fresh air flowing into the compartment through the opening. In a compartment with a large opening that allows a relatively abundant inflow of fresh air, flammable gas generated due to pyrolysis of combustibles burns completely inside the compartment. Such a mode of combustion is called the fuel-controlled fire or fuel-lean fire, where the fire intensity is associated with the supply rate of flammable gas due to pyrolysis. In contrast, in a compartment with a small opening that allows a limited inflow of fresh air, the fresh air reacts with the flammable gas immediately after its inflow into the compartment. The unburnt flammable gas is vented from the opening and reacts with fresh air outside the compartment. This is identified as an externally venting flame, often called as a window flame or window ejected flame. Such a mode of combustion is called the ventilation-controlled fire or under-ventilated fire, where the fire intensity is associated with the supply rate of fresh air (or oxygen) by ventilation. In either mode of combustion, the fully developed phase is the phase during which the fire in the compartment becomes most intense. Fire spread within the same building or to adjacent buildings occurs in this phase.

The transition of the combustion mode from the fully-developed phase to the decay phase occurs when the combustibles in the fire compartment start to burn out. In the decay phase, the fire intensity decays with a decrease in the HRR. The temperature inside the compartment decreases monotonically as the inflow of cold ambient air continues by ventilation. However, the progress of temperature reduction is slow as components of the compartment, such as walls and ceilings, have a large heat capacity inhibiting rapid heat losses.

4.4.3 Fully developed compartment fire

In outdoor fires, a fire that starts in a compartment becomes a heat source to the surrounding buildings and objects when it reaches the fully-developed phase. In the fully-developed phase, the thermal environment in a compartment is often compared to that of a well-stirred combustion vessel. All physical quantities, such as temperature and species concentration, can be assumed to be uniform in the compartment. A number of previous fire experiments support the validity of this assumption for practical fire behavior analysis.

Figure 4.9 shows a schematic representation of the mass and thermal energy transfer in a fully developed compartment fire. In the compartment, combustibles heated by the well-stirred hot gas (including the flaming portion) pyrolyze and release flammable gas into the gas phase. The flammable gas reacts with the fresh air (oxygen) that flows into the compartment through an opening, releasing heat to raise the compartment gas temperature. The conservation equations for the mass and thermal energy of the compartment gas in the fully-developed phase can be expressed as follows:

$$\frac{d}{dt}(\rho V) = \dot{m}_B - \dot{m}_S + \dot{m}_A \tag{4.29a}$$

$$\frac{d}{dt}(c_P \rho V T) = \dot{Q}_B - \dot{Q}_F - \dot{Q}_W - \dot{Q}_D + c_P \dot{m}_B T_P - c_P \dot{m}_S T + c_P \dot{m}_A T_\infty \tag{4.29b}$$

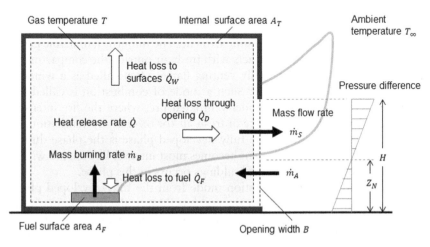

Figure 4.9 Schematic representation of mass and heat transfer in a fire compartment during the fully-developed phase.

where c_P is the specific heat of gas, ρ is the density, V is the compartment volume, T is the compartment gas temperature, $\dot{m}_B$ is the mass loss rate of the fuel, $\dot{m}_S$ is the mass outflow rate through the opening, $\dot{m}_A$ is the mass inflow rate through the opening, $\dot{Q}_B$ is the HRR, $\dot{Q}_F$ is the rate of heat loss to the fuel, $\dot{Q}_W$ is the rate of heat loss to the compartment surfaces, $\dot{Q}_D$ is the rate of radiative heat loss through the opening, T_P is the pyrolysis temperature, and T_∞ is the ambient gas temperature. Eq. (4.29) can be used to calculate the time evolution of the state of the compartment gas. However, to close the equations, additional formulations are required for the combustion of combustibles, mass transfer due to ventilation, and heat transfer at the compartment boundary.

(1) Mass loss rate (mass burning rate)

Under the ventilation-controlled condition, the following correlation for the MLR, $\dot{m}_B$, was empirically derived from fire experiments using wooden combustibles (Kawagoe, 1958):

$$\dot{m}_B \cong 0.1A\sqrt{H} \ \left(\mathrm{kg \cdot s^{-1}}\right) \tag{4.30}$$

where A is the opening area, and H is the opening height. $A\sqrt{H}$ is an important parameter when discussing the behavior of compartment fires called the ventilation factor. In Eq. (4.30), $\dot{m}_B$ is determined solely by the opening geometry and is not affected by the fire environment in the compartment. It is not easy to logically deduce the physical mechanism behind this relationship. Despite such a shortcoming, Eq. (4.30) applies to a wide range of compartment conditions, regardless of the compartment geometry, the material type, or the fuel type.

Later research demonstrated that Eq. (4.30) by Kawagoe applies only to ventilation-controlled fires that occur in compartments with small $A\sqrt{H}$ but not to fuel-controlled fires. Ohmiya et al. conducted a series of compartment fire experiments using wooden combustibles under conditions that covered a wider range of $A\sqrt{H}$. Figure 4.10 shows the regression equation of the experimental data (Ohmiya et al., 1996), which takes the form:

$$\frac{\dot{m}_B}{A_F} = \begin{cases} 0.1\chi & (\chi \le 0.07) \\ 0.007 & (0.07 \le \chi \le 0.1) \\ 0.003 + 0.12\chi\exp(-11\chi) & (0.1 \le \chi) \end{cases} \ \left(\mathrm{kg \cdot m^{-2} \cdot s^{-1}}\right) \tag{4.31}$$

where A_F is the surface area of combustible materials, and χ is the burn type factor given by:

$$\chi = \frac{A\sqrt{H}}{A_F} \tag{4.32}$$

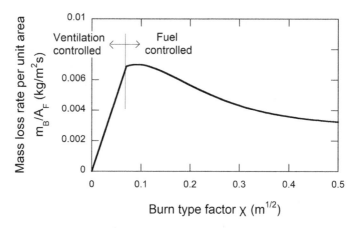

Figure 4.10 Relationship between the burn type factor and the mass loss rate per unit area in a compartment fire (after Ohmiya et al., 1996).

According to Eq. (4.31), the combustion mode of a compartment fire can be classified based on the burn type factor, χ; ventilation-controlled when $\chi \leq 0.07$ and fuel-controlled when $\chi \geq 0.07$. The models by Kawagoe (Eq. (4.30)) and Ohmiya (Eq. (4.31)) are identical in $\chi \leq 0.07$. For large values of χ, $\dot{m}_B$ asymptotically approaches a value obtained when a wooden combustible is burnt in an open environment outside an enclosure. While fresh air is abundantly supplied in an open environment, the heat generation by combustion is easily lost due to radiation and convection. This is the reason why $\dot{m}_B$ is smaller in an open environment than in a compartment. On the other hand, in a compartment with a small opening, $\dot{m}_B$ could become even smaller than that in an open environment. This is due to a limited supply of fresh air by ventilation that suppresses the combustion of flammable gas in the compartment. When χ takes a transitional value between the ventilation-controlled and fuel-controlled conditions, a relatively abundant supply of both fresh air and flammable gas enhances the combustion intensity. At the same time, the heat loss from the compartment is restrained as the fire source is enclosed with building members with high heat capacity. Thus, the $\dot{m}_B$ reaches the maximum in the transitional range of χ.

In evaluating the hazard of fire spread in outdoor fires, we need to identify the duration of the fully developed fire, t_{fire}, given that the hazard in the growth and decay phases is relatively minor. t_{fire} is often approximated by dividing the total amount of burnable combustibles in the compartment, W_F, by the MLR during the fully-developed phase, $\dot{m}_B$, which is:

$$t_{fire} \cong \frac{W_F}{\dot{m}_B} \tag{4.33}$$

This relationship ignores the mass loss of combustibles during the growth and decay phases. Consequently, Eq. (4.33) overestimates t_{fire} compared to actual fires. The use of Eq. (4.33) can be justified for cases when one needs to avoid underestimating the hazard of fire spread, such as determining a safe separation between buildings.

(2) Mass flow rate due to ventilation

When a fire starts in a compartment, the temperature rise causes a change in the profile of the hydrostatic pressure relative to the outdoors. This forms a vertical distribution of pressure difference at the opening facing the outdoors, causing the movement of gases in and out of the compartment. The neutral plane is the plane where the pressure difference at the opening is zero. The pressure difference above the neutral plane is positive, causing the hot gas to be vented from the compartment. The pressure difference below the neutral plane becomes negative, causing the fresh air to be introduced from the outdoors. If the state of the compartment gas can be considered uniform, the mass outflow and inflow rates through a rectangular opening, $\dot{m}_S$ and $\dot{m}_A$, can be respectively calculated with the following equations (Prahl et al., 1975; Steckler et al., 1982; Nakaya et al., 1986):

$$\dot{m}_S = \frac{2}{3}\alpha B(H - Z_N)\sqrt{2\Delta\rho \cdot \rho g(H - Z_N)} \tag{4.34a}$$

$$\dot{m}_A = \frac{2}{3}\alpha B Z_N \sqrt{2\Delta\rho \cdot \rho_\infty g Z_N} \tag{4.34b}$$

where α is an empirical coefficient called the mass flow rate coefficient that generally takes a value between 0.6 and 0.7, B is the width of the opening, H is the height of the opening, Z_N is the height of the neutral plane, ρ and ρ_∞ are the densities inside and outside of the compartment, respectively, $\Delta\rho = \rho_\infty - \rho$ is the density difference, and g is the acceleration due to gravity.

As discussed above, the mass conservation equation of the compartment gas is expressed by Eq. (4.29a). This relationship is valid through the entire period of a fire, including the unsteady period. However, for simplicity, we focus on the fully developed phase during which the condition inside the compartment is quasi-steady. We also assume that the MLR, $\dot{m}_B$, is sufficiently small compared to $\dot{m}_S$ and $\dot{m}_A$ ($\dot{m}_B \ll \dot{m}_S$ and $\dot{m}_B \ll \dot{m}_A$). In such a case, Eq. (4.29a) yields $\dot{m}_S \cong \dot{m}_A$. Substituting Eq. (4.34) into this relationship, we obtain the following equation for the height of the neutral plane:

$$\frac{Z_N}{H} = \frac{1}{1 + (\rho_\infty / \rho)^{1/3}} \tag{4.35}$$

We obtain the following expression by substituting Eq. (4.35) into Eq. (4.34) (Tanaka, 1993):

$$\dot{m}_S \cong \dot{m}_A = \frac{2}{3}\alpha BH\sqrt{2\Delta\rho \cdot \rho_\infty gH} \left\{ \frac{1}{1+\left(\rho_\infty / \rho\right)^{1/3}} \right\}^{3/2} \tag{4.36}$$

This can be viewed as an equation that expresses the $\dot{m}_S$ and $\dot{m}_A$ as a function of the compartment gas density, ρ, which is strongly associated with the gas temperature, T. Although it may appear complicated, the rate of change in $\dot{m}_S$ and $\dot{m}_A$ becomes small when ρ is sufficiently small, that is, T is sufficiently high. Thus, by substituting general values of the air, a simple equation applicable to the mass flow rate calculation in the fully developed phase can be obtained as follows (Rockett, 1976; Tanaka, 1993):

$$\dot{m}_S = \dot{m}_A \cong 0.5A\sqrt{H} \ \left(\text{kg}\cdot\text{s}^{-1}\right) \tag{4.37}$$

where $A = BH$ is the area of the opening. In deriving Eq. (4.37), the mass flow rate coefficient was assumed as $\alpha = 0.7$ in Eq. (4.36). Similar to the case of $\dot{m}_B$ in Eq. (4.30), $\dot{m}_S$ and $\dot{m}_A$ during the fully developed phase are dominantly controlled by the ventilation factor, $A\sqrt{H}$. Comparing $\dot{m}_S \left(\cong \dot{m}_A\right)$ in Eq. (4.37) with $\dot{m}_B$ in Eq. (4.30), $\dot{m}_S$ is approximately five times larger than $\dot{m}_B$ in ventilation-controlled fires ($\dot{m}_S / \dot{m}_B \cong 5$, and $\dot{m}_A / \dot{m}_B \cong 5$).

(3) Heat release rate (HRR)

According to the discussion in Section 4.2, the threshold for the ventilation-controlled and fuel-controlled fires can be expressed as follows, given that all fresh air introduced into the compartment is effectively consumed:

$$\left.\begin{aligned}\dot{m}_B &\leq \frac{\dot{m}_A}{r_{st,A}} \quad \left(fuel-controlled\right)\\[2mm] \dot{m}_B &\geq \frac{\dot{m}_A}{r_{st,A}} \ \left(ventilation-controlled\right)\end{aligned}\right\} \tag{4.38}$$

where $r_{st,A}$ is the stoichiometric air-to-fuel ratio defined in Eq. (4.19).

In a fuel-controlled fire, all flammable gas supplied due to pyrolysis burns completely in the compartment. Thus, the HRR can be expressed as:

$$\dot{Q}_B = \Delta H \dot{m}_B \tag{4.39}$$

where ΔH is the heat of combustion. Meanwhile, in a ventilation-controlled fire, only the stoichiometrically consumable amount of flammable gas, $\dot{m}_A / r_{st,A}$, contributes to the combustion. Thus, the HRR can be expressed as:

$$\dot{Q}_B = \Delta H \frac{\dot{m}_A}{r_{st,A}} = \Delta H_A \dot{m}_A \tag{4.40}$$

where $\Delta H_A = \Delta H / r_{st,A}$ is the heat release per unit consumption of fresh air, which is approximately $3\,kJ \cdot g^{-1}$ regardless of the fuel type (Huggett, 1980). Thus, by substituting the $\dot{m}_A$ in Eq. (4.37) into Eq. (4.40), the HRR in a ventilation-controlled condition can be further transformed as follows (Tanaka, 1993):

$$\dot{Q}_B \cong 3000\dot{m}_A \cong 1500A\sqrt{H} \; (kW) \tag{4.41}$$

This simple expression is often used to evaluate the HRR in a ventilation-controlled fire in the fire safety design practice of buildings. However, note that $\dot{Q}_B$ in Eq. (4.41) should be regarded as the HRR of the combustion completed inside the compartment, rather than the HRR of all the flammable gas generated due to pyrolysis. In ventilation-controlled fires, flammable gas that is not burnt inside the compartment burns outside the compartment as vented from the opening and mixed with fresh air. Thus, the HRR that corresponds to the combustion outside the compartment, $\dot{Q}_X$, can be obtained by subtracting the HRR inside the compartment from the HRR of all the generated flammable gas:

$$\dot{Q}_X \cong \Delta H \dot{m}_B - \Delta H_A \dot{m}_A \cong 1700A\sqrt{H} - 1500A\sqrt{H} = 200A\sqrt{H} \; (kW) \tag{4.42}$$

Note that in Eq. (4.42), we assumed $\Delta H = 17.0\,kJ \cdot g^{-1}$ of wooden fuels (refer to Table 4.2) and used Eq. (4.30) for the calculation of $\dot{m}_B$. Eq. (4.42) provides an estimate that the fraction of combustion outside the compartment is approximately 2/15 of that completed inside the compartment in ventilation-controlled fires. Given that we often encounter huge flames venting from the compartment in fire experiments, such a small fraction of combustion outside the compartment may appear unexpected.

(4) Heat loss rate

The heat released by the combustion of flammable gas inside the compartment causes an increase in the compartment gas temperature. The temperature difference between the compartment gas and the enclosure boundary or the ambient air induces heat transfer between them (ref. Chapter 3). The following is an overview of several modes of heat transfer involved in a compartment fire.

First, the heat loss rate through an opening by thermal radiation, $\dot{Q}_D$, can be expressed as follows:

$$\dot{Q}_D = \sigma\left(T^4 - T_\infty^4\right)A_D \tag{4.43}$$

where σ is the Stefan-Boltzmann constant, A_D is the area of the opening, and T and T_∞ are the absolute temperature of the compartment gas and the ambient air, respectively. For simplicity, the emissivity of the compartment gas is assumed to be unity. This assumption is reasonable in most compartment fires in the fully developed phase, where the compartment is filled with sooty flame and thick smoke.

Second, the heat loss rates to the fuel surfaces, $\dot{Q}_F$, and the enclosure surfaces, $\dot{Q}_W$, can be expressed as a combination of heat transfer by thermal radiation and convection, which respectively take the form:

$$\dot{Q}_F = \left\{ \sigma\left(T^4 - T_P^4\right) + h\left(T - T_P\right)\right\} A_F \tag{4.44a}$$

$$\dot{Q}_W = \left\{ \sigma\left(T^4 - T_W^4\right) + h\left(T - T_W\right)\right\} A_W \tag{4.44b}$$

where h is the heat transfer coefficient, T_P is the pyrolysis temperature, T_W is the surface temperature of the enclosure surface, and A_F and A_W are the exposed surface areas of the combustible and enclosure, respectively. Again, the emissivity of the compartment gas is assumed to be unity in Eq. (4.44).

The above equations for the heat loss rate cannot be readily solved as they involve terms proportional to the fourth power of the absolute temperatures, T, T_P, and T_W. Thus, a frequently used assumption in engineering calculations is to combine the fourth-power term on thermal radiation and the linear term on convection into a single linear term by introducing the effective heat transfer coefficient, h_k (McCaffrey et al., 1981). Assuming that the rate of radiative and convective heat transfer incident to the enclosure surface is equivalent to that of heat conduction inside the material subject to heating, Eq. (4.44b) can be approximated as:

$$\dot{Q}_W = h_k\left(T_W - T_\infty\right)A_W \cong h_k\left(T - T_\infty\right)A_W \tag{4.45}$$

In Eq. (4.45), the surface temperature, T_W, is approximated with the compartment gas temperature, T, for simplicity. A similar approximation is possible for Eq. (4.44a). In Eq. (4.45), McCaffrey et al. expressed the h_k as a function of the time elapsed after the heating has started, t, which takes the form (McCaffrey et al., 1981):

$$h_k = \begin{cases} \sqrt{\dfrac{k\rho c}{\pi t}} & \left(t \le \dfrac{\delta^2}{\pi(k/\rho c)}\right) \\[4mm] \dfrac{k}{\delta} & \left(t \ge \dfrac{\delta^2}{\pi(k/\rho c)}\right) \end{cases} \tag{4.46}$$

where k, ρ, c, and δ are the thermal conductivity, density, specific heat, and width of the compartment material. The conditional branch represents the shift of the heat conduction mode inside the material from unsteady to steady at the threshold, $\delta^2 / \pi(k / \rho c)$.

WORKED EXAMPLE 4.4

Assume a concrete compartment with a square base of 5 m × 5 m and a height of 3 m. The compartment has a square opening of 2 m × 2 m in one of its walls. The surface area of wooden combustibles stored inside the compartment is equivalent to that of the enclosure. The heat of combustion of wooden combustibles is $\Delta H = 17.0$ kJ·g^{-1}. Calculate the MLR and HRR in the fully-developed phase of a fire.

SUGGESTED SOLUTION

The experimental correlation in Eq. (4.31) can be used to calculate the MLR during the fully developed phase of a fire. The burn type factor, χ, which determines the combustion mode inside the compartment in Eq. (4.31), can be calculated using Eq. (4.32):

Ventilation factor: $A\sqrt{H} = (2 \times 2) \times \sqrt{2} = 5.66$ m$^{5/2}$

Fuel surface area: $A_F \cong A_T = (5 \times 5) \times 2 + (5 \times 2) \times 4 = 110$ m^2

Burn type factor: $\chi = \dfrac{A\sqrt{H}}{A_F} = \dfrac{5.66}{110} = 0.0514$ m$^{1/2}$

As χ is smaller than the threshold, which is 0.07 m$^{1/2}$, the combustion mode inside the compartment is ventilation-controlled according to Eq. (4.31). The MLR, $\dot{m}_B$, can be calculated using Eq. (4.31) as follows:

$$\dot{m}_B = 110 \times 0.1 \times 0.0514 = 0.566 \text{ kg / s}$$

The total HRR for the combustion of all the flammable gas released due to pyrolysis, $\dot{Q}_{B,T}$, can be obtained by multiplying $\dot{m}_B$ by the heat of combustion, ΔH, which is:

$$\dot{Q}_{B,T} = 0.566 \times 17.0 \times 1000 = 9622 \text{ kW}$$

Among the total HRR, the fraction completed inside the compartment, $\dot{Q}_B$, can be obtained by using Eq. (4.41) as follows:

$$\dot{Q}_B = 1500 \times 5.66 = 8490 \text{ kW}$$

Thus, the fraction burnt outside of the compartment, $\dot{Q}_X$, is the remaining, which is as follows:

$$\dot{Q}_X = 9622 - 8490 = 1132 \, \text{kW}$$

4.4.4 Compartment gas temperature

Together with the HRR, $\dot{Q}_B$, the compartment gas temperature, T, is an important physical quantity representing the potential hazard of a fire. T can be obtained by simultaneously solving the governing equations of the mass and thermal energy in Eq. (4.29) numerically. Alternatively, a tractable equation to predict the temperature rise above the ambient, $\Delta T (= T - T_\infty)$, has been derived by regressing data from various compartment fire experiments (McCaffrey et al., 1981):

$$\frac{\Delta T}{T_\infty} = 1.6 \left(\frac{\dot{Q}_B}{c_P \rho_\infty T_\infty g^{1/2} A\sqrt{H}} \right)^{2/3} \left(\frac{h_k A_T}{c_P \rho_\infty g^{1/2} A\sqrt{H}} \right)^{-1/3} \quad (4.47)$$

where A_T is the surface area of the enclosure. The expression consists of two dimensionless numbers: the HRR (the first term) and the heat loss rate to the enclosure surfaces (the second term), both normalized by the terms associated with the heat loss due to ventilation. As the power index of the first term is positive, the larger the first term is, the larger ΔT becomes. On the other hand, as the power index of the second term is negative, the larger the second term is, the smaller ΔT becomes.

The coefficient and power indices in Eq. (4.47) were determined based on the experiments of fuel-controlled compartment fires with ΔT below 600 K (McCaffrey et al., 1981). In general, the combustion in a compartment becomes more vigorous when it is ventilation-controlled. Thus, Eq. (4.47) has been extended to predict the compartment gas temperature during the fully developed phase of ventilation-controlled fires as follows (Tanaka et al., 1997; Matsuyama et al., 1998):

$$\frac{\Delta T}{T_\infty} \cong \begin{cases} 3.0 \left(\dfrac{A\sqrt{H}}{A_T} \right)^{1/3} \left(\dfrac{t}{k\rho c} \right)^{1/6} & \left(t \leq \dfrac{\delta^2}{\pi(k/\rho c)} \right) \\[3ex] 3.0 \left(\dfrac{A\sqrt{H}}{A_T} \right)^{1/3} \left(\dfrac{\delta}{k} \right)^{1/3} & \left(t > \dfrac{\delta^2}{\pi(k/\rho c)} \right) \end{cases} \quad (4.48)$$

This equation shows that the normalized temperature rise, $\Delta T / T_\infty$, is governed by the factors representing the geometry of the compartment, $A\sqrt{H} / A_T$, and the thermal property of the enclosure material, either $t/(k\rho c)$ or δ/k. The time evolution of ΔT corresponds to the change in the effective heat transfer coefficient of the enclosure material. ΔT is proportional to $t^{1/6}$ during the unsteady phase $(t \leq \delta^2 / \pi(k/\rho c))$, and constant during the steady phase $(t > \delta^2 / \pi(k/\rho c))$.

WORKED EXAMPLE 4.5

Assume a concrete compartment with a square base of 10 m × 10 m and a height of 3 m. The compartment has a rectangular opening of 5 m (width) ×2 m (height) in one of its walls. Calculate the rise in the compartment gas temperature, ΔT, 30 min following the ignition. The thermal conductivity of concrete is $k = 1.4 \times 10^{-3}$ kW·m^{-1}·K^{-1}, the density is $\rho = 2300$ kg·m^{-3}, and the heat capacity is $c = 0.88$ kJ·kg^{-1}·K^{-1}. The thickness of the compartment is $\delta = 0.1$ m. The ambient temperature is $T_\infty = 300$ K.

SUGGESTED SOLUTION

In the calculation of ΔT using Eq. (4.48), we first calculate the transition time of the compartment fire from unsteady to steady:

$$\frac{\delta^2}{\pi(k/\rho c)} = \frac{0.1^2}{3.14 \times 1.4 \times 10^{-3} \div (2300 \times 0.88)} = 46018.5\,\text{s} = 767 \text{ min}$$

Thus, the compartment fire is in the unsteady phase 30 min after the fire starts. For this specific compartment, the parameters involved in Eq. (4.48) can be obtained as follows:

Ventilation factor: $A\sqrt{H} = (5 \times 2) \times \sqrt{2} = 14.1$ m$^{5/2}$

Enclosure surface: $A_T = (10 \times 10) \times 2 + (10 \times 3) \times 4 = 320$ m^2

Thermal inertia: $k\rho c = (1.4 \times 10^{-3}) \times 2300 \times 0.88$
$$= 2.83 \text{ kJ}^2 \cdot \text{s}^{-1} \cdot \text{K}^{-2} \cdot \text{m}^{-4}$$

By substituting these parameters into Eq. (4.48), we obtain ΔT as follows:

$$\Delta T = 300 \times 3.0 \times \left(\frac{14.1}{320}\right)^{1/3} \left(\frac{30 \times 60}{2.83}\right)^{1/6} = 932.2 \text{ K}$$

Chapter 5

Fire plumes – quiescent environment

A fire plume is a hot updraft formed above a burning combustible. A fire plume begins as reacting plume (flame) followed by a non-reacting plume segment at higher elevations. The behavior of fire plumes is critical in evaluating the heating intensity against adjacent combustibles and their ignition. In our discussion, the reacting and non-reacting plumes are collectively called fire plumes. The term 'flame' is used only when the reacted segment is identified.

The behavior of a fire plume is dependent on its source geometry. A point and an infinite line are the fire source geometries often assumed for the theoretical analysis of the fire plume behavior. They are ideal fire sources, and in a strict sense, no such fire sources exist in actual fires. However, the overall behavior of fire plumes in actual fires can be reasonably represented by such ideal fire sources in many cases. For instance, the temperature rise around a burning object is often represented by that of a point fire source. The flame geometry of the fire-front of a wide-range burning area is often represented by that of a line fire source.

The behavior of fire sources is also affected by natural winds, especially in outdoor fires. The behavior of fire sources subject to crosswinds has also been the target of analysis for many years. However, analysis in a windy environment generally has had a basis on the theoretical considerations in a quiescent environment. Thus, in this chapter, we focus on the behaviors of ideal fire sources in a quiescent environment, and those in a windy environment will be discussed in the subsequent chapter.

The flame ejection from an opening of a fire-involved building is an important factor of fire spread in urban fires. A flame ejected from an opening, which can be termed as a window flame, is a special form of a flame that initially has a horizontal momentum but gains a vertical momentum due to buoyancy as it separates from the opening. The behavior of window flames is also discussed in this chapter.

DOI: 10.1201/9781003096689-5

5.1 BASIC CHARACTERISTICS OF FIRE PLUMES

Some important assumptions are introduced herein in modeling the behavior of fire plumes. The first is on the self-similarity of the mean temperature rise and velocity profiles in the direction perpendicular to the plume centerline. The second is on the periodic change of flame geometry and the corresponding segmentation of the plume domain.

5.1.1 Self-similarity

The mode of combustion mostly observed in outdoor fires is turbulent diffusion combustion. Physical quantities within and around a turbulent diffusion flame, such as the temperature rise and the flow velocity, fluctuate randomly in time and space. However, if they are averaged over time, they are the highest within the flame where the combustion reaction is in progress and decrease with distance from the flame. Although there is an exception in the fuel-rich core just above the solid fuel, the higher the elevation, z, and the further the distance from the plume centerline, r, the smaller the temperature rise, ΔT, and the flow velocity, w. Such a decrease in ΔT and w is caused by the fresh air entrainment from the surroundings, accompanied by the buoyancy-induced rise of the combustion gas. As the fire plume widens due to the fresh air entrainment, the domain of a fire plume above a point source resembles an inverted cone, and the one formed above a line source resembles an overturned triangular prism.

Figure 5.1 shows the measurement results of the plume characteristics above an alcohol lamp as an assumed point source (Yokoi, 1960). The middle and right panels of Figure 5.1 show the radial distributions of the temperature rise, $\Delta T(r,z)$, and the upward velocity, $w(r,z)$, normalized by the centerline values, $\Delta T_m(z)$, and, $w_m(z)$, respectively (Yokoi, 1960). In Figure 5.1, z is the height and r is the radial distance from the centerline. The result shows that both $\Delta T(r,z)/\Delta T_m(z)$ and $w(r,z)/w_m(z)$ take similar distributions regardless of the z as indicated by the overlap of the data points. This exhibits an important feature of fire plumes; the width, b, of the fire plume widens in proportion to the height, z, above the point source. Such a relationship is valid not only for point sources but also for line sources, as shown in Figure 5.2 (Yokoi, 1960). Generally, the normalized distributions are approximated with Gaussian profiles (Rouse et al., 1952; Yokoi, 1960):

$$\frac{\Delta T(r,z)}{\Delta T_m(z)} = \exp\left[-\beta\left(\frac{r}{b(z)}\right)^2\right] \tag{5.1a}$$

$$\frac{w(r,z)}{w_m(z)} = \exp\left[-\left(\frac{r}{b(z)}\right)^2\right] \tag{5.1b}$$

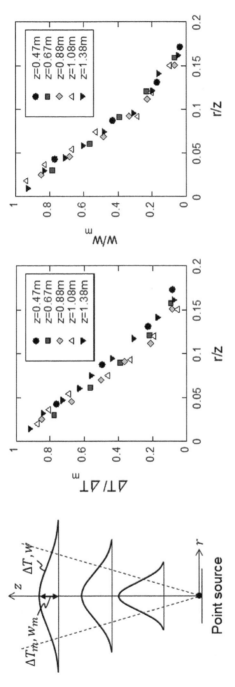

Figure 5.1 Radial distribution of temperature rise and upward velocity at different heights (point source) (adapted from Yokoi, 1960).

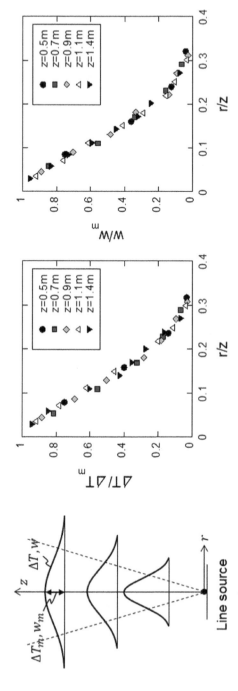

Figure 5.2 Radial distribution of temperature rise and upward velocity at different heights (line source) (adapted from Yokoi, 1960).

where b is the effective width of the normalized velocity distribution, $w(r,z)/w_m(z)$, and β is the coefficient that converts $w(r,z)/w_m(z)$ to $\Delta T(r,z)/\Delta T_m(z)$. However, as shown in Figures 5.1 and 5.2, the difference in the distributions is not that apparent between $\Delta T(r,z)/\Delta T_m(z)$ and $w(r,z)/w_m(z)$. According to the analysis of various experimental results, the values of β are 0.913 for point sources and 0.845 for linear sources (Quintiere et al., 1998).

Eq. (5.1) is an important relationship for the theoretical analysis of fire plumes. If the centerline values $(\Delta T_m(z)$ and $w_m(z))$ and the width, b, of the fire plume are identified, values at an arbitrary height z and radial distance r $(\Delta T(r,z)$ and $w(r,z))$ can be evaluated from Eq. (5.1).

5.1.2 Intermittency and domain segmentation

Figure 5.3 shows the morphological transition of a turbulent diffusion flame above a 30-cm diameter porous burner photographed continuously at a constant time interval of 1/24 s (McCaffrey, 1979). The flame periodically changes its shape, repeating the expansion and contraction alternately: In (a) and (b), the fuel supplied from the burner reacts with fresh air to form a flame in a relatively small zone near the burner outlet. In (c) and (d), the flame starts to elongate as it entrains the surrounding fresh air. At the same time, the flame forms a neck near the bottom. In (e) and (f), the neck becomes narrower and its position is raised as the flame stretches further. The upper portion of the flame is characterized by the formation of large toroidal vortices due to air entrainment. In (g) and (h), the flame eventually splits into the upper and lower portions at the neck. The upper portion disappears when all the fuel is consumed, while the lower portion remains

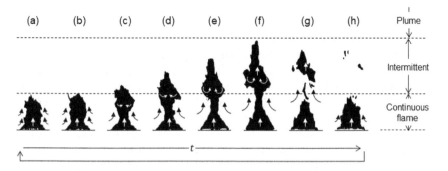

Figure 5.3 A buoyant diffusion flame on a 0.3-m porous burner photographed at a constant interval of 1/24 (s). Arrows qualitatively represent the local flow (adapted from McCaffrey, 1979).

as long as the fuel supply continues. At this stage, the process returns to (a), and the periodical oscillation of the turbulent diffusion flame continues.

As shown in Figure 5.3, the domain of a turbulent diffusion flame can be divided into three characteristic regimes depending on the elevation from the ground as follows.

- Continuous flame regime.
- Intermittent flame regime.
- Buoyant plume regime.

The continuous flame regime corresponds to the domain near the fuel outlet where the flame persistently exists. The intermittent flame regime corresponds to the domain above the continuous flame regime, where the presence and absence of the flame are repeated intermittently. The buoyant plume regime corresponds to the domain above the intermittent flame regime, where the flame is persistently absent as combustion of all supplied fuel has been completed. The behavior of a fire plume, represented by physical quantities such as the temperature rise, ΔT, and the flow velocity, w, is closely related to the domain segmentation. In a fire, the continuous and intermittent flame regimes become the dominant radiative heat sources causing ignition of the adjacent objects. On the other hand, the temperature rise in the buoyant plume regime may not be high enough to cause the ignition of the adjacent objects on its own. However, it is high enough to hamper human activity. Given that the effect of the buoyant plume regime reaches a broader range, one should consider its effect on the evacuation of residents and firefighting activity by fire services when assessing the risk of large outdoor fires.

5.2 POINT FIRE SOURCE

Figure 5.4 shows a schematic representation of a fire plume above a point source. The point source model is the most basic model for the theoretical consideration of fire plumes. Due to the air entrainment associated with the rise of the buoyancy-dominated flow, its temperature and velocity attenuate, and its flow rate increases at a higher elevation. Thus, the fire plume behavior is substantially affected by the fire source geometry at a lower elevation. However, the effect of the fire source geometry becomes weaker with a distance from the fire source. We can reasonably assume that the heat generation by the flame is confined within a single point for evaluating the fire plume behavior at a higher elevation. When discussing the fire plume behavior at a substantially high elevation, we often refer to it as 'a fire plume above a point fire source'. Despite such an assumption, the derived model is applicable to a wide range of conditions, even inside the flame regime in certain cases.

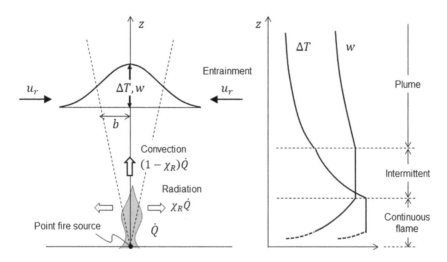

Figure 5.4 Schematic representation of a fire plume above a point source in a quiescent environment and the transition of the centerline behavior.

5.2.1 Governing equations

A fire plume is a buoyancy-driven flow above a fire source that entrains the surrounding fresh air while ascending. From the fire safety point of view, the key issue on a fire plume is to evaluate the changes in its temperature rise, ΔT, and flow velocity, w, above the fire source. The classical plume theory has often been applied to analyze the behavior of fire plumes (Morton et al., 1956). In this theory, the change in the behavior of fire plumes along the centerline (z-axis) has been analyzed by dimensional analysis. The dimensional analysis cleverly derives the dimensional relationships between the plume characteristics and the height, z, that are embedded in the conservational equations of mass, momentum, and thermal energy. Additionally, it analyzes the behavior at an arbitrary location above the fire source by assuming self-similar profiles for ΔT and w in the radial direction to the centerline (r-axis) (refer to Eq. (5.1)). The analytical procedure of the classical plume theory, which rigorously follows the conservational laws, is well established (Yokoi, 1960; Steward, 1964; McCaffrey, 1979; Cox et al., 1980; Zukoski, 1980/81; Baum et al., 1989; Quintiere et al., 1998). The analytical procedure rigorously follows the conservation laws of mass, momentum, and thermal energy.

On the other hand, one may find it less accessible if not familiar with analysis procedures commonly employed in fluid mechanics. As some good references are already available on the classical plume theory (Zukoski, 1995; Karlsson et al., 2000; Quintiere, 2006), we employ a more intuitive approach. The alternative approach also considers the conservational relationships between the characteristic quantities, but in a simplified manner by incorporating physical considerations of the flow behavior. It can derive the same dimensional relationships as the classical approach. In fact, the classical and alternative approaches are both dimensional analyses. The difference between the two approaches can be found in their strictness of formulation; the classical theory formulates relatively strict conservational relationships between physical quantities, by which a more theoretically supported model can be obtained. However, there is no complete analytical solution available for turbulent flows involving combustion reactions, such as fire plumes, in any case. The models derived from the classical plume theory are not exempted from introducing empirical parameters. In such a case, there is room to consider an approach that enables an intuitive understanding of the dimensional relationships between physical quantities.

In this section, characteristic quantities of a fire plume are modeled separately in the flame and buoyant plume regimes as described in Figure 5.5. The considered characteristic quantities are the centerline values of the temperature rise, ΔT_m, and upward velocity, w_m, and the effective width of the fire plume, b. In the following discussion, the subscript that denotes a centerline value, m, is omitted unless otherwise noted. Additionally, although the flow field in a turbulent fire plume randomly fluctuates by nature, we focus on

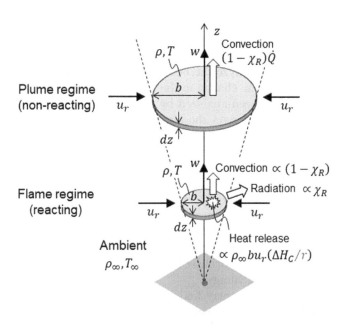

Figure 5.5 Behavior of characteristic quantities in flame and plume regimes above a point source.

the time-averaged values as they are the major concern for assessing the hazard of fire spread in outdoor fires.

(1) Plume regime

First, let us consider the behaviors of characteristic quantities in the buoyant plume regime. As shown in Figure 5.5, consider an infinitesimal control volume perpendicular to the plume centerline above a point source (effective width, b, flow velocity, w, and temperature rise, $\Delta T (= T - T_\infty)$. The geometry of the control volume is symmetrical to the plume centerline.

The b widens as the combustion gas rises due to buoyancy and entrains the surrounding fresh air. Given that the b is dependent on the cumulative amount of air entrainment, the mass conservation relationship for the plume can be expressed as:

$$\rho b^2 w \propto \int \rho_\infty b u_r dz \tag{5.2}$$

where ρ is the plume gas density, ρ_∞ is the ambient gas density, u_r is the radial velocity at the perimeter of the control volume (entrainment velocity), and z is the height above the point source. We assume that the scale and intensity of the plume turbulence are dominated by the degree of thermal instability. Additionally, u_r is proportional to w according to the classical plume theory (Morton et al., 1956):

$$u_r \propto w \tag{5.3}$$

By introducing Eq. (5.3) into Eq. (5.2), we obtain:

$$\rho b^2 w \propto \int \rho_\infty b w dz \tag{5.4}$$

The vertical momentum is controlled predominantly by the buoyancy force based on the density difference between the plume and the surrounding air, $\Delta \rho (= \rho_\infty - \rho)$. As with the conservation of mass, the momentum is determined by the cumulative effect of the buoyancy forces. Assuming that the density difference can be represented by the value at z, we obtain:

$$\rho w^2 \propto \int \Delta \rho g dz \cong \Delta \rho g z \tag{5.5}$$

where g is the acceleration due to gravity. In addition, ρT is approximately constant at atmospheric pressure according to the state equation of the perfect gas (Drysdale, 1998):

$$\rho T = \rho_\infty T_\infty \tag{5.6}$$

where T and T_∞ are the plume and ambient temperatures, respectively. Using Eq. (5.6), Eq. (5.5) can be transformed into:

$$w \propto \sqrt{\frac{\Delta T}{T_\infty} gz} \tag{5.7}$$

As for the conservation of thermal energy, we assume that all the heat is generated at the point source and none in the plume regime. Assuming that the change in the control volume is exclusively due to the increase in b by the air entrainment, the following equation can be obtained:

$$\frac{d}{dz}\left(c_P \rho b^2 w \Delta T\right) = 0 \tag{5.8}$$

where c_P is the specific heat of gas at constant pressure. A fraction of the thermal energy generated by combustion is lost by radiation (radiation fraction, χ_R). The remainder is used to raise the plume temperature causing the convective flow due to buoyancy (convection fraction, $1 - \chi_R$). The thermal energy retained by the plume is $(1 - \chi_R)\dot{Q}$ after excluding the thermal energy lost by radiation. By integrating Eq. (5.8) with respect to z, we obtain the conservation equation of thermal energy as follows:

$$(1 - \chi_R)\dot{Q} \propto c_P \rho b^2 w \Delta T \tag{5.9}$$

The χ_R is often considered as a constant depending on the type and size of the fuel (Mudan, 1984).

In summary, the conservation equations for the mass, momentum, and thermal energy take the form:

$$\text{Mass: } \rho b^2 w \propto \int \rho_\infty b w dz \tag{5.10a}$$

$$\text{Momentum: } w \propto \sqrt{\frac{\Delta T}{T_\infty} gz} \tag{5.10b}$$

$$\text{Energy: } (1 - \chi_R)\dot{Q} \propto c_P \rho b^2 w \Delta T \tag{5.10c}$$

These equations involve three unknown characteristic quantities, b, w, and ΔT.

(2) Flame regime

Second, let us consider the conservation equations in the flame regime following the same manner as in the plume regime. Among the three conservation equations in the plume regime, the relationships for the mass and momentum still hold in the flame regime. They change due to the air entrainment and the buoyancy force, respectively. Features of the flame regime are incorporated into the conservation equation of thermal energy.

Assume that the flammable gas supplied from the bottom and the oxygen supplied from the perimeter are uniformly mixed in the flame regime. The heat generated inside the control volume is proportional to the entrainment rate of oxygen through the perimeter:

$$\frac{d}{dz}\left(c_P \rho b^2 w \Delta T\right) \propto \left(1 - \chi_R\right) \Delta H_C \cdot \frac{\rho_\infty b u_r}{r} \tag{5.11}$$

where ΔH_C is the heat of combustion of the fuel, and r is the stoichiometric air-to-fuel ratio. In Eq. (5.11), the mass fraction of oxygen in the ambient air is assumed constant. Invoking Eq. (5.3) for the entrainment velocity, u_r, Eq. (5.11) can be transformed as:

$$\frac{d}{dz}\left(c_P \rho b^2 w \Delta T\right) \propto \left(1 - \chi_R\right) \Delta H_C \cdot \frac{\rho_\infty b w}{r} \tag{5.12}$$

In summary, the conservation equations for the mass, momentum, and thermal energy are as follows:

$$\text{Mass:} \quad \rho b^2 w \propto \int \rho_\infty b w \, dz \tag{5.13a}$$

$$\text{Momentum:} \quad w \propto \sqrt{\frac{\Delta T}{T_\infty} g z} \tag{5.13b}$$

$$\text{Energy:} \quad \frac{d}{dz}\left(c_P \rho b^2 w \Delta T\right) \propto \rho_\infty b w \cdot \left(1 - \chi_R\right) \frac{\Delta H_C}{r} \tag{5.13c}$$

These equations involve three unknown characteristic quantities, b, w, and ΔT.

The governing equations for fluid motion (Eq. (5.10) for the plume regime and (5.13) for the flame regime) do not represent exact balances of the characteristic quantities, but only their proportional relationships. However, the proportional relationships are often sufficient to provide correlations for analyzing the characteristic quantities required for practical applications.

5.2.2 Dimensional analysis

The governing equations are non-dimensionalized to reduce the number of variables and capture the physical relationships among the variables (Quintiere, 1989; Quintiere et al., 1998). We use the representative values

D for length, U for velocity, and T_∞ for temperature, to introduce the following dimensionless numbers:

$$Z^* = \frac{z}{D}, \quad B^* = \frac{b}{D}, \quad W^* = \frac{w}{U}, \quad \text{and} \quad T^* = \frac{\Delta T}{T_\infty} \tag{5.14}$$

By substituting them into Eqs. (5.10) and (5.13), the governing equations can be transformed into the following dimensionless forms:

$$\text{Mass: } B^* \propto Z^* \tag{5.15a}$$

$$\text{Momentum: } W^{*2} \propto \left(\frac{gD}{U^2}\right) T^* Z^* \tag{5.15b}$$

$$\text{Energy: } \begin{cases} B^* T^* \propto \left(1 - \chi_R\right)\left(\dfrac{\Delta H_C}{r c_P T_\infty}\right) Z^* & \text{(Flame)} \\[2em] B^{*2} W^* T^* \propto \left(1 - \chi_R\right)\left(\dfrac{\dot{Q}}{c_P \rho_\infty T_\infty D^2 U}\right) & \text{(Plume)} \end{cases} \tag{5.15c}$$

For simplicity, Boussinesq approximation is introduced, which assumes $\rho = \rho_\infty$ except for the buoyancy term. However, in these equations, the representative values for length, D, and velocity, U, are still not specified. To complete the non-dimensionalization, D and U are determined so that the values in parentheses in Eqs. (5.15b) and (5.15c) become unity:

$$D = \left(\frac{\dot{Q}}{c_P \rho_\infty T_\infty g^{1/2}}\right)^{2/5} \tag{5.16a}$$

$$U = g^{1/2}\left(\frac{\dot{Q}}{c_P \rho_\infty T_\infty g^{1/2}}\right)^{1/5} \tag{5.16b}$$

Although there is an arbitrariness whether or not to include $\left(1 - \chi_R\right)$ in the above representative values, it is excluded from the present formulation.

By performing dimensional analysis based on Eq. (5.15), the dependence of the characteristic quantities of the fire plume (B^*, W^*, and T^*) on the dimensionless height, Z^*, can be analyzed. Let us assume the following power dependencies.

$$B^* \propto \left(Z^*\right)^l, \quad W^* \propto \left(Z^*\right)^m, \quad \text{and} \quad T^* \propto \left(Z^*\right)^n \tag{5.17}$$

where l, m, and n are the power indices. Then, substituting Eq. (5.17) into Eq. (5.15), we obtain balance equations for Z^*. As the power indices on both sides of the equations need to be identical to ensure dimensional integrity, l, m, and n are specified in the flame and plume regimes as follows, respectively:

$$\text{Flame: } l = 1, \quad m = \frac{1}{2}, \quad \text{and} \quad n = 0 \tag{5.18a}$$

$$\text{Plume: } l = 1, \quad m = -\frac{1}{3}, \quad \text{and} \quad n = -\frac{5}{3} \tag{5.18b}$$

In summary, the following relationships are obtained for the dimensionless numbers B^*, W^*, and T^*, as follows:

$$\text{Width: } B^* \propto Z^* \tag{5.19a}$$

$$\text{Velocity: } W^* \propto \begin{cases} (1 - \chi_R)^{1/2} \left(\dfrac{\Delta H_C}{rc_P T_\infty} \right)^{1/2} (Z^*)^{1/2} & \text{(Flame)} \\ (1 - \chi_R)^{1/3} (Z^*)^{-1/3} & \text{(Plume)} \end{cases} \tag{5.19b}$$

$$\text{Temperature rise: } T^* \propto \begin{cases} (1 - \chi_R) \left(\dfrac{\Delta H_C}{rc_P T_\infty} \right) (Z^*)^0 & \text{(Flame)} \\ (1 - \chi_R)^{2/3} (Z^*)^{-5/3} & \text{(Plume)} \end{cases} \tag{5.19c}$$

In the flame regime, B^* and W^* increase as Z^* increases, while T^* remains constant. In the plume regime, while B^* continues to increase as Z^* vincreases, W^* and T^* turn to decrease.

Figures 5.6 and 5.7 show the measured ΔT and w along the centerline above a 0.3-m porous refractory burner operating on natural gas at several rates in the range of 14.4–57.5 kW (McCaffrey, 1979). Considering that $Z^* \propto z$, the experimental results in Figures 5.6 and 5.7 confirm that the relationships between the characteristic quantities (ΔT and w) and the elevation above a fire source, z, theoretically predicted by Eq. (5.19) is broadly valid for actual fire plumes. However, the theoretical model cannot sufficiently describe the deviations from the experimental results for both ΔT and w, especially in the flame regime close to the burner. This is because there is a fuel-rich core with a high concentration of low-temperature, low-speed fuel near the burner surface, which deviates from the assumptions introduced in the model derivation (e.g., the assumption that entrained air mixes uniformly with fuel supplied from the bottom in the flame regime). However, as such a deviation is limited to the close vicinity of the fuel outlet, the deviation effect is minor from the viewpoint of hazard evaluation of outdoor fires.

Eq. (5.19) only expresses proportional relationships among physical quantities and cannot be used to quantify the involved physical quantities

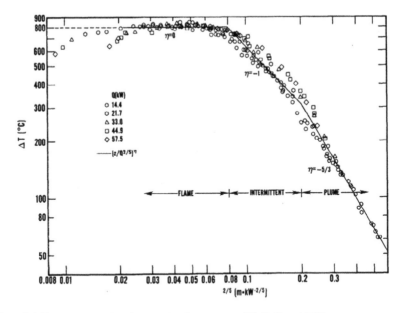

Figure 5.6 Temperature rise above point fire sources (McCaffrey, 1979).

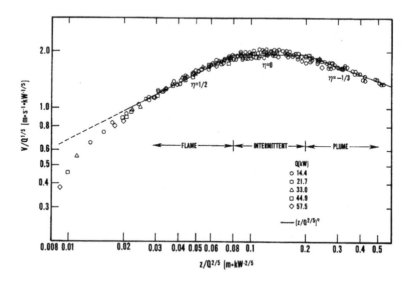

Figure 5.7 Upward velocity above point fire sources (McCaffrey, 1979).

as they are. Thus, Eq. (5.19) is generalized by incorporating coefficients, C_B, C_W, and C_T, and a constant, n, commonly involved in the power indices of Z^* as follows:

Width: $B^* = C_B \cdot Z^*$ (5.20a)

Velocity: $W^* = C_W \cdot \left(Z^*\right)^n$ (5.20b)

Temperature rise: $T^* = C_T \cdot \left(Z^*\right)^{2n-1}$ (5.20c)

Note that the $\left(1 - \chi_R\right)$ and $\Delta H_C Y_{O,\infty} / \left(rc_P T_\infty\right)$, could have been explicit terms in Eq. (5.20). However, they are both included in the coefficients C_W and C_T, as they can be regarded as constant over a wide range of conditions. Temperature rise, ΔT, and flow velocity, w, of fire plumes have been experimentally investigated by many researchers (Beyler, 1986). Although there is a variation in the values of C_T between researchers in the flame regime due to a difference in the temperature measurement accuracy in each experiment, the values for the other coefficients were generally similar. Table 5.1 shows a regression result by Cox et al. and Quintiere et al. (Cox et al., 1980; Quintiere et al., 1998). The values of n for the continuous flame and buoyancy plume regimes were identified based on the theoretical analysis described above. Although there was no theoretical ground, the value of n in the intermediate regime, the intermittent flame regime, was determined to be consistent with the theoretically supported value of n in the regimes at both ends.

Eq. (5.20) has been experimentally validated to hold for a wide range of conditions. However, it has been reported that ΔT in continuous and intermittent flame regimes could be underestimated for a fire source with a large diameter, W (Baum et al., 1989). This is because a large amount of smoke is produced from a fire source with a large W, which blocks radiative heat loss from the flame (Mudan, 1984). A fire source with a larger W has a smaller surface-area-to-volume ratio. This reduces the availability of fresh air from the perimeter, resulting in increased production of smoke. Such a blockage effect can be reflected in the above formulation as a reduction in the radiative fraction, χ_R, which in turn increases the ΔT. Figure 5.8 shows examples

Table 5.1 The plume parameters for point fire sources (Cox et al., 1980; Quintiere et al., 1998).

Regime	Range	n	C_B	C_W	C_T
Flame	$0 < Z^* \le 1.32$	$1/2$	0.179	2.18	3.00
Intermittent	$1.32 < Z^* \le 3.30$	0	–	2.40	3.94
Plume	$3.30 < Z^*$	$-1/3$	0.118	3.57	8.61

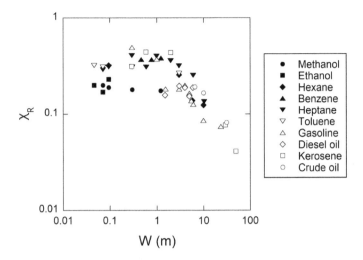

Figure 5.8 Radiative fraction of heat release as a function of pool fire diameter (Koseki, 1989; Hamins et al., 1991; Munoz et al., 2007; Zhou et al., 2017).

of χ_R measurements for several liquid fuels with various W (Koseki, 1989; Hamins et al., 1991; Muñoz et al., 2007; Zhou et al., 2017). As a whole, the χ_R of alcohol fuels are small, whereas those of heavy hydrocarbon fuels are large. It also shows that the change in χ_R is small for relatively small fire sources, but χ_R decreases as W increases for larger fire sources. There has been an attempt to model the behavior of fire plumes considering the variability of χ_R, which partially solves the underestimation of the ΔT (Ma et al., 2003).

The following dimensionless heat release rate (HRR), Q_z^*, has been widely used to describe various buoyancy-induced phenomena involved in a fire (Zukoski et al., 1980/81):

$$Q_z^* = \left(Z^*\right)^{-5/2} = \frac{\dot{Q}}{c_P \rho_\infty T_\infty g^{1/2} z^{5/2}} \tag{5.21}$$

Q_z^* is nothing but a variant of the dimensionless number, Z^*, defined earlier. When Q_z^* is used instead of Z^*, Eq. (5.20) can be transformed as:

$$\text{Width: } B^* = C_B \cdot \left(Q_z^*\right)^{-2/5} \tag{5.22a}$$

$$\text{Velocity: } W^* = C_W \cdot \left(Q_z^*\right)^{-2n/5} \tag{5.22b}$$

$$\text{Temperature rise: } T^* = C_T \cdot \left(Q_z^*\right)^{-2(2n-1)/5} \tag{5.22c}$$

The values of the coefficients, C_B, C_W, and C_T, and the constant, n, in Table 5.1 are applicable without modification.

In practical evaluations of the hazard of fire spread using Eq. (5.20) or Eq. (5.22), one needs to convert the calculated dimensionless values to finite dimensions. As some variables included in Z^* or Q_z^* can be regarded as approximately constant under a general outdoor environment, Eq. (5.20) or Eq. (5.22) can be transformed into the following tractable forms:

$$\text{Width: } b = \begin{cases} 0.179z & \left(z \le 0.08\dot{Q}^{2/5}\right) \\ 0.118z & \left(0.20\dot{Q}^{2/5} < z\right) \end{cases} (\text{m}) \tag{5.23a}$$

$$\text{Velocity: } w = \begin{cases} 6.83\dot{Q}^{1/5}\left(\dfrac{z}{\dot{Q}^{2/5}}\right)^{1/2} & \left(z \le 0.08\dot{Q}^{2/5}\right) \\[2mm] 1.85\dot{Q}^{1/5} & \left(0.08\dot{Q}^{2/5} < z \le 0.20\dot{Q}^{2/5}\right) \\[2mm] 1.08\dot{Q}^{1/5}\left(\dfrac{z}{\dot{Q}^{2/5}}\right)^{-1/3} & \left(0.20\dot{Q}^{2/5} < z\right) \end{cases} (\text{m}/\text{s}) \tag{5.23b}$$

$$\text{Temperature rise: } \Delta T = \begin{cases} 880 & \left(z \le 0.08\dot{Q}^{2/5}\right) \\[2mm] 70.0\left(\dfrac{z}{\dot{Q}^{2/5}}\right)^{-1} & \left(0.08\dot{Q}^{2/5} < z \le 0.20\dot{Q}^{2/5}\right) \\[2mm] 23.6\left(\dfrac{z}{\dot{Q}^{2/5}}\right)^{-5/3} & \left(0.20\dot{Q}^{2/5} < z\right) \end{cases} (\text{K})$$

$$\tag{5.23c}$$

The only variables included in Eq. (5.23) are the HRR, $\dot{Q}$ (kW), and the elevation, z (m). Figure 5.9 shows the calculated results of Eq. (5.23) with different $\dot{Q}$. There are contrasting changes in ΔT and w in the intermittent flame and buoyancy plume regimes; the reduction rate is considerably larger for ΔT than for w.

5.2.3 Virtual origin

The concept of virtual origin is often introduced to improve the prediction accuracy of point source plume models. In point source plume models, the source of the heat generation is represented by a point on the ground. However, such an assumption was introduced to allow for theoretical considerations, but ignores the spatial extent of actual fire sources. The virtual origin is aimed at correcting the error associated with the point source

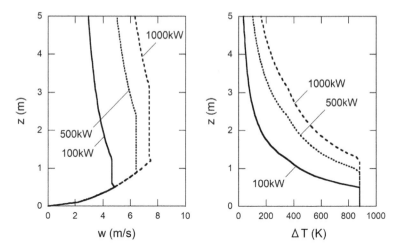

Figure 5.9 Calculated upward velocity and temperature rise along the plume centerline with different HRR (point source).

assumption by adjusting the position of the point source in the vertical direction (Figure 5.10). By introducing a correction, Δz, the height, z, in the plume model is replaced by $z + \Delta z$:

$$z \rightarrow z + \Delta z \qquad (5.24)$$

Heskestad obtained the following regression equation from the results of several fire experiments, deriving Δz in a manner consistent with the plume theory (Heskestad, 1983b):

$$\Delta z = 1.02W - 0.083\dot{Q}^{2/5} \,(\mathrm{m}) \qquad (5.25)$$

where W is the diameter or side length of the fire source. In Figure 5.10, the virtual origin is located under the ground. However, according to Eq. (5.25), the virtual origin can be either under or above the floor. Δz is positive when $W > 0.081\dot{Q}^{2/5}$ whereas Δz is negative when $W < 0.081\dot{Q}^{2/5}$.

5.2.4 Flame height

The equation for the temperature rise, ΔT, allows us to evaluate the ignition probability or the time to ignition of combustibles immersed in a fire plume. However, if a combustible is at a distant location from the fire source, one needs to calculate the radiative heat transfer from the fire source to the combustible, $\dot{Q}_R$, in a form:

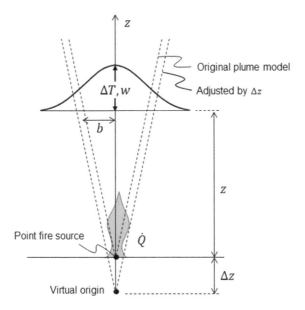

Figure 5.10 Virtual origin of a fire plume.

$$\dot{Q}_R = AF \cdot E \qquad (5.26)$$

where A is the exposed surface area of the combustible, F is the view factor of the flame as viewed from the combustible, and E is the emissive power. E can be calculated using Stefan-Boltzmann's law. The calculation procedure could be complicated if the spatial non-uniformity of temperature within the flame is considered. To avoid this complexity, the flame is often regarded as a pseudo-solid object with uniform radiation characteristics. The radiative portion of the HRR, $\chi_R \dot{Q}$, is regarded as the radiation energy, E. In this case, it is more important to identify the fire source geometry (flame height L_{fl} in no-wind conditions) for the calculation of the F rather than ΔT as the latter can be considered uniform in the continuous flame regime.

Various researchers have proposed models for the average flame height, L_{fl} (Thomas et al., 1961; Thomas, 1963a; Zukoski, 1980/81; Heskestad, 1981; Heskestad, 1983a; Delichatsios, 1987; Quintiere et al., 1998). In this subsection, we discuss L_{fl} based on the relationships of the characteristic quantities as derived above. Assuming infinite reaction kinetics in the diffusion flame, the height of the flame tip can be considered as the height at which all the flammable gas supplied from the source on the ground is consumed by the reaction with the oxygen contained in the entrained air (Heskestad, 1981). In other words, $\dot{Q}$ can be viewed as the HRR due to the

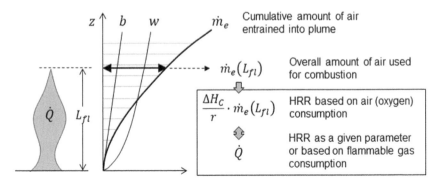

Figure 5.11 Relationship between the rate of air entrainment and the flame height.

consumption of fresh air entrained by the fire plume before it reaches the height, L_{fl} (Figure 5.11). Such a relationship can be expressed as follows:

$$\dot{Q} \propto \frac{\Delta H_C}{r} \cdot \dot{m}_e\left(L_{fl}\right) \tag{5.27}$$

where ΔH_C is the heat of combustion of the combustible, r is the stoichiometric air-to-fuel ratio, and $\dot{m}_e(z)$ is the cumulative amount of air entrainment at a height, z. For most fuels, $\Delta H_C/r$ is approximately constant ($\cong 3.0\,\text{MJ}/\text{kg}$) (Huggett, 1980). Thus, the relationship between $\dot{Q}$ and L_{fl} can be derived by quantifying $\dot{m}_e\left(L_{fl}\right)$.

In general, it is not straightforward to theoretically formulate $\dot{m}_e(z)$. However, recall that the mass flow rate of the plume formed above a fire source, $\dot{m}_z$, is the sum of the flammable gas supply rate and the $\dot{m}_e(z)$. Among the two components, the $\dot{m}_e(z)$ is predominant as the flammable gas is supplied only from the bottom of the fire plume. In other words, the $\dot{m}_e(z)$ can be approximated using the conservation equation of mass (Eq. (5.4)) as follows:

$$\dot{m}_e(z) \cong \dot{m}_z \propto \rho b^2 w_m \tag{5.28}$$

By introducing the relationships in the plume regime obtained earlier (Eq. (5.20a) for b and Eq. (5.20b) for w_m), Eq. (5.28) can be transformed into:

$$\dot{m}_e(z) \propto \rho_\infty \cdot z^2 \cdot \sqrt{gD}\left(z/D\right)^{-1/3} \tag{5.29}$$

However, note that we assumed $\rho \cong \rho_\infty$ for simplicity. $\dot{m}_e\left(L_{fl}\right)$ calculated from Eq. (5.29) is further substituted in Eq. (5.27) to yield:

$$\left(\frac{\dot{Q}}{c_P \rho_\infty T_\infty g^{1/2} L_{fl}^{5/2}} \right)^{2/3} \propto \frac{\Delta H_C}{r \cdot c_P T_\infty} \tag{5.30}$$

By isolating L_{fl} on the left-hand side, we obtain:

$$\frac{L_{fl}}{W} \propto \left(\frac{\dot{Q}}{c_P \rho_\infty T_\infty g^{1/2} W^{5/2}} \right)^{2/5} \left(\frac{r \cdot c_P T_\infty}{\Delta H_C} \right)^{3/5} \tag{5.31}$$

where W is the diameter or the side length of the fire source. This is a physical relationship that can be used to predict the flame length, L_{fl}. Note that the first term of the right-hand side of this equation is nothing but the dimensionless HRR, which is a modified version of the one in Eq. (5.21):

$$Q_W^* = \frac{\dot{Q}}{c_P \rho_\infty T_\infty g^{1/2} W^{5/2}} \tag{5.32}$$

As discussed above, intermittency is an important characteristic of turbulent diffusion flames. Therefore, the height of flames fluctuates randomly along with their periodical expansion and contraction in shape. However, the average value of flame heights is generally sufficient to evaluate the hazard of fire spread in outdoor fires. Heskestad used the relationship in Eq. (5.31) to regress the average flame height of various burning fuels as shown in Figure 5.12 (Heskestad, 1983a):

$$\frac{L_{fl}}{W} = 15.6 Q_W^{*2/5} \left(\frac{r \cdot c_P T_\infty}{\Delta H_C} \right)^{3/5} - 1.02 \tag{5.33}$$

The average height is defined as the height at which the frequency of flame presence is 50% as shown in Figure 5.13 (Heskestad, 1983a). Recalling that the amount of heat release per unit consumption of air contributing to combustion is approximately constant for a wide variety of fuels, $\Delta H_C / r \cong 3.0$ MJ / kg (Huggett, 1980), Eq. (5.33) can be transformed into a form (Heskestad, 1984):

$$\frac{L_{fl}}{W} = 3.7 Q_W^{*2/5} - 1.02 \tag{5.34a}$$

$$L_{fl} = 0.23 \dot{Q}^{2/5} - 1.02W \, (\text{m}) \tag{5.34b}$$

It is noteworthy that Eq. (5.34b) is an equation with a finite dimension. The variables included in Eq. (5.24) are the flame length, L_{fl} (m), the HRR, $\dot{Q}$ (kW), and the fire source diameter W (m).

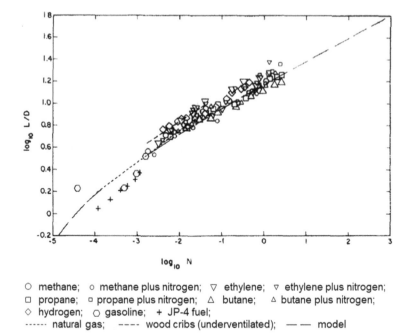

○ methane;   ○ methane plus nitrogen;   ▽ ethylene;   ▽ ethylene plus nitrogen;
□ propane;   ▫ propane plus nitrogen;   △ butane;   △ butane plus nitrogen;
◇ hydrogen;   ○ gasoline;   + JP-4 fuel;
------ natural gas;   ---- wood cribs (underventilated);   — — model

Figure 5.12 Average flame heights of various burning fuels (Heskestad, 1983a).

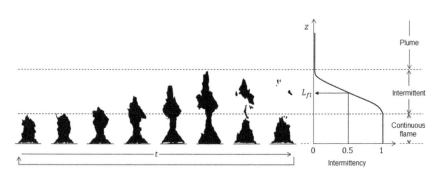

Figure 5.13 Definition of the average flame height based on the visible flame structure (adapted from McCaffrey, 1979).

Zukoski et al. derived a similar relationship for the average flame height, L_{fl} (Zukoski, 1980/81; Zukoski, 1986). The following equation is a regression of the results of experiments using various fuels with different combustion characteristics, including gaseous fuels, alcohol fuels, and woods:

$$\frac{L_{fl}}{W} = \begin{cases} 3.3Q_W^{*2/3} & (Q_W^* < 1.0) \\ 3.3Q_W^{*2/5} & (Q_W^* \geq 1.0) \end{cases}$$ (5.35)

This can be transformed into the form with a finite dimension (Tanaka, 1993):

$$L_{fl} = \begin{cases} 0.03(\dot{Q}/W)^{2/3} & (\dot{Q}/W^{5/2} < 1120) \\ 0.20\dot{Q}^{2/5} & (\dot{Q}/W^{5/2} \geq 1120) \end{cases} \text{(m)}$$ (5.36)

The Heskestad and Zukoski models give similar predictions on flame heights. However, the Heskestad model can give negative predictions for fire sources with a relatively large W, whereas the Zukoski model always gives positive predictions. Such a difference may give the Zukoski model prevalence in practical applications.

WORKED EXAMPLE 5.1

Assume an object above a fire source with a diameter $W = 1$ m and a constant HRR $\dot{Q} = 2$ MW. The location of the object is $z = 5$ m above the fire source and $r = 0.5$ m off the centerline. Calculate the temperature of the plume around the object.

SUGGESTED SOLUTION

We first need to determine which regime the object is included in (continuous flame regime/intermittent flame regime/buoyant plume regime). According to Eq. (5.23), the interface heights between the regimes can be estimated as follows:

continuous flame/intermittent flame: $z = 0.08\dot{Q}^{2/5} = 1.67$ m

intermittent flame/buoyant plume: $z = 0.20\dot{Q}^{2/5} = 4.18$ m

In the given case, the target object at $z = 5$ m is in the buoyant plume regime.

We then calculate the temperature rise along the centerline, ΔT_m, using Eq. (5.23c). We do not simply substitute $z = 5$ m into Eq. (5.23c), but consider an adjustment by the virtual origin. The position of the virtual origin, Δz, can be calculated using Eq. (5.25):

$$\Delta z = 1.02W - 0.083\dot{Q}^{2/5} = -0.72 \text{ m}$$

By substituting Δz into Eq. (5.23c), ΔT_m can be obtained as:

$$\Delta T_m = 23.6\left((z+\Delta z)/\dot{Q}^{2/5}\right)^{-5/3} = 331.5 \text{ K}$$

Since the target object deviates $r = 0.5$ m from the centerline, b also needs to be determined. By using Eq. (5.23a), we obtain:

$$b = 0.118(z+\Delta z) = 0.51 \text{ m}$$

Assuming that the temperature decay from the centerline follows a Gaussian distribution in Eq. (5.1a), the temperature rise of an object at $r = 0.5$ m can be calculated as:

$$\Delta T = \Delta T_m \exp\left[-\beta(r/b)^2\right] = 135.7 \text{ K}$$

Note that we assumed $\beta = 0.913$ (Quintiere et al., 1998).

WORKED EXAMPLE 5.2

Assume three fire sources with side lengths $W = 1, 2$, and 3 m, respectively. The HRR of all the fire sources is $\dot{Q} = 1$ MW. Compare the segmentation of the flame structure (continuous flame, intermittent flame, and buoyant plume regimes) and the average flame height, L_{fl}, of each fire source.

SUGGESTED SOLUTION

According to Eq. (5.23), the interface heights between the regimes can be estimated as follows:

continuous flame/intermittent flame: $z = 0.08\dot{Q}^{2/5} = 1.27$ m

intermittent flame/buoyant plume: $z = 0.20\dot{Q}^{2/5} = 3.17$ m

The heights of the regime interfaces can be determined solely by $\dot{Q}$, and are independent of W. As we assume that $\dot{Q}$ are identical, the interface heights are common for all the fire sources.

Eq. (5.34b) can be used to calculate the average flame height, L_{fl}. In contrast to the regime interfaces, the calculation of L_{fl} requires values of two variables, $\dot{Q}$ and W. Thus, L_{fl} is calculated separately for each fire source, which are:

$$W = 1\text{m}: L_{fl} = 0.23\times(1000)^{2/5} - 1.02\times1 = 2.63 \text{ m}$$

$$W = 2\text{m}: L_{fl} = 0.23\times(1000)^{2/5} - 1.02\times2 = 1.61 \text{ m}$$

$$W = 3\,\text{m} : L_{fl} = 0.23 \times (1000)^{2/5} - 1.02 \times 3 = 0.59\ \text{m}$$

If $\dot{Q}$ is constant, L_{fl} decreases as W increases.

By definition, the average flame height L_{fl} should be larger than the continuous flame height and smaller than the intermittent flame height. Among the three W cases tested, the obtained L_{fl} are consistent with the definition for $W = 1$ m and 2 m, but are out of the range for $W = 3$ m. This is partly because the effect of source geometry becomes more pronounced and the applicability of the point source model diminishes with an increase in the W.

WORKED EXAMPLE 5.3

Assume a square burner with a side length $W = 0.5$ m is fueled with methane at a rate, $\dot{m}_B = 0.0025$ kg $\cdot$ s^{-1}. Calculate the height of the flame, L_{fl}, and its emissivity, ε.

SUGGESTED SOLUTION

Table 4.2 in Chapter 4 shows that the heat of combustion of methane is $\Delta H = 50.1$ kJ $\cdot$ g^{-1}. The HRR, $\dot{Q}$, can be calculated as follows:

$$\dot{Q} = \Delta H \dot{m}_B = (50.1 \times 10^3) \times 0.0025 = 125\ \text{kW}$$

The obtained $\dot{Q}$ is substituted into Eq. (5.34b) to calculate L_{fl} as follows:

$$L_{fl} = 0.23\dot{Q}^{2/5} - 1.02W = 0.23 \times (125)^{2/5} - 1.02 \times 0.5 = 1.08\ \text{m}$$

In the calculation of ε, we need to identify the mean optical length of the flame as a radiator, $\bar{L}$. We consider the flame geometry as a square cylinder with the base side length $W = 0.5$ m and the height $L_{fl} = 1.08$ m. Eq. (3.86) discussed in Chapter 3 can be used to calculate $\bar{L}$ as follows:

$$\bar{L} \cong \frac{3.6V}{A} = \frac{3.6 \times (0.5 \times 0.5 \times 1.08)}{2 \times 0.5 \times 0.5 + 4 \times 0.5 \times 1.08} = 0.365\ \text{m}$$

Substituting the obtained $\bar{L}$ and the effective absorption coefficient of methane, $\kappa = 6.45$ m^{-1}, shown in Table 3.3 in Eq. (3.85), ε can be calculated as follows:

$$\varepsilon = 1 - \exp(-\kappa\bar{L}) = 1 - \exp(-6.45 \times 0.365) = 0.905$$

5.3 LINE FIRE SOURCE

The behavior of line fire sources also has been well investigated, as has the case of point fire sources. A linear fire source is an infinitely long fire source

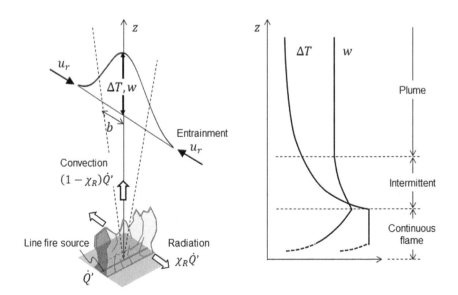

Figure 5.14 Schematic representation of the fire plume above a line source in a quiescent environment and the transition of the centerline behavior.

with infinitesimal thickness, as schematically illustrated in Figure 5.14. Actual fire sources have finite lengths and thicknesses. Thus, a linear fire source is an ideal fire source. However, a fire source with a sufficiently long side length relative to its thickness can be considered as a line fire source, especially when the thickness effect becomes weaker at a higher elevation. In outdoor fires, several types of fire sources can be considered as line sources. For example, the earlier the combustibles get ignited, the earlier they bun out in outdoor fires. This forms a linear fire-front at the perimeter of the burning area.

5.3.1 Governing equations

Similar to the case of point fire sources, the behavior of line fire sources has also been modeled based on the classical plume theory (Rouse et al., 1952; Yokoi, 1960; Lee et al., 1961; Yuan et al., 1996; Quintiere et al., 1998). However, to maintain the consistency of the discussion from the previous section, we alternatively employ the approach that incorporates the physical considerations of the flow behavior into the conservational relationships between characteristic quantities (Figure 5.15).

The derivation process of the governing equations for the behavior of a line fire source is omitted as there is no procedural difference from that of a

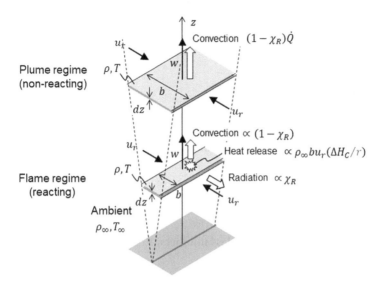

Figure 5.15 Behavior of characteristic parameters in flame and plume regimes above a line source.

point fire source. The final forms of the conservation equation for the mass, momentum, and thermal energy are summarized as follows:

$$\text{Mass:}\ \rho b w \propto \int \rho_\infty w dz \tag{5.37a}$$

$$\text{Momentum:}\ w \propto \sqrt{\frac{\Delta T}{T_\infty} g z} \tag{5.37b}$$

$$\text{Energy:}\ \begin{cases} \dfrac{d}{dz}(c_P \rho b w \Delta T) \propto \rho_\infty w \cdot (1 - \chi_R) \dfrac{\Delta H_C}{r} & \text{(Flame)} \\[2mm] (1 - \chi_R)\dot{Q}' \propto c_P \rho b w \Delta T & \text{(Plume)} \end{cases} \tag{5.37c}$$

Here, b is the cross-sectional half-width, $\dot{Q}'$ is the HRR per unit length, c_P is the heat capacity of the gas, g is the gravitational acceleration, and s and u are the distances and velocities along the trajectory, respectively. w is the vertical velocity, ρ and ρ_∞ are the gas density and ambient gas density, respectively, and $\Delta\rho$ is the density difference.

A major difference between the two types of ideal fire sources is the way they entrain air from the surroundings. It is radial to the center for the point source, whereas it is perpendicular to the linear center for the line source. The differences appear in the final forms of the conservation equations in

Eqs. (5.11) and (5.13) for the point source and Eq. (5.37) for the line source. In Eq. (5.37), the conservation equations for the mass and momentum are common between the flame and buoyancy plume regimes. These equations involve three unknown characteristic quantities, b, w, and ΔT.

5.3.2 Dimensional analysis

The governing equations are simplified for elucidating the relationships between the variables by non-dimensionalization. We use the representative values D for length, U for velocity, and T_∞ for temperature, to introduce the following dimensionless numbers:

$$Z^* = \frac{z}{D}, \quad B^* = \frac{b}{D}, \quad W^* = \frac{w}{U}, \quad \text{and} \quad T^* = \frac{\Delta T}{T_\infty} \tag{5.38}$$

By substituting these dimensionless numbers, Eq. (5.37) can be transformed as follows:

Mass: $B^* \propto Z^*$ (5.39a)

Momentum: $W^{*2} \propto \left(\frac{gD}{U^2}\right) T^* Z^*$ (5.39b)

$$\text{Energy:} \begin{cases} B^* T^* \propto \left(1 - \chi_R\right)\left(\dfrac{\Delta H_C}{r c_P T_\infty}\right) Z^* & \text{(Flame)} \\[3mm] B^* W^* T^* \propto \left(1 - \chi_R\right)\left(\dfrac{\dot{Q}'}{c_P \rho_\infty T_\infty D U}\right) & \text{(Plume)} \end{cases} \tag{5.39c}$$

For simplicity, Boussinesq approximation is introduced, which assumes $\rho = \rho_\infty$ except for the buoyancy term. However, in these equations, the representative values for length, D, and velocity, U, are still not specified. To complete the non-dimensionalization, D and U are determined so that the values in the parentheses in Eqs. (5.39b) and (5.39c) become unity:

$$D = \left(\frac{\dot{Q}'}{c_P \rho_\infty T_\infty g^{1/2}}\right)^{2/3} \tag{5.40a}$$

$$U = g^{1/2}\left(\frac{\dot{Q}'}{c_P \rho_\infty T_\infty g^{1/2}}\right)^{1/3} \tag{5.40b}$$

The expressions for D and U are similar to Eq. (5.16) for the point source. As for D, $\dot{Q}$ has been replaced by the one per unit length $\dot{Q}'$, and the power index has been changed to 2/3 from 2/5. As for U, $\dot{Q}$ has also been replaced by $\dot{Q}'$, and the power index has been changed to 1/3 from 1/5. Again, the $(1 - \chi_R)$ is excluded from the above definition.

By performing dimensional analysis based on Eq. (5.39), the dependence of the characteristic quantities of the fire plume (B^*, W^*, and T^*) on the dimensionless height, Z^*, can be analyzed. Let us introduce the following power dependencies:

$$B^* \propto \left(Z^*\right)^l, \quad W^* \propto \left(Z^*\right)^m, \quad \text{and} \quad T^* \propto \left(Z^*\right)^n \tag{5.41}$$

where l, m, and n are the exponents. Then, substituting Eq. (5.41) into Eq. (5.39), we obtain balance equations for Z^*. As the power indices on both sides of the equations need to be identical to ensure dimensional integrity, l, m, and n are specified in the flame and plume regimes as follows, respectively:

$$\text{Flame:} \, l = 1, \, m = \frac{1}{2}, \, \text{and} \, n = 0 \tag{5.42a}$$

$$\text{Plume:} \, l = 1, \, m = 0, \, \text{and} \, n = -1 \tag{5.42b}$$

In summary, the following relationships are obtained for the dimensionless numbers B^*, W^*, and T^* :

$$\text{Width:} \, B^* \propto Z^* \tag{5.43a}$$

$$\text{Velocity:} \, W^* \propto \begin{cases} (1 - \chi_R)^{1/2} \left(\dfrac{\Delta H_C}{r c_p T_\infty} \right)^{1/2} \left(Z^*\right)^{1/2} & \text{(Flame)} \\ (1 - \chi_R)^{1/3} \left(Z^*\right)^0 & \text{(Plume)} \end{cases} \tag{5.43b}$$

$$\text{Temperature rise:} \, T^* \propto \begin{cases} (1 - \chi_R) \left(\dfrac{\Delta H_C}{r c_p T_\infty} \right) \left(Z^*\right)^0 & \text{(Flame)} \\ (1 - \chi_R)^{2/3} \left(Z^*\right)^{-1} & \text{(Plume)} \end{cases} \tag{5.43c}$$

B^* increases in proportion to Z^* in both the buoyant plume and flame regimes, which is consistent with the relationship for the point source. W^* increases in proportion to Z^* to the power of 1/2 in the flame regime and remains constant in the buoyant plume regime. T^* is constant in the flame regime and decreases inversely proportional to Z^* in the buoyant plume regime. In Eq. (5.43), the power indices are identical to the point source in the flame regime. The differences appear in the power indices of T^* and W^* in the buoyant plume regime. Among these, the constancy of

W^* in the buoyant plume regime implies that the influence of a line source extends to infinity without attenuation, which is not realistic in actual fires. This is because the length of the interface between the plume and the ambient, through which the surrounding air is entrained, does not change with height z in the model. One should be mindful of the applicability range of the model.

Figures 5.16 and 5.17 show the measured temperature rise, ΔT, and upward flow velocity, w, along the centerline above different fire sources (Hasemi et al., 1989; Yuan et al., 1996). The source was a 1.0 m long × 0.1 m wide porous diffusion burner operating on propane in the range of 5–350 kW for Hasemi et al. (Hasemi et al., 1989). The sources were 0.2 m long × 0.015 m wide and 0.5 m long × 0.015 m wide porous refractory burners, and a 0.5 m long × 0.05 m wide sandbox burner, all operating on natural gas at several rates in the range of 2–110 kW for Yuan et al. (Yuan et al., 1996). The solid lines are the regressed results based on the relationships obtained by the dimensional analysis, which reasonably correlates with the data points. A relatively large variation in the flow velocity results could be attributed to the difficulty in the measurement without disturbing the flow.

As Eq. (5.43) only expresses proportional relationships between physical quantities, we generalize the equations by introducing coefficients, C_B, C_W, and C_T, and a common constant, n, for the power indices of Z^* in each equation:

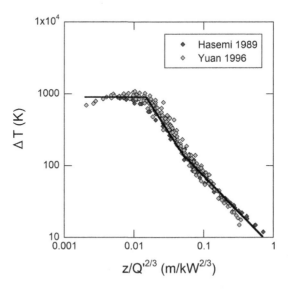

Figure 5.16 Temperature rise above line fire sources (Hasemi et al., 1989; Yuan et al., 1996).

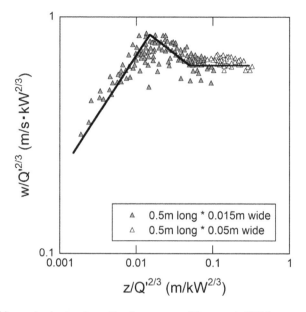

Figure 5.17 Upward velocity above line fire sources (Yuan et al., 1996).

Width: $B^* = C_B \cdot Z^*$ (5.44a)

Velocity: $W^* = C_W \cdot \left(Z^*\right)^n$ (5.44b)

Temperature rise: $T^* = C_T \cdot \left(Z^*\right)^{2n-1}$ (5.44c)

Note that the terms, $(1 - \chi_R)$ and $\Delta H_C Y_{O,\infty} / (r c_P T_\infty)$, could have been given explicitly. However, they were both included in the coefficients C_W and C_T, as they can be regarded as constant over a wide range of conditions. By referring to some experimental results (Hasemi et al., 1989; Yuan et al., 1996; Quintiere et al., 1998), the coefficients, C_B, C_W, and C_T, and the common constant, n, can be determined as shown in Table 5.2. The power indices for the continuous flame and buoyant plume regimes were given based on the theoretical analysis described above. However, for the intermittent flame regime, there was arbitrariness in determining the power index as it was not determined theoretically. Although Yuan et al. suggested $n = 0$ (Yuan et al., 1996), this would result in discontinuous changes in ΔT and w at the regime interfaces. Thus, $n = -1/4$ was selected to interpolate the regimes at both ends.

The dimensionless HRR for line fire sources, $Q_l^{\prime*}$, can be derived similarly to that for point fire sources, Q_z^*, as defined in Eq. (5.21):

Table 5.2 The plume parameters for line fire sources (Hasemi et al., 1989; Yuan et al., 1996; Quintiere et al., 1998).

Regime	Range	n	C_B	C_W	C_T
Flame	$0 < Z^* \leq 1.60$	$1/2$	0.444	2.20	3.06
Intermittent	$1.60 < Z^* \leq 5.34$	$-1/4$	–	3.13	6.22
Plume	$5.34 < Z^*$	0	0.103	2.06	2.63

$$Q_z^{'*} = \left(Z^*\right)^{-3/2} = \frac{\dot{Q}'}{c_P \rho_\infty T_\infty g^{1/2} z^{3/2}} \tag{5.45}$$

By substituting Z^* with $Q_z^{'*}$, Eq. (5.44) can be transformed as follows:

Width: $B^* = C_B$ (5.46a)

Velocity: $W^* = C_W \cdot \left(Q_z^*\right)^{-2n/3}$ (5.46b)

Temperature rise: $T^* = C_T \cdot \left(Q_z^*\right)^{-2(2n-1)/3}$ (5.46c)

The values for the coefficients, C_B, C_W, and C_T, and the constant, n, in Table 5.2 are applicable to Eq. (5.46) without modification.

In a practical evaluation of the hazard of fire spread using Eq. (5.44) or Eq. (5.46), one needs to convert the calculated dimensionless values to finite dimensions. As some variables included in Z^* or Q_z^* can be regarded as approximately constant under a general outdoor environment, Eq. (5.44) or Eq. (5.46) can be transformed into the following tractable forms:

$$\text{Width: } b = \begin{cases} 0.444z & \left(z \leq 0.015\dot{Q}'^{\,2/3}\right) \\ 0.103z & \left(0.05\,\dot{Q}'^{\,2/3} < z\right) \end{cases} \text{(m)} \tag{5.47a}$$

$$\text{Velocity: } w = \begin{cases} 6.88\dot{Q}'^{1/3}\left(\dfrac{z}{\dot{Q}'^{2/3}}\right)^{1/2} & \left(z \leq 0.015\dot{Q}'^{2/3}\right) \\ 0.295\dot{Q}'^{1/3}\left(\dfrac{z}{\dot{Q}'^{2/3}}\right)^{-1/4} & \left(0.015\dot{Q}'^{2/3} < z \leq 0.05\dot{Q}'^{2/3}\right) \\ 0.624\dot{Q}'^{1/3} & \left(0.05\dot{Q}'^{2/3} < z\right) \end{cases} \text{(m/s)}$$

(5.47b)

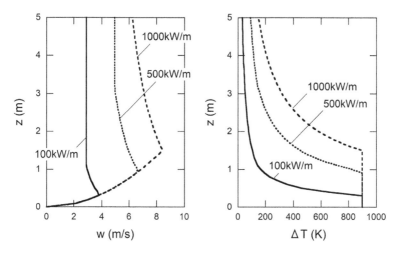

Figure 5.18 Calculated upward velocity and temperature rise along the plume centerline with different HRR (line source).

$$
\text{Temperature rise: } \Delta T = \begin{cases} 898 & \left(z \leq 0.015\,\dot{Q}'^{\,2/3}\right) \\[3mm] 1.65\left(\dfrac{z}{\dot{Q}'^{\,2/3}}\right)^{-3/2} & \left(0.015\,\dot{Q}'^{\,2/3} < z \leq 0.05\,\dot{Q}'^{\,2/3}\right) \\[3mm] 7.20\left(\dfrac{z}{\dot{Q}'^{\,2/3}}\right)^{-1} & \left(0.05\,\dot{Q}'^{\,2/3} < z\right) \end{cases} \text{(K)}
$$

$$(5.47c)$$

Note that the variables included in Eq. (5.47) are the HRR per unit length, $\dot{Q}'$ (kW·m^{-1}), and the height, z (m). ΔT in the continuous flame regime is 898°C for the line source, which is similar to 880°C for the point source (Eq. (5.23)). Figure 5.18 shows the calculated w and ΔT with varying $\dot{Q}'$.

5.3.3 Flame height

For linear sources, several investigators have also proposed models for the average flame height, L_{fl} (Thomas et al., 1961; Thomas, 1963a; Steward, 1964; Yuan et al., 1996; Quintiere et al., 1998). We can use the relationships between the characteristic quantities discussed above to derive the L_{fl}. The derivation procedure follows the case of the point source model. Assuming infinite reaction kinetics in the diffusion flame, the height of the flame tip can be considered as the height at which all the flammable gas supplied from the source on the ground is consumed by the reaction with the oxygen contained in the entrained air (Heskestad, 1981). Such a relationship can be expressed as:

$$\dot{Q}' \propto \frac{\Delta H_C}{r} \cdot \dot{m}_e\left(L_{fl}\right) \tag{5.48}$$

where ΔH_C is the heat of combustion, and r is the stoichiometric air-to-fuel ratio. In Eq. (5.48), $\Delta H_C / r$ can be considered as constant for most fuels ($\cong$ 3.0 MJ / kg) (Huggett, 1980).

As in the case of a point fire plume, the mass flow rate per unit length of a line fire plume, $\dot{m}_e$, mostly attributed to the air entrained from the perimeter. The cumulative amount of air entrainment by the line fire plume at a height z, which is $\dot{m}_e(z)$, can be approximated using the conservation equation of mass (Eq. (5.37a)) as:

$$\dot{m}_e(z) \cong \dot{m}_z \propto \rho b w_m \tag{5.49}$$

By introducing correlations in the plume regime (Eq. (5.43a) for b and Eq. (5.43b) for w_m), Eq. (5.49) is transformed into:

$$\dot{m}'_e(z) \propto \rho_\infty g^{1/2} z^{3/2}\left(Z^*\right)^{-1/2} \tag{5.50}$$

However, $\rho \cong \rho_\infty$ was assumed for simplicity. The $\dot{m}_e\left(L_{fl}\right)$ calculated from Eq. (5.50) is further substituted in Eq. (5.48) to yield:

$$\frac{L_{fl}}{W} \propto \left(\frac{\dot{Q}'}{c_P \rho_\infty T_\infty g^{1/2} W^{3/2}}\right)^{2/3}\left(\frac{r \cdot c_P T_\infty}{\Delta H_C}\right) \tag{5.51}$$

where W is the width of the line source. Note that the first term of the right-hand side of Eq. (5.51) is the dimensionless HRR of the line source, which is:

$$Q_W^{'*} = \frac{\dot{Q}'}{c_P \rho_\infty T_\infty g^{1/2} W^{3/2}} \tag{5.52}$$

The second term on the right-hand side, $r \cdot c_P T_\infty / \Delta H_C$, can be regarded as approximately constant over a wide range of conditions. Thus, Eq. (5.51) can be simplified to be a proportional relationship between the normalized flame height, L_{fl} / W, and the dimensionless HRR, $Q_W^{'*}$. The validity of the derived equation can be confirmed by the measurement results of several investigators, as plotted in Figure 5.19 (Hasemi et al., 1989; Sugawa et al.,

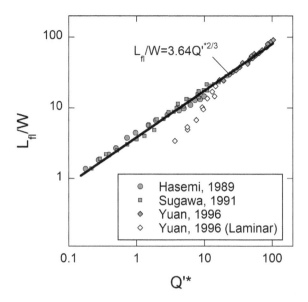

Figure 5.19 Correlation of flame height data above various line sources (Hasemi et al., 1989; Sugawa et al., 1991; Yuan et al., 1996).

1991; Yuan et al., 1996). Figure 5.19 also includes the following regression equation proposed by Yuan et al. (Yuan et al., 1996):

$$\frac{L_{fl}}{W} = 3.64 Q_W^{\prime*2/3} \qquad (5.53)$$

The open symbols in Figure 5.19 represent the measurement results for laminar diffusion flames, which show a different trend from the others. However, the remaining data points for turbulent diffusion flames are successfully regressed by Eq. (5.53). As most of the flames expected in outdoor fires are turbulent diffusion flames, Eq. (5.53) is applicable to a wide variety of practical problems. The dimensionless equation can be transformed into a form with a finite dimension as follows (Yuan et al., 1996):

$$L_{fl} = 0.034 \dot{Q}^{\prime 2/3} \, (\text{m}) \qquad (5.54)$$

The variables included in Eq. (5.54) are $\dot{Q}'$ (kW · m⁻¹) and L_{fl} (m).

WORKED EXAMPLE 5.4

Assume that there are two burners, one is square with a side length of 1 m, and the other is rectangular with a long side of 5 m and a short side of

0.2 m, both with an area of 1 m². Compare the average lengths of the flames formed above the square and rectangular burners, $L_{fl,1}$ and $L_{fl,2}$, respectively. The fuel supply rate of each burner is controlled so that the HRR becomes $\dot{Q} = 1$ MW.

SUGGESTED SOLUTION

The $L_{fl,1}$ for the square burner can be calculated using Eq. (5.34b) as follows:

$$L_{fl,1} = 0.23 \times (1000)^{2/5} - 1.02 \times 5 = 2.63 \text{ m}$$

The rectangular burner has an aspect ratio of 1/25, which can be regarded as a line source. The HRR per unit length, $\dot{Q}'$, is as follows:

$$\dot{Q}' = 1000 \div 5 = 200 \text{ kW / m}$$

By substituting the $\dot{Q}$ in Eq. (5.54), $L_{fl,2}$ can be calculated as follows:

$$L_{fl,2} = 0.034 \dot{Q}'^{2/3} = 1.16 \text{ m}$$

Even though the HRR were the same at 1 MW, the flame lengths above the two burners $L_{fl,1}$ and $L_{fl,2}$ were significantly different. As discussed above, the flame length depends on the rate of fresh air entrainment. In general, the larger the rate of fresh air entrainment is, the shorter the flame length is. The rate of air entrainment depends on the perimeter of the burner, d_{fl}, which is the representative length of the plume–ambient interface. Comparing the perimeter, d_{fl}, of the two burners, $d_{fl,1} = 4$ m for the square burner and $d_{fl,2} = 10.4$ m for the rectangular burner. The relationship between the perimeters, which is $d_{fl} < d_{fl,2}$, corresponds to the relationship between the flame lengths, which is $L_{fl,1} > L_{fl,2}$.

5.4 FLAME EJECTION FROM AN OPENING (WINDOW FLAME)

When a building is involved in a fire, the generation of hot combustion product gas inside the building causes the breakage of window panes. The unburnt flammable gas that could not be consumed inside is then vented from the opening together with combustion product gas. As they mix with the fresh air outside, the flammable gas reacts with the ambient fresh air to release thermal energy. Such an ejection of the flammable and combustion product gases forms an external flame above the opening (Figure 5.20). In this section, we call such a flame as a window flame.

In Chapter 4, we discussed that a fully developed compartment fire can be categorized either into a fuel-controlled or ventilation-controlled fire. In a fuel-controlled fire, the formation of a window flame is basically avoided as

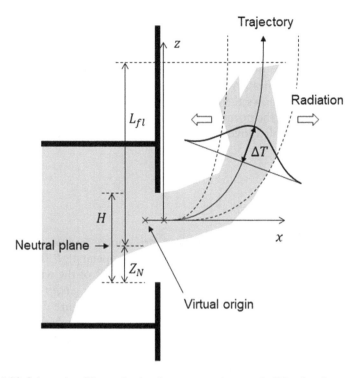

Figure 5.20 Schematic of flame ejection from an opening on a building façade.

all of the flammable gas is consumed inside the compartment. Meanwhile, in a ventilation-controlled fire, a portion of the flammable gas that could not be consumed inside the compartment is vented and reacts with the ambient fresh air to form a window flame. Window flames are a major factor in fire spread to the upper floors of the burning compartment and combustibles (vegetation and structures) adjacent to the fire-involved building in outdoor fires.

5.4.1 Thermal behavior along the trajectory

Window flame is formed as unburnt flammable gas, together with combustion product gas, vented from an opening reacts with oxygen outside the compartment. Therefore, unlike the point and line fire sources discussed above, window flame is already at a high thermal energy state and has an initial velocity in the horizontal direction at the stage of ejection. The hot gas vented from the opening gains the vertical momentum due to buoyancy, while the horizontal momentum gradually decays due to diffusion. Although the effect of buoyancy decreases as the distance from the venting point increases, the hot gas eventually rises almost vertically upward.

The formation process of window flames considerably differs from that of point and line fire sources that rise vertically upward. However, Yokoi demonstrated that the scaling relationship obtained for the point fire sources could be extended to regress the measured temperature rise along the curved trajectory of window flames (including both combusting and non-combusting regimes), ΔT, ejected from a small compartment (Yokoi, 1960). Figure 5.21 shows the correlations between the dimensionless temperature rise, Θ, and the normalized height, z / r_0, for the test cases with and without a vertical wall above an opening. Θ is defined as follows:

$$\Theta = \frac{\Delta T r_0^{5/3}}{\left(\dfrac{\dot{Q}^2 T_\infty}{c_P^2 \rho^2 g}\right)^{1/3}} \tag{5.55}$$

where $r_0 = \sqrt{BH / 2\pi}$ is the equivalent fire source radius (radius of a circle whose area is equal to the upper half of the opening), B is the opening width, H is the opening height, $\dot{Q}$ is the virtual HRR of the window flame, T_∞ is the ambient temperature, c_P is the specific heat of gas, ρ is the plume density, and g is the acceleration due to gravity. In Figure 5.21, Θ takes almost a constant value when z / r_0 is small, regardless of whether there is a vertical wall above the opening or not. However, if the vertical wall above the opening is not present, the attenuation of Θ varied depending on the aspect ratio of the opening, B / H. The degree of attenuation of Θ is greater for wide openings than for narrow openings. On the other hand, if the vertical wall above the opening is present, the effect of the B / H was insignificant.

Several attempts have been made to formulate the temperature rise along the trajectory of window flames since Yokoi (Ohmiya et al., 2001; Yamaguchi et al., 2005; Himoto et al., 2009a; Lee et al., 2012; Tang et al., 2012; Hu et al., 2013). However, the one proposed by Yokoi is still the most commonly used model for evaluating the hazard of fire spread to the upper floors or adjacent buildings due to window flames. In order to evaluate the hazard of fire spread by window flames, the relationship between the dimensionless temperature, Θ, and the dimensionless height, z / r_0, needs to be quantified. The data points for the case with a vertical wall above the opening, which can be considered as a more general case, can be regressed as follows:

$$\Theta = \begin{cases} 0.455 & \left(\dfrac{z}{r_0} < 1\right) \\[2ex] 0.455\left(\dfrac{z}{r_0}\right)^{-1/2} & \left(1 \leq \dfrac{z}{r_0} < 4.83\right) \\[2ex] \left(\dfrac{z}{r_0}\right)^{-1} & \left(4.83 \leq \dfrac{z}{r_0}\right) \end{cases} \tag{5.56}$$

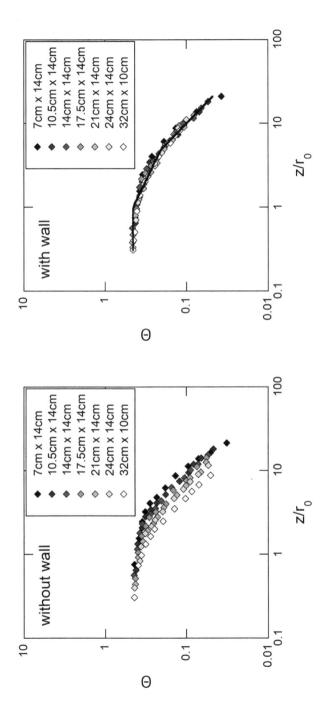

Figure 5.21 Dimensionless temperature rises along the trajectory of window flames with and without a vertical wall above the opening (adapted from Yokoi, 1960).

In Eq. (5.56), the power indices on z/r_0 were determined empirically. With this equation, Θ can be calculated at an arbitrary height z. By substituting Θ into the following transformed form of Eq. (5.55), the temperature rise, ΔT, at an arbitrary height, z, can also be evaluated as follows:

$$\Delta T = \Theta \frac{\left(\dfrac{\dot{Q}^2 T_\infty}{c_P^2 \rho^2 g}\right)^{1/3}}{r_0^{5/3}} \tag{5.57}$$

However, it should be noted that ρ involved in Θ is the plume density and is supposed to take a lower value from that of the ambient density, ρ_∞. To highlight this difference, we can further transform Eq. (5.57) into:

$$\frac{\Delta T}{T_\infty} = \Theta \cdot \left(\frac{\dot{Q}}{c_P \rho_\infty T_\infty g^{1/2} r_0^{5/2}}\right)^{2/3} \left(\frac{\rho_\infty}{\rho}\right)^{2/3} \tag{5.58}$$

In Eq. (5.58), the dimensionless temperature rise, $\Delta T / T_\infty$, is formulated as a product of three dimensionless parameters, Θ and $\dot{Q}/c_P\rho_\infty T_\infty g^{1/2}r_0^{5/2}$, and ρ_∞/ρ. The values of Θ and $\dot{Q}/c_P\rho_\infty T_\infty g^{1/2}r_0^{5/2}$ can be identified relatively easily, assuming that variables required for the calculation are available as given fire source conditions. However, ρ_∞/ρ is associated with ΔT on the left-hand side of this equation. Thus, we need to solve a nonlinear equation for either ΔT or ρ to obtain a solution. We may solve this equation numerically. However, in practice, ΔT is often determined by assuming a presumable value for ρ or by trial-and-error calculations.

To evaluate the thermal energy possessed by a window flame (virtual HRR), $\dot{Q}$, we consider a fire occurring in a compartment with a single opening with an area A and height H (Figure 5.20). The thermal energy is generated by the combustion of solid combustibles installed in the compartment. However, as some portion of the generated heat is lost through its perimeter walls, only a portion of it is used to drive the window flame. Thus, the product of the mass loss rate (MLR), $\dot{m}_B$, and the heat of combustion, ΔH_C, would be an overestimation for the virtual HRR of the window flame. $\dot{Q}$ can be regarded as the sum of the thermal energy possessed by the flow at the stage of ejection, $\dot{Q}_E$, and the thermal energy generated by the combustion of unburned flammable gas after ejection, $\dot{Q}_F$, which is:

$$\dot{Q} = \dot{Q}_E + \dot{Q}_F \tag{5.59}$$

Considering that the mass flow rate through the opening can be approximated by $0.5A\sqrt{H}\,(\text{kg}/\text{s})$ during the fully developed phase of a fire (Kawagoe, 1958), $\dot{Q}_E$ can be expressed as follows:

$$\dot{Q}_E = 0.5A\sqrt{H} \cdot c_P (T - T_\infty) \tag{5.60}$$

where c_P is the specific heat of gas at constant pressure, T is the compartment gas temperature, and T_∞ is the ambient temperature.

On the other hand, $\dot{Q}_F$ can be obtained by subtracting the HRR completed inside the compartment, $1500A\sqrt{H}\,(kW)$, (Tanaka, 1993) from the total HRR, $\Delta H_C \dot{m}_B$. Considering that the MLR of solid combustibles can be approximated as $\dot{m}_B \cong 0.1A\sqrt{H}\,(kg/s)$ in ventilation-controlled fires (Kawagoe, 1958), $\dot{Q}_F$ can be calculated from the following equation:

$$\dot{Q}_F = (1 - \chi_R)(0.1\Delta H_C A\sqrt{H} - 1500A\sqrt{H}) \tag{5.61}$$

where χ_R is the radiative fraction of the heat generation outside the compartment. For point and line fire sources in the open environment, the value of χ_R depends on the type and scale of the fire source. However, it is expected to be smaller for window flames, because a large amount of smoke that mitigates radiative heat loss is generally produced during combustion in a compartment. Assuming $\Delta H_C \cong 16000 \, kJ/kg$ for wood, which is a common combustible in buildings, Eq. (5.61) can be simplified to take the form:

$$\dot{Q}_F = (1 - \chi_R) \cdot 100A\sqrt{H} \tag{5.62}$$

Although there is an arbitrariness in the value of ΔH_C which varies with the fuel type, Eq. (5.62) infers that $\dot{Q}_F$ is substantially smaller than the amount completed inside the compartment, which is $1500A\sqrt{H}$.

5.4.2 Trajectory of the centerline

As Eq. (5.57) or (5.58) for ΔT evaluates the temperature rise along the plume centerline, the trajectory of the plume centerline also needs to be evaluated. The trajectory is a curve connecting points of the maximum temperature rise at different heights. The trajectory of the plume centerline changes depending on conditions such as the virtual HRR, $\dot{Q}$, and the geometry of the opening. Compared to the case of ΔT, there are a few examples that have investigated trajectories of the plume centerline (Yokoi, 1960; Himoto et al., 2009b). Here, we compare two alternative approaches.

Considering the relationship between the horizontal velocity at the ejection point and the buoyancy-induced acceleration in the vertical direction, Yokoi derived the following equation for window plumes when there is no vertical wall above the opening (Yokoi, 1960):

$$\frac{z}{H - Z_N} = \frac{1}{9\beta T_\infty} \frac{\left\{ \left[(x + x_0)/(H - Z_N) \right]^{3/2} - \left[x_0/(H - Z_N) \right]^{3/2} \right\}^2}{x_0/(H - Z_N)} \tag{5.63}$$

In this equation, z and x are the vertical and horizontal coordinates of the plume centerline, respectively, x_0 is the distance from the virtual origin of the fire source, H is the height of the opening, Z_N is the height of the neutral plane, and β is the coefficient of thermal expansion of the air ($0.00341 \, \text{K}^{-1}$ at $293 \, \text{K}$). The values of some variables involved in Eq. (5.63) are not readily available, but require extra calculations. However, the following approximations based on compartment fire experiments are often introduced to simplify Eq. (5.63) (Yokoi, 1960; Nakaya et al., 1986):

$$\frac{x_0}{H - Z_N} \cong 0.0558 \tag{5.64a}$$

$$\frac{H - Z_N}{H} \cong 0.64 \tag{5.64b}$$

By substituting these approximations into Eq. (5.63), the following equation can be obtained (Tanaka, 1993):

$$\frac{z}{H} = 4.9 \left(\frac{x}{H} \right)^3 \tag{5.65}$$

In Eq. (5.65), z is expressed as a monotonically increasing function of x. This indicates that once the hot gas is vented from the opening, it departs from the wall surface (i.e., the building facade) and does not approach the wall surface again.

Figure 5.22 shows the temperature contour map of a window flame ejected from a cubic compartment ($0.9 \, \text{m} \times 0.9 \, \text{m} \times 0.9 \, \text{m}$) (Himoto et al., 2009b). Connected lines of data points represent the plume trajectory, which are the locations of the maximum temperature at each measurement height. In Figure 5.22, the difference between the two panels is in the width of the opening; $B = 0.3$ m on the left and $B = 0.5$ m on the right. The heights of the opening are both $H = 0.5$ m. When $B = 0.3$ m, the window flame rises almost vertically upward after being separated from the wall above the opening. However, when $B = 0.5 \, \text{(m)}$, the window flame approaches the wall above the opening forming a curved trajectory.

The attachment of a window flame to the wall is caused by the Coanda effect. The Coanda effect is a phenomenon that occurs when a flow pulls the surrounding fluid due to its viscous effect. The air entrainment by a window flame occurs symmetrically along the centerline when there is no object that obstructs the flow. However, if there is an object that obstructs the air entrainment of a window flame, a pressure difference is caused to pull the window flame toward the object. Yokoi demonstrated in his experiments that the larger the opening width, B, the more likely the window flame to be pulled to the wall (Yokoi, 1960). This is because the larger B, the larger the interface area between the window flame and the wall, causing a larger pressure difference across the window flame.

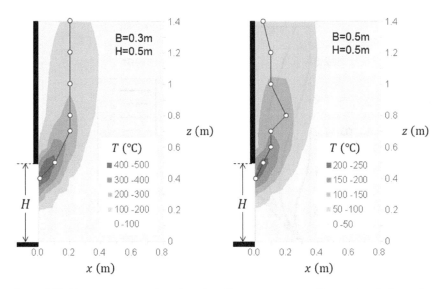

Figure 5.22 Measurement results of window flame temperature. Data points indicate the locations of the maximum temperature at each measurement height (adapted from Himoto et al., 2009b).

When a window flame attaches to the wall, building components above the opening are exposed to a higher intensity of heating, thus increasing the hazard of vertical fire spread. On the other hand, when a window flame does not attach to the wall, objects facing the opening are exposed to a higher intensity of heating, thus increasing the hazard of horizontal fire spread. Eq. (5.65) by Yokoi has a simple form. However, disregard for the flow attachment may lead to an underestimation of the hazard of vertical fire spread in certain cases. A unified trajectory model that considers both the attached and detached flows takes the form of cubic curves as follows (Himoto et al., 2009b):

$$\frac{x}{B} = \begin{cases} Attached\,flame\,F^* \le 3.33: \\ \begin{cases} 0.603\dfrac{z}{B}\left(1.33-\dfrac{z}{B}\dfrac{1}{F^*}\right)^2 & \left(0\le\dfrac{z}{B}\dfrac{1}{F^*}\le1.33\right) \\ 0 & \left(1.33\le\dfrac{z}{B}\dfrac{1}{F^*}\right) \end{cases} \\ Detached\,flame\,F^* \ge 3.33: \\ \begin{cases} 0.603\dfrac{z}{B}\left(1.33-\dfrac{z}{B}\dfrac{1}{F^*}\right)^2 & \left(0\le\dfrac{z}{B}\dfrac{1}{F^*}\le0.44\right) \\ 0.21F^* & \left(0.44\le\dfrac{z}{B}\dfrac{1}{F^*}\right) \end{cases} \end{cases} \qquad (5.66)$$

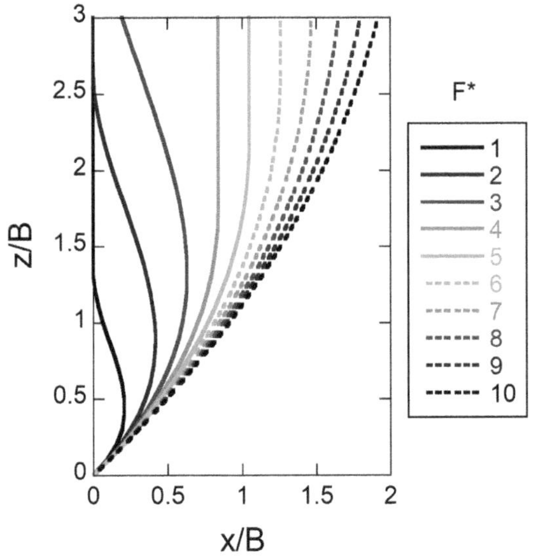

Figure 5.23 Calculated trajectory of window plume as a function of the dimensionless parameter F^*.

where F^* is the dimensionless parameter defined by:

$$F^* = \left(\frac{u_0}{\sqrt{gB}}\right)^2 \left(\frac{\dot{Q}}{c_p \rho_\infty T_\infty g^{1/2} B (H - Z_N)^{3/2}}\right)^{-2/3} \tag{5.67}$$

In Eq. (5.67), u_0 is the maximum ejection velocity of the flow at the opening. The coefficients in Eq. (5.66) are experimentally determined to satisfy the momentum conservation of the flow. In Eq. (5.67), the first term represents the effect of initial horizontal momentum, and the second term represents the effect of vertical momentum gained by buoyancy. Thus, F^* represents the dimensional relationship between the horizontal and vertical components of the flow momentum. Figure 5.23 shows the trajectories when F^* is changed in a stepwise manner. The window flame attaches to the wall when F^* is small, whereas the flame detaches from the wall when F^* is large, as the effect of the horizontal component becomes greater than that of the vertical component.

5.4.3 Flame geometry

Evaluation of the temperature rise along the trajectory, ΔT, may not be enough to evaluate the hazard of fire spread by window flames. It is valid when the heat-receiving object is in the proximity of the window flame where the effect of convective heat transfer is prominent. However, when the target object is separated from the window flame, the effect of radiative heat transfer becomes dominant. The radiative energy transferred to a target object, $\dot{Q}_R$, can be calculated by assuming the window flame as a solid mass of homogeneous radiative properties, as expressed in Eq. (5.26).

For radiative heat transfer calculation from a window flame to a target object, the view factor of the window flame as viewed from the target object needs to be calculated. The geometry of the window flame that determines the geometric relationship between the two is the key to the view factor calculation. Tang et al. conducted a series of fire experiments measuring the average height of window flames above the neutral plane, L_{fl} (Tang et al., 2012). Figure 5.24 shows the relationship between L_{fl} and the dimensionless HRR, Q^*. The following equation is the regression of data points (Tang et al., 2012):

$$\frac{L_{fl}}{\left(A\sqrt{H}\right)^{2/5}} = 2\left(\frac{\dot{Q}}{c_p \rho_\infty T_\infty g^{1/2} A\sqrt{H}}\right)^{0.44} \qquad (5.68)$$

In Eq. (5.68), the power index is 0.44, which is close to 2/5 for fire plumes with three-dimensional air entrainment above a point source. However, in Figure 5.24, data points vary depending on the opening geometry. It

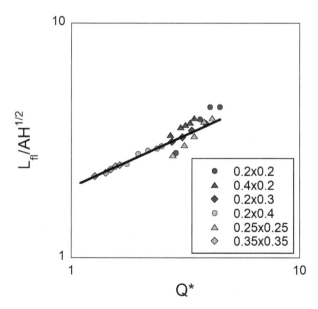

Figure 5.24 Relationship between the average flame height normalized by the ventilation factor and the dimensionless HRR (Tang et al., 2012).

is noteworthy that the behavior of window flames is affected by several factors that have not been considered in modeling idealistic point and line fire sources. These include the possession of initial momentum and thermal energy at the stage of ejection, and the asymmetry of air entrainment due to curvature of the trajectory and interference with the wall above the opening.

WORKED EXAMPLE 5.5

Assume a window flame ejected from an opening ($B = 2$ m and $H = 2$ m) with a virtual HRR of $\dot{Q} = 1$ MW. Calculate the temperature rise along the trajectory, ΔT, at $z = 2$ m from the top of the opening.

SUGGESTED SOLUTION

Prior to the calculation of ΔT using Eq. (5.58), we calculate the equivalent radius of the fire source, r_0:

$$r_0 = \sqrt{\frac{BH}{2\pi}} = 0.798 \text{ m}$$

Thus, the normalized height, z / r_0, becomes:

$$\frac{z}{r_0} = \frac{2}{0.798} = 2.51$$

Substituting z / r_0 into the regression model given by Eq. (5.56), the dimensionless temperature rise, Θ, can be obtained as follows:

$$\Theta = 0.455 \left(\frac{z}{r_0} \right)^{-1/2} = 0.287$$

Further substituting Θ into Eq. (5.58), we obtain:

$$\Delta T = T_\infty \cdot \Theta \cdot \left(\frac{\dot{Q}}{c_P \rho_\infty T_\infty g^{1/2} r_0^{5/2}} \right)^{2/3} \cdot \left(\frac{\rho_\infty}{\rho} \right)^{2/3} = 114.8 \left(\frac{\rho_\infty}{\rho} \right)^{2/3}$$

where $c_P = 1 \text{ kJ} \cdot \text{kg}^{-1} \cdot \text{K}^{-1}$, $T_\infty = 293 \text{ K} (20°C)$, and $g = 9.8 \text{ m} \cdot \text{s}^{-2}$. Note that in the above calculation, $\rho_\infty T_\infty = 353 \text{ kg} \cdot \text{K} \cdot \text{m}^{-3}$ is assumed from the equation of state.

The difficulty of this equation is that the plume density, ρ, depends on the temperature rise, ΔT. If we assume $\rho = \rho_\infty = 1.20 \text{ kg} \cdot \text{m}^{-3}$ in the above equation, we obtain $\Delta T = 114.8 \text{ K}$. However, the plume density needs to be $\rho = 0.87 \text{ kg} \cdot \text{m}^{-3}$ to satisfy the equation of state with this temperature rise.

This is inconsistent with the initial assumption on ρ. We need to adjust the plume density substituted into the equation, $\rho_{(0)}$, so that the plume density calculated from the equation of state, $\rho_{(1)} = 353 / (\Delta T_{(0)} + T_{\infty})$, is equivalent to $\rho_{(0)}$. However, if we assume the value of $\rho_{(0)}$ randomly, there is no guarantee that we can reach a solution. Thus, we take an iterative approach; $\rho_{(1)}$ is used for the next assumed value of $\rho_{(0)}$, until we reach $\rho_{(0)} = \rho_{(1)}$. This is nothing but the classical numerical procedure called the successive substitution method.

Let us start with $\rho_{(0)} = 0.87$ kg·m^{-3} from the currently available values, the calculated temperature rise is $\Delta T = 142.6$ K. This yields $\rho_{(1)} = 0.81$ kg·m^{-3} from the equation of state. Next, assuming $\rho_{(0)} = 0.81$ kg·m^{-3}, we obtain $\Delta T = 149.6$ K. This yields $\rho_{(1)} = 0.80$ kg·m^{-3}. Further, assuming $\rho_{(0)} = 0.80$ kg·m^{-3}, we obtain $\Delta T = 150.8$ K. This yields $\rho_{(1)} = 0.80$ kg·m^{-3}, thus reaching the convergence requirement, $\rho_{(0)} = \rho_{(1)}$.

One might find it cumbersome to seek ΔT in a trial-and-error manner. However, if we use spreadsheet software, we can easily replicate the above calculation procedure and obtain the desired results on ΔT and $\rho_{(1)}$.

WORKED EXAMPLE 5.6

Assume a compartment with an opening $(B = 4$ m, $H = 1.5$ m, and $A\sqrt{H} = 7.35$ m$^{5/2}$). The fire in the compartment is in the fully developed phase. The temperature inside the compartment is approximately uniform at 1000°C. Calculate the average height of the flame ejected from the opening, L_{fl}.

SUGGESTED SOLUTION

L_{fl} can be calculated using Eq. (5.68). Among the variables included in Eq. (5.68), the virtual HRR, $\dot{Q}$, is the sum of the energy already possessed at the stage of ejection, $\dot{Q}_E$, and the energy generated by the combustion of unburned flammable gas outside the compartment, $\dot{Q}_F$. $\dot{Q}_E$ and $\dot{Q}_F$ can be calculated using Eqs. (5.60) and (5.62), respectively:

$$\dot{Q}_E = 0.5A\sqrt{H} \cdot c_P (T - T_{\infty}) = 3600.8 \text{ kW}$$

$$\dot{Q}_F = (1 - \chi_R) \cdot 100A\sqrt{H} = 587.9 \text{ kW}$$

in which $\chi_R = 0.2$ was assumed. Thus, $\dot{Q}$ becomes:

$$\dot{Q} = \dot{Q}_E + \dot{Q}_F = 4188.6 \text{ kW}$$

Although the ratios of $\dot{Q}_E$ and $\dot{Q}_F$ to $\dot{Q}$ vary depending on the fire conditions in the compartment, about 14.0% of $\dot{Q}$ comes from the combustion of unburned flammable gas outside the compartment, $\dot{Q}_F$, in this specific case. Substituting $\dot{Q}$ into Eq. (5.68), we obtain L_{fl} as follows:

$$L_{fl} = 2\left(A\sqrt{H}\right)^{2/5}\left(\frac{\dot{Q}}{c_P\rho_\infty T_\infty g^{1/2}A\sqrt{H}}\right)^{0.44} = 3.32 \text{ m}$$

5.5 OTHER FIRE SOURCES

In the previous sections we discussed the behavior of fire plumes above sources of relatively simple configurations. The assumption of ideal fire source geometries, such as the point and line sources, is applicable to the analysis of various source geometries that exist in actual fires. However, there are certain cases when this assumption is not applicable as is, but requires a modification for better applicability. Even though, the fire plume models based on the ideal source assumption are still the basis for analyzing the behavior of fire plumes above non-ideal sources.

5.5.1 Rectangular fire sources

Many investigators have demonstrated that the point and line source models can adequately evaluate the temperature rise and flow velocity in the regime far from the fire source of various geometries. However, the effect of fire source geometry becomes significant in the regime near the fire source. The point source model is applicable when the aspect ratio of the fire source, r, is close to unity. In contrast, the line source model is applicable when r is sufficiently large. The r of actual fire sources often falls between them.

The fire plume behavior of rectangular fire sources is expected to have intermediate properties between those of point and line sources. To analyze the fire plume behavior above rectangular fire sources, we consider applying the model that assumes a line fire source, one end of the two extreme geometries. To apply this model to a rectangular fire source, we obtain the HRR per unit length, $\dot{Q}'$, by dividing the HRR, $\dot{Q}$, by the longer side length of the rectangular fire source, L:

$$\dot{Q}' = \dot{Q}/L \tag{5.69}$$

Thus, the dimensionless HRR of the line source, $Q_W'^*$, given by Eq. (5.52), can be rewritten as:

$$Q_{mod}^* = \frac{\dot{Q}}{c_P\rho_\infty T_\infty g^{1/2}LW^{3/2}} \tag{5.70}$$

where W is the shorter side length of the rectangular fire source. This is a modified dimensionless HRR for the line source. However, comparing Eq. (5.70) for the line fire source and Eq. (5.32) for the point fire source, the two equations become identical when $L = W$, i.e., when the fire source geometry is square. In other words, the dimensionless HRR, Q^*_{mod}, in Eq. (5.70) is commonly applicable to the two extreme ends of fire source geometries.

Figure 5.25 shows the measurement results of the average flame height, L_{fl}, above fire sources of the aspect ratio, r, ranging from 1 to 10 (Hasemi et al., 1989). When the aspect ratio is $r = 1$, L_{fl} is proportional to Q^*_{mod} to the power of 5/2, which is equivalent to Eq. (5.33). When the aspect ratio is $r = 10$, L_{fl} is proportional to Q^*_{mod} to the power of 3/2, which is equivalent to Eq. (5.53). The power relationship between L_{fl} and Q^*_{mod} is not necessarily apparent from Figure 5.25. However, given that data points of the intermediate aspect ratios are bounded by the two regressed lines of the point source ($r = 1$) and the line source ($r = 10$), L_{fl} above a rectangular fire source of an arbitrary r can be generalized with the following equation:

$$\frac{L_{fl}}{W} = C \cdot \left(Q^*_{mod} \right)^N + D \tag{5.71}$$

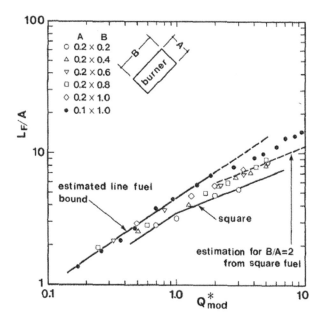

Figure 5.25 Flame height above burners of various aspect ratios (Hasemi et al., 1989).

where C, D, and N are the parameters specific to r. These parameters were identified only for the point and line sources given by Eqs. (5.33) and (5.53), respectively. For the other r, one needs to determine the parameters experimentally.

5.5.2 Group fires

In outdoor fires, multiple fire sources could occur simultaneously. When these fire sources are located close together, they could merge to form a pseudo-single flame, as described in Figure 5.26. When multiple fire sources burn in an integrated manner, oxygen is depleted around each fire source. Thus, the rate of oxygen supply to individual fire sources decreases compared to the case they burn independently. Accordingly, the height of the pseudo-single flame extends (Sugawa et al., 1993; Fukuda et al., 2004; Delichatsios, 2007; Zhang et al., 2018). This may increase the hazard of fire spread to adjacent combustible objects.

Let us consider that multiple fire sources are arranged in a grid $(n \times n)$. All the conditions of each fire source, including the HRR, $\dot{Q}$, side length, d, and separation from the others, s, are identical. Assuming that a pseudo-single flame is formed above the fire sources, the representative length of the fire source arrangement, W, is given by:

$$W = nd + (n-1)s \tag{5.72}$$

Referring to the dimensionless HRR of a single fire source in Eq. (5.32), that of the pseudo-single fire source can be expressed as follows:

$$Q_W^* = \frac{n^2 \dot{Q}}{c_P \rho_\infty T_\infty g^{1/2} \left[nd + (n-1)s \right]^{5/2}} = \frac{n^2 \dot{Q}}{c_P \rho_\infty T_\infty g^{1/2} W^{5/2}} \tag{5.73}$$

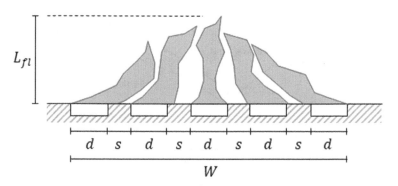

Figure 5.26 Schematic of flames merged from multiple fire sources.

When $n = 1$, Eq. (5.73) is equivalent to Eq. (5.32).

Figure 5.27 shows the measurement results of average flame length, L_{fl}, above a group of propane-operated burners with a side length of $d = 5.6\,\text{cm}$ (Zhang et al., 2018). The side-length-to-separation ratio, s / d, varied between 0.07 and 2. The ratio, s / d, was greater than 0 due to the 2 mm thickness of the burner equipment. When $s / d = 0.07$, the burners were aligned without gaps given that $(0.2 + 0.2) / 5.6 \cong 0.07$. The tested burners, n^2, was either 4, 16, or 36. In Figure 5.27, the correlation between Q_W^* and L_{fl} / W is compared with the calculation result by the Heskestad model in Eq. (5.34a). When $s / d = 0.07$, L_{fl} was in good agreement with the model. However, the L_{fl} deviated from the model as the s / d increased. The degree of deviation was greater with smaller Q_W^*. Based on their experimental results, Zhang et al. suggested that L_{fl} above gridded fire sources can be generalized with the following expression (Zhang et al., 2018):

$$\frac{L_{fl}}{W} = 3.7Q_W^{*2/5} - c \tag{5.74}$$

where $c(\geq 1.02)$ is a constant. Eq. (5.74) is equivalent to Eq. (5.34a) by Heskestad when $n = 1$.

If the separation between fire sources, s, increases, the mutual influence of fire plumes decreases. There is a minimum separation of fire sources by which each fire plume can be regarded as independent. Delichatsios estimated the

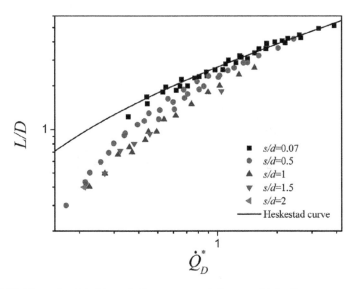

Figure 5.27 Normalized heights of flames merged from multiple fire sources (Zhang et al., 2018).

critical condition for the occurrence of the flame merge between two identical fire sources as follows (Delichatsios, 2007):

$$L_{fl} > 3s \tag{5.75}$$

The relevant critical condition for gridded fire sources has not been provided. However, we can expect that the minimum separation would be smaller than that given in Eq. (5.75), because each plume is deflected by the combined field of the adjacent plumes when they do not act symmetrically (Delichatsios, 2007).

WORKED EXAMPLE 5.7

Consider two fire cases: a house burning independently and four houses in a grid pattern burning simultaneously. $L_{fl,1}$ and $L_{fl,2}$ are the average flame heights of each fire.. Compare $L_{fl,1}$ and $L_{fl,2}$ when the HRR of each house, $\dot{Q}$, changes from 100 to 200 MW in a stepwise manner. However, the footprint of the houses is square with a side length of $d = 10$ m, and the side-length-to-separation ratio for the latter fire case is $s / d = 0.25$. Assume the value of the constant as $c = 1.2$ in Eq. (5.74) based on the experimental results by Zhang et al. (Zhang et al., 2018).

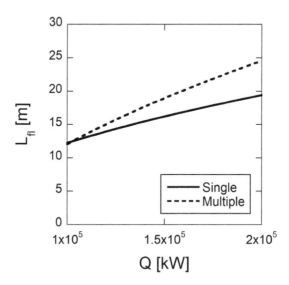

Figure 5.28 Calculated average flame heights above single and multiple fire sources.

SUGGESTED SOLUTION

We use Eq. (5.34a) to calculate $L_{fl,1}$ for the single house case. When $\dot{Q}$ ranges from 100 to 200 MW, the dimensionless HRR, Q_W^*, in Eq. (5.32) ranges from 0.29 to 0.57. Substituting the Q_W^* into Eq. (5.34a), we obtain the $L_{fl,1}$ ranging from 12.2 to 19.4 m.

By referring the above result to the merging condition in Eq. (5.75), it is inferred that the flame merge consistently occurs under the given separation condition, $s / d = 0.25$. Thus, $L_{fl,2}$ for the multiple fire sources can be calculated using the dimensionless HRR, Q_W^*, defined by Eq. (5.73). Substituting this into Eq. (5.74), we obtain the $L_{fl,2}$ ranging from 12.1 to 24.5 m.

The above results are plotted in Figure 5.28. The obtained average flame lengths were $L_{fl,1} < L_{fl,2}$ for most of $\dot{Q}$. In general, the larger $\dot{Q}$ and the smaller s, the greater the flame extension due to merging.

Chapter 6

Fire plumes – windy environment

Following the discussion on the characteristics of fire plumes in a quiescent environment in Chapter 5, we now discuss those in a windy environment. Fire plumes begin as reacting plumes (flames) followed by non-reacting plume segments at higher elevations. As in Chapter 5, the reacting and non-reacting plumes are collectively called fire plumes in this chapter.

The damage caused by outdoor fires is significantly affected by the wind condition. Without wind, fire plumes above burning fuels rise vertically upward due to buoyancy. Thus, the temperature rise is the highest within the zone right above the burning fuel, whereas, in a windy environment where fire plumes are blown down, the zone that involves a large temperature rise shifts to the leeward side closer to the ground surface. As fuels that may be involved in fire spread are distributed horizontally on the ground, the hazard of fire plumes is generally more pronounced in a windy environment.

In the theoretical analysis of fire plumes in a quiescent environment in Chapter 5, we assumed ideal fire source geometries, such as a point and a line. Although this is purposed for the convenience of analysis, it is widely acknowledged that the models of ideal fire source geometries can reasonably represent the characteristics of various fire plumes formed above real fire sources. This is also true in windy conditions. We also consider fire sources of ideal geometries in this chapter.

Theoretical analysis of fire plumes in a windy environment is based extensively on that in a quiescent environment, given that buoyancy is the common factor majorly affecting the behavior of fire plumes. However, in contrast to the cases in a quiescent environment, few attempts have been made to analyze the two continuous regimes, the flame and buoyant plume regimes, in an integrated manner. Thus, in this chapter, we discuss the characteristics of the flame regime (Section 6.4) and the buoyant plume regime (Sections 6.2 and 6.3) in separate sections. We precede our discussion on the buoyant plume regime as it is the basis of discussion on the flame regime.

DOI: 10.1201/9781003096689-6

6.1 BASIC CHARACTERISTICS OF FIRE PLUMES

Figure 6.1 illustrates the difference in physical processes associated with the combustion of liquid fuels in no-wind and windy conditions (Hu, 2017). Most fuels in outdoor fires are solids. Despite the difference in the form of gasification, either evaporation or thermal degradation, and an exception in the surface oxidation of solid fuels, both liquid and solid fuels react with oxygen in the gas phase to form diffusion flames. In diffusion combustion of liquid and solid fuels, the flame is formed right above the fuels where the evaporated or decomposed flammable gases mix with the fresh ambient air. In a windy environment, the zone of flame formation shifts to the leeward side due to the advection of flammable gases.

In modeling the behavior of fire plumes in a quiescent environment, we focused on the self-similarity of temperature and velocity profiles, and the regime segmentation associated with the cyclic change in the flame morphology. In addition, the wind-induced tilt and extension of flames are considered in modeling the behavior of fire plumes in a windy environment.

The flame tilt is a readily identifiable morphological change of flames in a crosswind. A fraction of heat generation within the flame regime is transferred to its surrounding object by thermal radiation. The flame tilt reduces the separation between the flame and fuels on the leeward side, increasing the rate of radiative heat transfer and raising the possibility of fire spread. On the other hand, the flame tilt extends the separation between the flame and fuels on the upwind side, decreasing the rate of radiative heat transfer and reducing the possibility of fire spread. The remaining fraction of heat generation within the flame regime raises the gas temperature and creates a buoyancy-induced updraft above the burning fuel. This is advected in the crosswind heating fuels on the leeward side. However, the heated flow mixes with the ambient air during advection and gradually decreases its temperature and buoyancy.

The flame extension and shortening is another type of morphological change of flames in a crosswind. Given that the damage caused by outdoor fires is often extensive in windy conditions, one may consider that a crosswind unexceptionally extends the flame length. However, the results of some relevant experiments exhibit that it is not generally the case (Thomas, 1967; Pitts, 1991; Lin et al., 2019). Changes in the conditions of the fuel heating, fresh air supply, and advection of gas mixture are the predominant factors affecting the flame length extension and shortening in a windy environment. These factors have opposing effects on flame length extension and shortening.

6.1.1 Fuel heating

The flame tilt due to crosswind changes the heat flux incident on the fuel, which in turn changes the supply rate of flammable gas into the gas phase.

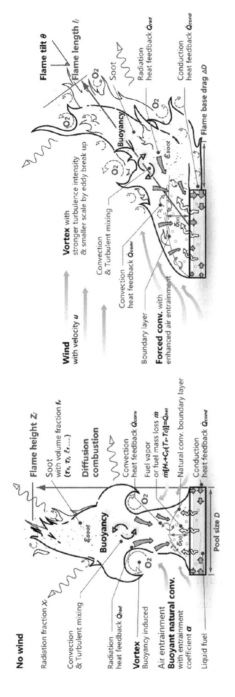

Figure 6.1 Physical processes involved in flame formation in no-wind and windy conditions (Hu, 2017).

The flame extends if the supply rate of flammable gas increases, while it shortens if the supply rate decreases. As illustrated in Figure 6.1, the heat transfer to liquid fuels comprises radiative and convective heat transfer from the flame, and heat conduction from the fuel container. It has been demonstrated that the dominant mode of heat transfer depends on the diameter of the fuel container (Hottel, 1959; Blinov et al., 1961). Namely, if the diameter is small (up to approximately 0.1–1 m), heat conduction through the fuel container is dominant. If the diameter is large, radiative and convective heat transfer from the flame becomes dominant. Nevertheless, liquid fuels are not common fuels in outdoor fires. We would focus on radiative and convective heat transfer, rather than heat conduction through the fuel container, as the dominant mode of heat transfer from the flame to the fuel.

As discussed in Chapter 3, the rate of heat transfer from the flame is proportional to the fourth power of its absolute temperature by radiation and proportional to its absolute temperature by convection. Thus, the radiative effect generally prevails the convective effect in the heat transfer from hot heat sources such as flames. However, at the same time, the emissive power of flame is proportional to its emissivity. The emissivity, ε, is dependent on the effective flame thickness, δ; the ε decreases as the δ decreases, and it increases and approaches unity as δ increases. In a windy environment, the thickness of the flame that covers the fuel becomes thinner than the vertical flame in a quiescent environment due to the tilt. Thus, the radiative heat flux to the fuel is generally reduced in a windy environment than in a quiescent environment. However, the reduction in the rate of radiative heat transfer would be minimized for large diameter fuels, as the flame maintains sufficient thickness even when tilted.

In a crosswind, the mode of convective heat transfer from fire plumes to fuel surfaces changes from natural convection to mixed convection, a combination of forced and natural convection. The heat transfer coefficient is generally enhanced in mixed convection compared with natural convection. This increases the rate of convective heat transfer, provided the heat source temperature is unchanged. However, the fire plume temperature decreases in a crosswind due to fresh air entrainment. As a consequence, the rate of convective heat transfer from fire plumes does not necessarily increase in a crosswind.

6.1.2 Fresh air supply

Figure 6.2 shows a schematic of the flow around a fire source in a crosswind (Church et al., 1980). Near the ground surface, a horizontal shear was originally formed corresponding to the velocity gradient of the crosswind. The onset of a fire source in the flow field results in a clustering of vortex lines with a strong component parallel to the plume axis along each downwind flank (Church et al., 1980). The clustering of vortex lines can be

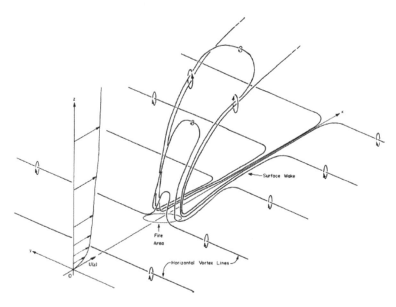

Figure 6.2 Interaction of a rising plume with a crosswind (Church et al., 1980).

identified as the formation of a counter vortex pair, by which fresh ambient air is entrained into the fire plume. Direct measurement of the air entrainment rate of a fire plume in a windy environment is not straightforward. However, according to some experimental observations, the rate of temperature reduction along the plume axis is greater in a windy environment than in a quiescent environment (Himoto, 2019), inferring a greater air entrainment rate in the former.

Applying crosswind to a fire source enhances the mixing of flammable gases and fresh air, which may increase the reaction rate of the two. However, unlike combustion in an enclosure, ample fresh air is generally available outdoors, independent of crosswind conditions. Thus, the heat release rate (HRR) due to the combustion of flammable gases outdoors is dominantly controlled by the supply rate of flammable gases rather than that of fresh air. As discussed in Chapter 5, the flame length can be viewed as the path length of flammable gases supplied at the ground level until fully consumed by reacting with the fresh air. If the crosswind increases the entrainment rate of fresh air, the combustion of flammable gases will be completed at an earlier stage. Thus, an oversupply of fresh air has a shortening effect on the flame length. The flame shortening further reduces the heat flux incident on the burning fuel and the supply rate of flammable gases into the gas phase. An increase in the fresh air supply may also reduce the temperature and concentration of the flammable gas mixture. This may lead to the flammable gases being transported out of the combustion zone without completing the reaction.

On the other hand, flames can be extended when the fresh air entrainment is restricted due to the flame attachment to the ground. The flame attachment occurs in strong crosswind conditions where the advection effect in the horizontal direction prevails the buoyancy effect. In such a situation, the formation of a counter vortex pair, as described in Figure 6.2, would be physically prevented. As a result, the entrainment of fresh air into the fire plume is restricted, and a longer path length is required to complete the reaction of flammable gases.

6.1.3 Advection of gas mixture

Given that the flame length is the path length of flammable gases supplied at the ground level until fully consumed by reacting with the fresh air, flames extend as the advection velocity of the flammable gas mixture increases. In a quiescent environment, the upward velocity of the gas mixture is determined predominantly by buoyancy, whereas, in a windy environment, the advection velocity of the gas mixture is affected by the crosswind velocity. Depending on the crosswind velocity, the advection velocity exceeds the buoyancy-induced upward velocity, resulting in the flame extension. However, if the advection velocity increases, the temperature and concentration of the gas mixture decrease, which may cause a deviation from the combustion limit of flammable gases. In such a situation, some flammable gases are no longer involved in the combustion, and the flame shortens. In other words, the relationship between crosswind velocity and flame length is not monotonous. In the range of low wind velocities, the flame extends as the wind velocity increases. Meanwhile, in the range of high wind velocities, the flame may be shortened despite an increase in the wind velocity.

6.2 NON-REACTING FIRE PLUMES DOWNWIND OF A POINT FIRE SOURCE

Point and line sources are the two ideal fire source geometries that have been widely employed to analyze the behavior of fire plumes in crosswinds. In this section, we discuss the characteristics of non-reacting fire plumes formed downwind of a point fire source.

6.2.1 Governing equations

The classical plume theory has been one of the common approaches to the analysis of fire plumes above a point fire source in a windy environment (Briggs, 1965; Slawson et al., 1967; Csanady, 1971; Davidson et al., 1982; Fisher et al., 2001; Himoto, 2019). An advantage of this approach is that the relationships among physical parameters affecting the behavior of fire plumes are formulated in relatively simple forms. An explicit formulation of physical mechanisms helps focus on the primary features of fire plumes when analyzing the behavior. Simple forms of equations enable the use of the relationships in numerical simulations of outdoor fires involving a large number of fire sources. On the other hand, the drawback lies in the difficulty of incorporating the effect of interactions between multiple fire plumes. To address this issue, the integral models that extend the buoyant plume theory have been developed (Hoult et al., 1969; Hirst, 1971; Mercer et al., 2001; Houf et al., 2008; Freitas et al., 2010; Marro et al., 2014; Tohidi et al., 2016). As with the plume theory models, the integral models assume certain profiles for the temperature rise and flow velocity along the trajectory. However, the plume characteristics downstream of the fire source are calculated by integrating the governing equations of fluid motion along the trajectory numerically. The integral models possess higher generality than the plume theory models. However, in this chapter, the fire plume behavior is analyzed using one of the plume theory models, considering its simplicity and the continuity of the discussion from Chapter 5. Readers interested in the integral models are referred to the literature mentioned above.

The analysis approach to the fire plume behavior follows that of the no wind case. Instead of discussing the strict conservational relationships of mass, momentum, and thermal energy, we use a dimensional analysis focusing on the transition of the characteristic properties along the trajectory, as schematically described in Figure 6.3 (Himoto, 2019). The x-axis is

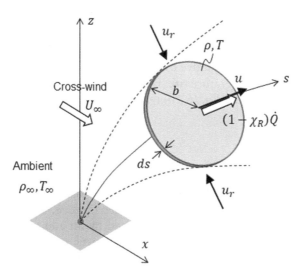

Figure 6.3 Behavior of characteristic parameters in the plume regime downwind of a point fire source.

parallel to the airflow, the z-axis is vertically upward, and the s-axis is along the trajectory. A control volume of a thickness ds is taken along the s-axis. The model assumptions are listed below:

1) The fire source geometry negligibly affects the behavior of a fire plume relatively far downwind of the fire source. Therefore, the fire source can be represented as a point source on the ground.
2) The scale and intensity of turbulence are governed mainly by the thermal instability of the fire plume, and the rate of ambient air entrainment is proportional to the vertical velocity driven by buoyancy (Morton et al., 1956).
3) The vertical momentum is predominantly driven by the buoyancy force. The viscous force can be ignored.
4) The heat loss from the wind-blown fire plume to the ground is negligible at all tilt angles.

Based on these assumptions, the conservational relationships of mass, vertical momentum, and thermal energy are, respectively, given by:

$$\text{Mass:} \ \rho b^2 u \propto \int \rho_\infty b w \ ds \tag{6.1a}$$

$$\text{Momentum:} \ \rho u w \propto \Delta \rho g s \tag{6.1b}$$

$$\text{Energy:} \ (1 - \chi_R)\dot{Q} \propto c_P \rho b^2 u \Delta T \tag{6.1c}$$

where ρ is the gas density, ρ_∞ is the ambient gas density, $\Delta \rho$ is the density difference, b is the effective cross-sectional width, u and w are the velocities along the trajectory and in the vertical direction, respectively, g is the acceleration due to gravity, s is the distance along the trajectory, $\dot{Q}$ is the HRR of the fire source, χ_R is the radiative fraction of $\dot{Q}$, c_P is the heat capacity of the gas, and ΔT is the temperature rise. Under atmospheric pressure, ρT can be considered constant, and Eq. (6.1b) reduces to:

$$u w \propto \frac{\Delta T}{T_\infty} g s \tag{6.2}$$

where T_∞ is the ambient gas temperature.

The three conservational relationships, Eqs. (6.1a), (6.1b), and (6.2), involve four unknown characteristic parameters: b, u, w, and ΔT. Thus, these simultaneous equations must be closed by an additional model. We focus on the relationship of the three characteristic parameters, u, w, and U_∞. When U_∞ is sufficiently small, u can be considered equivalent to w. On the other hand, when U_∞ is sufficiently large, u can be considered equivalent to U_∞. Exploiting this antipodal relationship, u can be approximated by (Himoto, 2019):

$$u \sim w^{1-N}U_\infty^N \tag{6.3}$$

where N is a real number between 0 and 1; the crosswind effect is negligible if $N = 0$, and prevailing if $N = 1$. Thus, N can be considered as the fractional contribution of crosswind velocity on velocity along trajectory.

6.2.2 Dimensional analysis

In conducting dimensional analysis, we introduce the following dimensionless parameters:

$$S^* = \frac{s}{D}, \quad B^* = \frac{b}{D}, \quad W^* = \frac{w}{U}, \quad U^* = \frac{U_\infty}{U}, \quad \text{and} \quad T^* = \frac{\Delta T}{T_\infty} \tag{6.4}$$

where D is the reference length and U is the reference velocity. By substituting Eq. (6.3) and the above dimensionless parameters in Eqs. (6.1a), (6.1b), and (6.2), the conservational relationships can be transformed into the following forms:

$$\text{Mass: } B^{*2}W^{*1-N}U^{*N} \propto B^*W^*S^* \tag{6.5a}$$

$$\text{Momentum: } W^{*2-N}U^{*N} \propto \left(\frac{gD}{U^2}\right)T^*S^* \tag{6.5b}$$

$$\text{Energy: } B^{*2}W^{*1-N}U^{*N}T^* \propto \left(1 - \chi_R\right)\left(\frac{\dot{Q}}{c_P\rho_\infty T_\infty D^2 U}\right) \tag{6.5c}$$

For simplicity, we assumed $\rho = \rho_\infty$ except in the buoyancy term in Eq. (6.5b) (Boussinesq approximation). In the dimensionless parameters defined by Eq. (6.4), D and U can be determined such that the parameters in the parentheses in Eqs. (6.5b) and (6.5c) become unity, namely:

$$D = \left(\frac{\dot{Q}}{c_P\rho_\infty T_\infty g^{1/2}}\right)^{2/5} \tag{6.6a}$$

$$U = g^{1/2}\left(\frac{\dot{Q}}{c_P\rho_\infty T_\infty g^{1/2}}\right)^{1/5} \tag{6.6b}$$

Dimensional analysis can be used to elucidate how the dimensionless characteristic parameters of the fire plume (the effective width, B^*, the vertical velocity, W^*, and the temperature rise, T^*) are influenced by the distance along the trajectory, S^*, and the crosswind velocity, U^*. Thus, we express B^*, W^*, and T^* as the power-law models of S^* as follows:

$$B^* = C_l \left(S^*\right)^l, W^* = C_m \left(S^*\right)^m, \text{and } T^* = C_n \left(S^*\right)^n \qquad (6.7)$$

Note that U^* is indifferent to S^*, i.e., $U^* \propto \left(S^*\right)^0$. The power indices, l, m, and n, and the coefficients, C_l, C_m, and C_n can be determined by introducing Eq. (6.7) into Eq. (6.5) as follows:

$$l = 1 - \frac{N}{3}, \quad m = -\frac{1}{3}, \quad \text{and} \quad n = \frac{N}{3} - \frac{5}{3} \qquad (6.8a)$$

$$C_l \propto \left(U^*\right)^{-N}, \quad C_m \propto \left(U^*\right)^0, \text{ and } C_n \propto \left(U^*\right)^N \qquad (6.8b)$$

Summarizing the above considerations, the transitions of the characteristic parameters along the trajectory under the crosswind effect are given by the following dimensionless expressions:

$$B^* \propto \left(S^*\right)^{1-N/3} \left(U^*\right)^{-N} \qquad (6.9a)$$

$$W^* \propto \left(S^*\right)^{-1/3} \qquad (6.9b)$$

$$T^* \propto \left(S^*\right)^{N/3-5/3} \left(U^*\right)^N \qquad (6.9c)$$

When $N = 0$, i.e., the effect of the crosswind velocity is negligible, the above relationships are equivalent to those in a quiescent environment (refer to Chapter 5).

In Eq. (6.9), N is left undetermined. The value of N needs to be determined through comparison with experimental data so that the model can adequately represent actual phenomena. Himoto conducted a series of wind tunnel experiments to investigate the temperature rise on the leeward side of a square fire source with a side length of 0.4 m (Himoto, 2019). $\dot{Q}$ and U_∞ were varied from 38.3 to 89.4 kW and from 0.59 to 1.44 m/s, respectively. Through comparison with the experimental data, ΔT was best correlated when $N = 0$ (Himoto, 2019). This indicates that there is no difference in the dimensional relationship between the characteristic features along the trajectory between quiescent and windy environments for point fire sources. Such an agreement in the dimensional relationship does not necessarily signify that the fire plume characteristics are identical between quiescent and windy environments. In the case of a windy environment, the characteristic features are evaluated for their transitions along the trajectory bent by the crosswind. In contrast, in the case of a quiescent environment, they are evaluated for their transition along the vertical axis. Differences appear in the proportionality coefficients of experimentally correlated relationships.

Assuming that $N = 0$, B^* and T^* along the trajectory were correlated with S^* raised to the powers of 1 and $-5/3$, respectively. The correlation obtained

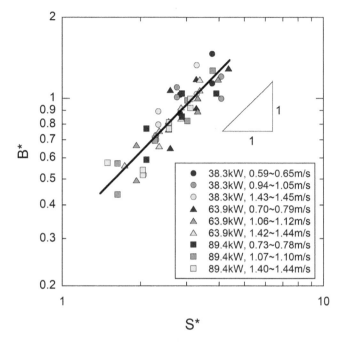

Figure 6.4 Dimensionless effective width of the cross-section along the trajectory of a fire plume (Himoto, 2019).

from the results of the wind tunnel experiment is plotted in Figures 6.4 and 6.5. The data points are almost linearly distributed and satisfy the following regression equations (Himoto, 2019):

Width: $B^* = 0.315S^* \ (S^* \ge 1.5)$ (6.10a)

Temperature rise: $T^* = 2.08S^{*-5/3} \ (S^* \ge 1.5)$ (6.10b)

For comparison, the expressions for B^* and T^* in a quiescent environment take the following expressions (refer to Chapter 5):

Width: $B^* = 0.118Z^* \ (Z^* \ge 3.3)$

Temperature rise: $T^* = 8.61Z^{*-5/3} \ (Z^* \ge 3.3)$

The basic structure of the expressions for a windy environment is equivalent to that for a quiescent environment. However, the definition of the distance from the fire source differs; it is the distance along the curved trajectory for a windy environment, whereas it is the height from the fire source in a

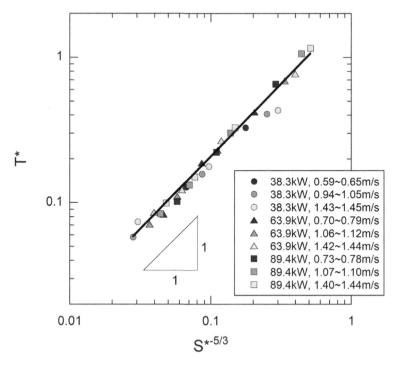

Figure 6.5 Dimensionless temperature rise along the trajectory of a flame plume (Himoto, 2019).

quiescent environment. However, the proportionality coefficient for B^* was larger and for T^* was smaller in a windy environment than those in a quiescent environment. This can be attributed to the difference in the entrainment rate of ambient air into fire plumes between the two environments; the entrainment rate is greater in a windy environment than in a quiescent environment, resulting in wider width (larger coefficient for B^*) and lower temperature rise (smaller coefficient for T^*) in a windy environment. As the flow velocity was not measured, the proportionality coefficient of W^* was not determined.

The above expressions can take another form using the following dimensionless heat release rate, Q_s^*, which has been conventionally used in fire research:

$$Q_s^* = \left(S^*\right)^{-5/2} = \frac{\dot{Q}}{c_p \rho_\infty T_\infty g^{1/2} s^{5/2}} \qquad (6.11)$$

By substituting S^* with Q_s^*, Eq. (6.10) can be expressed as follows:

Width: $B^* = 0.315 Q_s^{*-2/5}$ $\left(Q_s^* \le 0.363 \right)$ (6.12a)

Temperature rise: $T^* = 2.08 Q_s^{*2/3}$ $\left(Q_s^* \le 0.363 \right)$ (6.12b)

In general, the hazard of fire spread is finally evaluated with finite dimensions. As some variables in the dimensionless parameters, S^* and Q_s^*, can be regarded as approximately constant under ambient temperature and pressure conditions, Eq. (6.12) can be transformed into the following tractable forms:

Width: $b = 0.315 s$ (m) $\left(s \ge 0.09 \dot{Q}^{2/5} \right)$ (6.13a)

Temperature rise: $\Delta T = 5.70 \dfrac{\dot{Q}^{2/3}}{s^{5/3}}$ (K) $\left(s \ge 0.09 \dot{Q}^{2/5} \right)$ (6.13b)

In the transformation, the ambient temperature was assumed to be $T_\infty = 293\mathrm{K}$. In Eq. (6.13), the units of the variables are $\dot{Q}$ (kW) and s (m).

WORKED EXAMPLE 6.1

Assume a fire source with an HRR of $\dot{Q} = 10,000\ \mathrm{kW}$ in a crosswind. Calculate the temperature rise, ΔT, at distances $s = 10, 20$, and $30\ \mathrm{m}$ from the fire source along the trajectory.

SUGGESTED SOLUTION

ΔT along the trajectory in the plume regime can be calculated using Eq. (6.13b). The range of applicability of Eq. (6.13b) is:

$$s \ge 0.09 \dot{Q}^{2/5} = 0.09 \times \left(10000 \right)^{2/5} = 3.58\ \mathrm{m}$$

Thus, all the given distances $s = 10, 20$, and $30\ \mathrm{m}$ are within the range of applicability. We substitute the given values of $\dot{Q}$ and s into Eq. (6.13b) for evaluating the ΔT. The ΔT at $s = 10\ \mathrm{m}$ can be obtained as follows:

$$\Delta T = 5.70 \times \frac{10000^{2/3}}{10^{5/3}} = 57.0\ \mathrm{K}$$

Similarly, we obtain $\Delta T = 18.0\ \mathrm{K}$ at $s = 20\ \mathrm{m}$ and $\Delta T = 9.1\ \mathrm{K}$ at $s = 30\ \mathrm{m}$.

6.2.3 Trajectory

The temperature rise caused by a wind-blown fire plume is maximized along the trajectory and decreases with distance from the trajectory. However, because the fire plume rarely travels along the ground, the trajectory must

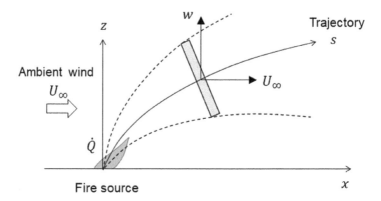

Figure 6.6 Cross-sectional view of a wind-blown fire plume downwind of a point fire source.

be predicted to evaluate the hazards posed to objects and human activity in the downwind of fire sources. Figure 6.6 shows the cross-sectional view of a wind-blown fire plume downwind of a point fire source. Based on the arguments in the previous subsection, the trajectory of a fire plume can be obtained by solving the following equation:

$$z = \int \frac{dz}{dx} dx = \int \frac{w}{U_\infty} dx \qquad (6.14)$$

where the horizontal velocity of the fire plume is assumed to be equivalent to U_∞. Using Eq. (6.9b), w of the fire plume is given by:

$$w \propto g^{1/2} \left(\frac{\dot{Q}}{c_P \rho_\infty T_\infty g^{1/2}} \right)^{1/3} s^{-1/3} \qquad (6.15)$$

If the distance along the trajectory is considered proportional to the horizontal distance from a wide-range view, i.e., $s \propto x$, we can substitute Eq. (6.15) into Eq. (6.14), which yields the following expression:

$$z \propto \frac{g^{1/2}}{U_\infty} \left(\frac{\dot{Q}}{c_P \rho_\infty T_\infty g^{1/2}} \right)^{1/3} x^{2/3} \qquad (6.16)$$

The above equation is further nondimensionalized by incorporating the side length of the fire source, W, as the reference length:

$$\frac{z}{W} \propto \dot{Q}_W^{*1/3} U_W^{*-1} \left(\frac{x}{W} \right)^{2/3} \qquad (6.17)$$

where $\dot{Q}_W^*$ and U_W^* are the dimensionless parameters, respectively, defined as follows:

$$\dot{Q}_W^* = \frac{\dot{Q}}{c_p \rho_\infty T_\infty g^{1/2} W^{5/2}} \tag{6.18a}$$

$$U_W^* = \frac{U_\infty}{\sqrt{gW}} \tag{6.18b}$$

In Eq. (6.18), the fire plume continuously loses buoyancy by entraining ambient air as it separates from the fire source. By this mechanism, it gradually becomes parallel to the crosswind. However, the trajectory cannot converge onto a horizontal line because the decay of upward momentum by viscous forces is not considered in the model.

Figure 6.7 displays the measured trajectory with the dimensionless height, z/W, and the dimensionless distance, $\dot{Q}_W^{*1/3} U_W^{*-1} (x/W)^{2/3}$, which was estimated by Eq. (6.17). The standard deviations are plotted as the error bars in Figure 6.7. The data points are linearly distributed, indicating a proportionality relationship between the two parameters. Setting the intercept

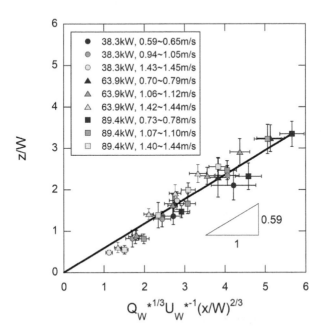

Figure 6.7 Regression result of the trajectory model. The error bars are the standard deviations (Himoto, 2019).

to zero (because the trajectory starts from the origin), the regression line of these data points is given by:

$$\frac{z}{W} = 0.59\dot{Q}_W^{*1/3}U_W^{*-1}\left(\frac{x}{W}\right)^{2/3} \left(\dot{Q}_W^{*1/3}U_W^{*-1}\left(\frac{x}{W}\right)^{2/3} < 5.7\right) \tag{6.19}$$

For calculating the trajectory of actual fire plumes, Eq. (6.19) needs to be converted to a form with finite dimensions. As some variables can be approximately constant under the ambient pressure and temperature conditions, Eq. (6.19) can be transformed into the following tractable form:

$$z = 0.179\left(\frac{\dot{Q}^{1/3}}{U_\infty}\right)x^{2/3}(\text{m}) \left(\left(\frac{\dot{Q}^{1/3}}{U_\infty}\right)x^{2/3} < 18.8W\right) \tag{6.20}$$

Note that in Eq. (6.20), the units of the variables are $\dot{Q}(\text{kW})$, U_∞ $(\text{m}\cdot\text{s}^{-1})$, $W(\text{m})$, and $x(\text{m})$.

WORKED EXAMPLE 6.2

Assume a fire source with a diameter of $W = 5$ m and a HRR of $\dot{Q} = 10,000$ kW in a crosswind with a velocity of $U_\infty = 2$ m / s. Calculate the height of the fire plume trajectory, z, at $x = 20$ m downwind from the fire source.

SUGGESTED SOLUTION

The height of the fire plume trajectory, z, can be calculated using Eq. (6.20). We first test the range of applicability of Eq. (6.20) as follows:

$$\left(\frac{\dot{Q}^{1/3}}{U_\infty}\right)\times 20^{2/3} = \left(\frac{10000^{1/3}}{2}\right)\times 20^{2/3} = 79.4 < 18.8\times 5 = 94$$

Thus, Eq. (6.20) can be used to solve this problem. The z at $x = 20$ m can be calculated by substituting the values of $\dot{Q}$ and U_∞ into Eq. (6.20) as follows:

$$z = 0.179\times\left(\frac{10000^{1/3}}{2}\right)\times 20^{2/3} = 14.2 \text{ m}$$

6.3 NON-REACTING FIRE PLUMES DOWNWIND OF A LINE FIRE SOURCE

A line fire source is a band of burning area that often appears at the front edge of spreading fires outdoors. Given that they have a certain width and curvature in actual outdoor fires, the line fire source is not an exact representation. Although such an assumption is introduced for analytical purposes, the line fire source model has wide applicability to various actual fire situations. In this section, we discuss the characteristics of non-reacting fire plumes formed downwind of a line fire source.

6.3.1 Governing equations

The classical plume theory has also been widely applied to the analysis of line fire sources (Thomas, 1964; Yokoi, 1965; Albini, 1981; Saga, 1990; Himoto et al., 2020). The approach to the analysis follows that of the point fire source model discussed in the previous section. Instead of considering the strict conservational relationships of mass, momentum, and thermal energy, we employ a dimensional analysis focusing on the transition of the characteristic properties of the fire plume along the trajectory, as schematically illustrated in Figure 6.8. The assumptions adopted for the point fire source model are also introduced in the line fire source model.

The conservational relationships of mass, vertical momentum, and thermal energy for a line fire plume are respectively given by:

$$\text{Mass:} \ \rho b u \propto \int \rho_\infty w \, ds \tag{6.21a}$$

$$\text{Momentum:} \ \rho u w \propto \Delta \rho g s \tag{6.21b}$$

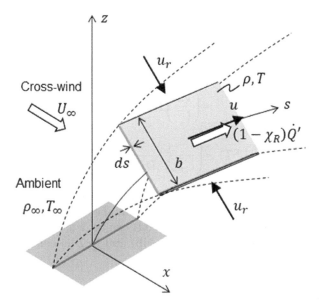

Figure 6.8 Behavior of characteristic parameters in the plume regime downwind of a line fire source.

$$\text{Energy}: \dot{Q}' \propto c_P \rho b u \Delta T \tag{6.21c}$$

where b is the cross-sectional half-width, $\dot{Q}'$ is the HRR per unit length, c_P is the heat capacity of the gas, g is the acceleration due to gravity, and s and u are the distance and velocity along the trajectory, respectively. w is the vertical velocity; ρ is the gas density, ρ_∞ is the ambient gas density, and $\Delta\rho$ is the difference in density. Under atmospheric pressure, ρT can be considered constant, and Eq. (6.21b) can be reduced to:

$$uw \propto \frac{\Delta T}{T_\infty} gs \tag{6.22}$$

where T_∞ is the ambient gas temperature.

6.3.2 Dimensional analysis

In the previous section, a similarity model was introduced to describe the transition of the characteristic properties along the trajectory of a fire plume from a point fire source. On deriving this model, we assumed that u can be expressed as a composed function of w and U_∞, as given by Eq. (6.3). By substituting u in Eq. (6.3) and introducing the dimensionless parameters:

$$S^* = \frac{s}{D}, \quad B^* = \frac{b}{D}, \quad W^* = \frac{w}{U}, \quad U^* = \frac{U_\infty}{U}, \quad \text{and} \quad T^* = \frac{\Delta T}{T_\infty} \tag{6.23}$$

Eqs. (6.21a), (6.21c), and (6.22) can be transformed into the following forms, respectively:

$$\text{Mass}: B^* W^{*-N} U^{*N} \propto S^* \tag{6.24a}$$

$$\text{Momentum}: W^{*2-N} U^{*N} \propto \left(\frac{gD}{U^2}\right) T^* S^* \tag{6.24b}$$

$$\text{Energy}: B^* W^{*1-N} U^{*N} T^* \propto \left(\frac{\dot{Q}'}{c_P \rho_\infty T_\infty D U}\right) \tag{6.24c}$$

where D is the reference length, and U is the reference velocity. In the above transformation, we assumed $\rho = \rho_\infty$ for simplicity. In the dimensionless parameters defined by Eq. (6.23), D and U can be determined such that the parameters in parentheses in Eqs. (6.24b) and (6.24c) can become unity, namely:

$$D = \left(\frac{\dot{Q}'}{c_P \rho_\infty T_\infty g^{1/2}}\right)^{2/3} \tag{6.25a}$$

$$U = g^{1/2} \left(\frac{\dot{Q}'}{c_P \rho_\infty T_\infty g^{1/2}} \right)^{1/3} \tag{6.25b}$$

Note that we also had the reference scales for a point fire source as defined by Eq. (6.6). They are not identical to those for a line fire source in Eq. (6.25).

To elucidate how the dimensionless characteristic parameters of the fire plume (B^*, W^*, and T^*) are influenced by the distance along the trajectory, S^*, and the crosswind velocity, U^*, the following expressions are substituted into Eqs. (6.24):

$$B^* = C_l \left(S^* \right)^l, \; W^* = C_m \left(S^* \right)^m, \text{ and, } T^* = C_n \left(S^* \right)^n \tag{6.26}$$

Considering that U^* is indifferent to S^*, i.e., $U^* \propto \left(S^* \right)^0$, the power indices, l, m, and n, and the coefficients, C_l, C_m, and C_n, can be determined as follows:

$$l = 1, \, m = 0, \text{and } n = -1 \tag{6.27a}$$

$$C_l \propto \left(U^* \right)^{\frac{3N}{N-3}}, \, C_m \propto \left(U^* \right)^{\frac{N}{N-3}}, \text{ and } C_n \propto \left(U^* \right)^{-\frac{N}{N-3}} \tag{6.27b}$$

After summarizing the above considerations, the transitions of the characteristic parameters along the trajectory under the crosswind effect are given by the following dimensionless expressions:

$$B^* \propto \left(S^* \right) \left(U^* \right)^{\frac{3N}{N-3}} \tag{6.28a}$$

$$W^* \propto \left(U^* \right)^{\frac{N}{N-3}} \tag{6.28b}$$

$$T^* \propto \left(S^* \right)^{-1} \left(U^* \right)^{-\frac{N}{N-3}} \tag{6.28c}$$

When $N = 0$, i.e., when the effect of the crosswind velocity is negligible, the above relationships are equivalent to those in a quiescent environment.

In Eq. (6.28), N is left undetermined. The value of N needs to be determined through comparison with experimental data so that the model can adequately represent actual phenomena. Himoto et al. used a rectangular burner with an aspect ratio of $r = 10$ (side lengths parallel and orthogonal to the wind were 0.2 m and 2.0 m, respectively) as the fire source in a series of wind tunnel experiments. They investigated the temperature rise on the leeward side of the fire source under different $\dot{Q}$ and U_∞ (Himoto et al., 2020). $\dot{Q}$ and U_∞ were varied from 127.8 to 253.0 kW and from 0.65 to 1.28 m/s, respectively. Through comparison with the experimental data, ΔT was best correlated when $N = -1.39$ (Himoto et al., 2020). The obtained value deviates from the originally assumed range of N, which was between 0 and 1. If N takes a negative value, Eq. (6.28) yields an estimation of a lower T^* under higher U^*. The deviation is attributed to the attachment of the wind-blown fire plume to the ground, restricting the air entrainment from the ambient under high crosswind conditions. However, the interaction between the fire plume and the ground is ignored in the present formulation of the model. Although there is room for discussion on the theoretical validity of the assumptions made in Eq. (6.28), we will proceed with the analysis assuming $N = -1.39$ in view of the accountability of the experimental results. In this case, the power index over U^* is $3N / (N - 3) = 0.950$ for B^* in Eq. (6.28a), $N / (N - 3) = 0.317$ for W^* in Eq. (6.28b), and $-N / (N - 3) = -0.317$ for T^* in Eq. (6.28c).

Based on the relation given by Eq. (6.28), Figure 6.9 displays T^* when $N = -1.39$. The data points are almost linearly distributed, satisfying the following regression equation (Himoto et al., 2020):

$$T^* = 1.98(U^*)^{-0.32}(S^*)^{-1} \quad (S^* \geq 3.9) \tag{6.29}$$

A positive value of the power index leads to the prediction of a higher T^* under a higher U^*, whereas a negative value of the power index leads to the prediction of a lower T^* under a higher U^*. Note that the dimensionless

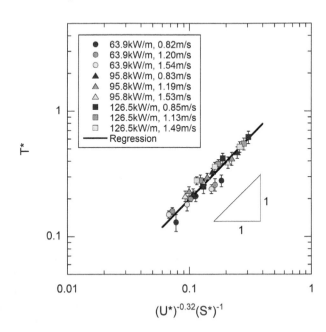

Figure 6.9 Dimensionless temperature rise for a fire source with an aspect ratio of 10 ($r = 10$) (Himoto et al., 2020).

distance, S^*, can be converted to another dimensionless parameter, Q'^*_s, which has been conventionally used in fire research:

$$Q'^*_s = \left(S^*\right)^{-3/2} = \frac{\dot{Q}'}{c_P\rho_\infty T_\infty g^{1/2} s^{3/2}} \tag{6.30}$$

where $\dot{Q}'$ is the HRR per unit length. By substituting S^* with Q'^*_s, Eq. (6.29) can be converted to an equation which takes the form:

Temperature rise: $T^* = 1.98\left(U^*\right)^{-0.32} Q'^{*2/3}_s$ (6.31)

As discussed later, the trajectory of a line fire source has a constant gradient with respect to the ground, i.e., $S^* \propto X^*$. Thus, S^* can be replaced by X^* in Eq. (6.28c). In this case, the following equation is obtained by regressing the experimental results:

$$T^* = 1.91\left(U^*\right)^{-0.27}\left(X^*\right)^{-1} \quad \left(X^* \geq 3.8\right) \tag{6.32}$$

Given that fire sources and heat-receiving objects are gathered near the ground in outdoor fires, it would be more practical to use Eq. (6.32) with X^* as a variable due to its ease of computation relative to Eq. (6.29) with S^* as a variable. For practical convenience, Eq. (6.32) can be converted to a form with finite dimensions so that the ΔT is readily obtained:

$$\Delta T = 5.73\left(\frac{\dot{Q}'^{0.67}}{U_\infty^{0.27}}\right)x^{-1}\,(\text{K}) \quad \left(x \geq 0.035\dot{Q}'^{2/3}\right) \tag{6.33}$$

In the conversion, the ambient temperature was assumed to be $T_\infty = 293\,\text{K}$. In Eq. (6.33), the units of the variables are $\dot{Q}'$ $(\text{kW}\cdot\text{m}^{-1})$, U_∞ $\left(\text{m}\cdot\text{s}^{-1}\right)$, and x (m).

WORKED EXAMPLE 6.3

Assume a line fire source with a HRR per unit length of $\dot{Q}' = 1,000\,\text{kW}\cdot\text{m}^{-1}$ in a crosswind with a velocity of $U_\infty = 2\,\text{m}\cdot\text{s}^{-1}$. Calculate the temperature rise, ΔT, along the trajectory at $x = 10, 20,$ and $30\,\text{m}$ downwind from the fire source.

SUGGESTED SOLUTION

ΔT along the trajectory can be calculated using Eq. (6.33). We first test the range of applicability of Eq. (6.33):

$$x \geq 0.035 \dot{Q}'^{2/3} = 3.5 \text{ m}$$

Thus, the given distances from the fire source, $x = 10, 20$, and 30 m, are all within the range of applicability. By substituting the values of $\dot{Q}'$ and U_∞ in Eq. (6.33), ΔT at $x = 10$ m can be obtained as follows:

$$\Delta T = 5.73 \left(\frac{\dot{Q}'^{0.67}}{U_\infty^{0.27}} \right) x^{-1} = 5.73 \times \left(\frac{1000^{0.67}}{2^{0.27}} \right) \times 10^{-1} = 48.6 \text{ K}$$

We can similarly obtain the temperature rises at the further downwind locations, which are $\Delta T = 24.3$ K at $s = 20$ m and $\Delta T = 16.2$ K at $s = 30$ m.

6.3.3 Trajectory

The temperature rise caused by a wind-blown fire plume is maximized along the trajectory and decreases with the distance from the trajectory. Figure 6.10 shows the cross-sectional view of a wind-blown fire plume downwind of a line fire source. Based on the arguments in the previous subsection, the trajectory of a fire plume can be obtained by solving the following equation:

$$z = \int \frac{dz}{dx} dx = \int \frac{w}{U_\infty} dx \tag{6.34}$$

where the horizontal velocity of the fire plume is assumed to be equal to U_∞. Using Eq. (6.28b), w of the fire plume is given by:

$$\frac{w}{U_\infty} \propto \left(U^* \right)^{\frac{3}{N-3}} \tag{6.35}$$

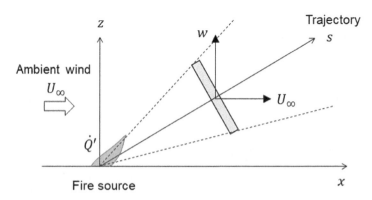

Figure 6.10 Cross-sectional view of a wind-blown fire plume downwind of a line fire source.

If the distance along the trajectory is considered proportional to the horizontal distance from a wide-range view, i.e., $s \propto x$, we can substitute Eq. (6.35) into Eq. (6.34), which yields:

$$\frac{z}{W} \propto (U^*)^{\frac{3}{N-3}} \left(\frac{x}{W} \right) \tag{6.36}$$

According to the obtained expression, the height, z, of the trajectory is proportional to the horizontal distance from the fire source, x. It should be noted that the line fire source is an ideal fire source with an infinite length by definition. However, the above equation is nondimensionalized by incorporating the side length orthogonal to the wind, W, of the actual fire source as the reference length.

The coefficient of Eq. (6.36) for the trajectory of a wind-blown fire plume is determined by regressing the experimental data, which is as follows (Himoto et al., 2020):

$$\frac{z}{W} = 0.24 (U^*)^{-0.68} \left(\frac{x}{W} \right) \left((U^*)^{-0.68} \left(\frac{x}{W} \right) < 14.9 \right) \tag{6.37}$$

Note that the value of N is unchanged from the case of the temperature rise, which was -1.39. The measured z/W are compared with the prediction by Eq. (6.37) as shown in Figure 6.11. Although the overall agreements were reasonable, the predicted z/W were comparatively lower than the measured data in the vicinity of the fire sources. This is attributed to the attachment of the fire plume to the ground, which is not explicitly considered in the present formulation. Also, although the present model predicts a linear relationship between z and x, an actual fire plume is expected to lose its vertical momentum along the attenuation of the buoyancy with distance from the fire source.

Eq. (6.37) takes a dimensionless form. To calculate the trajectory of the fire plume and to predict ΔT around the trajectory in an actual outdoor fire, it would be practical to have Eq. (6.37) in a form with finite dimensions. By substituting the values of the physical parameters that can be regarded as approximately constant under the atmospheric temperature and pressure, Eq. (6.37) can be converted as follows:

$$z = 0.107 \left(\frac{\dot{Q}'^{1/3}}{U_\infty} \right)^{0.68} x \, (\text{m}) \left(\frac{x}{W} < 33.6 \left(\frac{U_\infty}{\dot{Q}'^{1/3}} \right)^{0.68} \right) \tag{6.38}$$

where the units of the variables are $\dot{Q}'$ ($\text{kW} \cdot \text{m}^{-1}$), U_∞ ($\text{m} \cdot \text{s}^{-1}$), and x (m). Eq. (6.38) shows a relationship that the smaller $\dot{Q}'$ and the larger U_∞, the larger the tilt angle of the trajectory (the angle between the vertical axis and the trajectory).

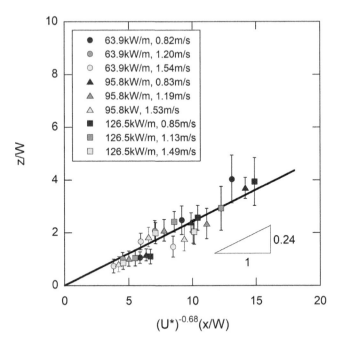

Figure 6.11 Regression result of the trajectory model. The error bars are the standard deviations (Himoto et al., 2020).

WORKED EXAMPLE 6.4

Assume a fire source with a HRR per unit length of $\dot{Q}' = 1,000 \ \text{kW} \cdot \text{m}^{-1}$ in a crosswind with a velocity of $U_\infty = 2 \ \text{m} / \text{s}$. The geometry of the fire source is an elongated rectangle with its width in the direction perpendicular to the wind direction is $W = 10$ m. Calculate the height, z, of the trajectory at $x = 20$ m downwind from the fire source.

SUGGESTED SOLUTION

The height of the trajectory, z, of the line fire source can be calculated using Eq. (6.38). The range of applicability of Eq. (6.38) can be tested as follows:

$$\frac{x}{W} = \frac{20}{10} = 2 < 33.6 \left(\frac{U_\infty}{\dot{Q}'^{1/3}} \right)^{0.68} = 33.6 \times \left(\frac{2}{1000^{1/3}} \right)^{0.68} = 11.2$$

Thus, Eq, (6.38) is applicable to solve this problem. The z at $x = 20$ m can be obtained by substituting the values of $\dot{Q}'$ and U_∞ into Eq. (6.38):

$$z = 0.107\left(\frac{1000^{1/3}}{2}\right)^{0.68} \times 20 = 6.4 \text{ m}$$

6.4 FLAME GEOMETRY

In this section, we discuss the characteristics of the reacting (flaming) segment of fire plumes in a windy environment. In no-wind conditions, the distributions of temperature and flow velocity within flames were the major subjects of investigations. However, relatively few studies have examined wind-blown flames from such a perspective, partly because of difficulty specifying the trajectory prior to measurement. Rather, the subjects of existing studies have been its geometry, including flame length, flame base drag, and tilt angle, as shown in Figure 6.12 (Lam et al., 2015; Hu, 2017). This is because the geometric relationship between the flame and the heat-receiving objects is the key to radiative heat transfer calculations.

6.4.1 Flame length

Expressions for the flame length, L_{fl}, in a windy environment are generally obtained by regressing experimental data with composite functions of multiple governing dimensionless numbers. One of the basic forms of such composite functions was presented by Thomas as follows (Thomas, 1963a; Thomas et al., 1963b):

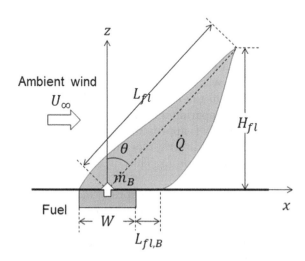

Figure 6.12 Cross-sectional view of a diffusion flame in a crosswind.

$$\frac{L_{fl}}{W} = a \cdot \left(M_F^*\right)^p \left(U_W^*\right)^q \tag{6.39}$$

where M_F^* is the dimensionless mass loss rate (MLR), and U_F^* is the dimensionless wind velocity. These dimensionless numbers are a Froude number representing the ratio of inertial force to gravity. They were respectively defined as follows (Thomas, 1963a; Thomas et al., 1963b):

$$M_W^* = \frac{\dot{m}_B''}{\rho_\infty \sqrt{gW}} \tag{6.40a}$$

$$U_W^* = \frac{U_\infty}{\sqrt{gW}} \tag{6.40b}$$

where $\dot{m}_B''$ is the MLR per unit area, ρ_∞ is the ambient gas density, g is the acceleration due to gravity, U_∞ is the crosswind velocity, and W is the width of the fire source. Note that Eq. (6.40b) is a restatement of Eq. (6.18b).

The values of the proportionality coefficient, a, and the power indices, p and q, in Eq. (6.40) have been determined experimentally by several researchers (Thomas, 1963a; Thomas et al., 1963b; Atallah et al., 1974; Moorhouse, 1982; Ferrero et el., 2007). However, these model parameters vary among researchers. The one by Thomas is based on a series of experiments burning rectangular wood cribs of different aspect ratios (Thomas, 1963a, 1967; Thomas et al., 1963b):

$$\frac{L_{fl}}{W} = 70\left(M_W^*\right)^{0.86} \left(U_W^*\right)^{-0.22} \tag{6.41}$$

Figure 6.13 shows the relationship between the dimensionless flame length, L_{fl} / W, and the dimensionless wind velocity, U_W^*, formed above a rectangular fire source. M_W^* was varied in three steps, 0.01, 0.05, and 0.1, by referring to the range in the experiment by Thomas (Thomas, 1963a; Thomas et al., 1963b). Comparing the absolute values of the power indices in Eq. (6.41), the one for M_W^* is approximately four times greater than that for U_W^*. The absolute values directly represent the influential scale of the relevant dimensionless number on L_{fl}. Thus, in Eq. (6.41), L_{fl} is more sensitive to M_W^* than to U_W^*.

The negative value of the power index for U_W^* indicates that the stronger the crosswind, the shorter L_{fl}. This relationship may contradict an intuitive but common understanding that the stronger the wind, the more intense the fire becomes. As discussed in Section 6.1, the flame tilt in a crosswind may reduce the burning intensity. This reduces the radiative heat flux incident on the burning fuel and shortens the flame length. On the other hand, the flame tilt in a crosswind may reduce the air entrainment rate as the flame

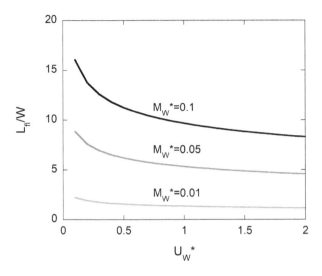

Figure 6.13 Dimensionless flame length above a rectangular fire source in a crosswind.

approaches the ground surface, restricting the supply path of fresh air to the flame. This leads to an extension of the flame length, given that it is the path length of flammable gases supplied at the ground level until fully consumed by reacting with the fresh air. The regression result in Eq. (6.41) exhibits the predominance of the shortening effect on the flame length. Contrastively, the predominance of the extending effect was demonstrated in experiments of some other investigators. An example is an experiment using a line fire source where the air entrainment is more likely to be restricted due to its elongated fire source geometry (Kolb et al., 1997). It should be noted that the crosswind effect is also dependent on the state of the fuel (solid or liquid) due to the difference in the gasification process under external heating. The evaporation rate of liquid fuels is controlled not only by the radiative and convective heating from the flame, but also by the heat conduction through the fuel container. Depending on how the fuel container is exposed to or covered by the tilted flame, stronger crosswinds may lead to an extension of the flame above liquid fuels (Hu et al., 2011).

In Eq. (6.41), the effect of buoyancy due to density difference is represented by the MLR, $\dot{m}_B''$. However, it is based solely on the measurement of L_{fl} when burning wood cribs. As the heat of combustion varies among fuels, burns of an identical mass of fuel may not create a similar buoyancy effect if the fuel type is different. It would be more appropriate to represent the buoyancy effect by the HRR, $\dot{Q}$, rather than the MLR, $\dot{m}_B''$. There were attempts to regress the experimentally obtained flame lengths by the dimensionless HRR, $\dot{Q}_W^*$, as defined in Eq. (6.18a), instead of M_W^* (Oka et al., 2000; Oka et al., 2003). Given that $\dot{m}_B'' W^2 \propto \dot{Q}$ and that the change in specific heat, c_P, is minor, M_W^* defined in Eq. (6.40a) can take the form:

$$M_W^* = \frac{\dot{m}_B''}{\rho_\infty \sqrt{gW}} \propto \frac{\dot{Q}}{c_P \rho_\infty T_\infty g^{1/2} W^{5/2}} = \dot{Q}_W^* \tag{6.42}$$

Based on this relationship, Eq. (6.39) can be expressed in an alternative form as follows:

$$\frac{L_{fl}}{W} = a \cdot \left(\dot{Q}_W^*\right)^p \left(U_W^*\right)^q \tag{6.43}$$

Oka et al. obtained the following equation for the flame height, H_{fl}, instead of the flame length, L_{fl}, based on the experiment using a 0.1 m × 0.1 m propane gas burner (Oka et al., 2000):

$$\frac{H_{fl}}{W} = 0.88\left(\dot{Q}_W^*\right)^{3/4}\left(U_W^*\right)^{-1}\left(0.11 \le \left(\dot{Q}_W^*\right)^{3/4}\left(U_W^*\right)^{-1} \le 3.34\right) \tag{6.44}$$

Figure 6.14 shows the results of the calculation of H_{fl}/W using Eq. (6.44) with U_W^* on the horizontal axis. To examine the sensitivity of H_{fl}/W to the dimensionless HRR, $\dot{Q}_W^*$ is varied in three steps, 0.5, 1, and 2. Note that the tilt angle of the flame, θ, is additionally required to obtain L_{fl} from this equation.

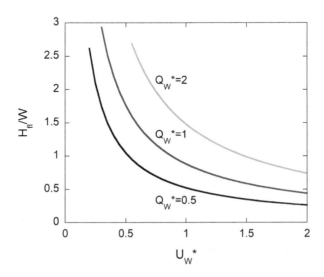

Figure 6.14 Dimensionless flame height above square fire source in a crosswind.

6.4.2 Flame base drag

Flame base drag is a part of the wind-blown flame that attaches to the ground surface on the leeward side of the fuel (Figure 6.12). The part of the ground surface covered by the flame base drag is subject to a higher intensity of heating than the other part.

The formation mechanism of the flame base drag differs between low and high crosswind conditions. In low crosswind conditions, the flame base drag appears when the density of the flammable gas is greater than that of the ambient air. This is because the dense flammable gas remains near the ground surface until it is heated and begins to rise by gaining buoyancy (Welker et al., 1966). Meanwhile, in high crosswind conditions, the flame base drag appears when the momentum of the crosswind prevails over the buoyancy force (Lam et al., 2015). Thus, these two effects need to be accounted for in the modeling of the length of the flame base drag, $L_{fl,B}$. The basic structure of the model was presented by Welker et al., and can be expressed in a generalized form as follows (Welker et al., 1966):

$$\frac{L_{fl,B} + W}{W} = a \cdot \left(\frac{\rho}{\rho_\infty}\right)^p (U_W^*)^q \tag{6.45}$$

where ρ and ρ_∞ are the densities of fuel vapor at boiling point and ambient air, respectively, and U_W^* is the dimensionless wind speed as defined in Eq. (6.40b).

Several research groups have experimentally investigated the values of the proportionality coefficients, a, and the power indices, p and q, in Eq. (6.45) (Welker et al., 1966; Moorhouse, 1982; Mudan, 1984; Lautkaski, 1992; Raj, 2010). Lam et al. compared the predictions by several existing models with their own experiment of flame base drag, which were obtained for 2 m diameter Jet A fires in crosswinds of 3–10 m/s. They reported that their experimental results were well correlated with a modified version of the models presented by Welker et al. and Lautkaski, which takes the form (Lam et al., 2015):

$$\frac{L_{fl,B} + W}{W} = 1.2 \left(\frac{\rho}{\rho_\infty}\right)^{0.48} (U_W^*)^{0.42} \tag{6.46}$$

Figure 6.15 shows the calculation results of the dimensionless flame base drag, $\left(L_{fl,B} + W\right)/W$, for LNG ($\rho/\rho_\infty = 1.6$ (Lautkaski, 1992)) and Jet A fuel ($\rho/\rho_\infty = 3.6$ (Lam, 2015)) under different U_W^* conditions. The $\left(L_{fl,B} + W\right)/W$ of Jet A fuel was consistently larger than that of LNG due to its high density ratio, ρ/ρ_∞.

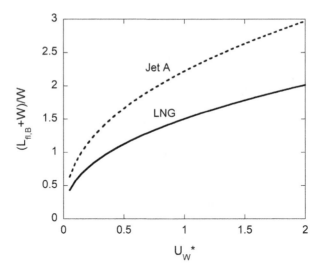

Figure 6.15 Dimensionless flame base drag downwind of a fire source in a crosswind.

6.4.3 Tilt angle

As in the case of the flame length, L_{fl}, one of the earliest models for the flame tilt angle, θ, was proposed by Thomas. Thomas introduced the following dimensionless number, focusing on the relationship between the crosswind velocity, U_∞, and the minimum wind velocity at which the flame tilt occurs, $U_C = \left(\dot{m}_B'' g W / \rho_\infty\right)^{1/3}$ (Thomas, 1965):

$$U_F^* = \frac{U_\infty}{U_C} = \frac{U_\infty}{\left(\dot{m}_B'' g W / \rho_\infty\right)^{1/3}} \tag{6.47}$$

By this definition of U_F^*, the flame tilt only appears when $U_F^* \geq 1$. Using U_F^*, the flame tilt angle is often generalized in a form (Thomas, 1965; Atallah et al., 1974; Moorhouse, 1982; Ferrero et al., 2007; Tang et al., 2015):

$$\cos\theta = \begin{cases} 1 & (U_F^* < 1) \\ a \cdot (U_F^*)^q & (U_F^* \geq 1) \end{cases} \tag{6.48}$$

where a is the coefficient and q is the power index, both to be determined experimentally. a and q obtained for correlating θ also vary among researchers. The most frequently referenced model is by Thomas, who conducted a series of burn experiments using wood cribs and identified θ from visual images (Thomas, 1965). This model takes the form:

$$\cos\theta = \begin{cases} 1 & (U_F^* < 1) \\ 0.7\left(U_F^*\right)^{-0.49} & (U_F^* \geq 1) \end{cases} \tag{6.49}$$

As in the case of L_{fl}, the buoyancy effect associated with the density difference is represented by the MLR, $\dot{m}_B''$, in Eq. (6.47). By replacing $\dot{m}_B''$ with $\dot{Q}$ in the same manner as in Eq. (6.42), U_F^* can be transformed into:

$$U_F^* = \frac{U_\infty}{\left(\dot{m}_B'' g W / \rho_\infty\right)^{1/3}} \propto \frac{U_\infty}{\left(\dot{Q}g / (c_P \rho_\infty T_\infty W)\right)^{1/3}} \tag{6.50}$$

Furthermore, the numerator and denominator of this dimensionless number can be separated into two dimensionless numbers by normalizing them by $\sqrt{gW}$:

$$U_F^* = \frac{U_\infty}{\left(\dot{m}_B'' g W / \rho_\infty\right)^{1/3}} \propto \frac{U_\infty / \sqrt{gW}}{\left(\dot{Q}/\left(c_P \rho_\infty T_\infty g^{1/2} W^{5/2}\right)\right)^{1/3}} = \frac{U_W^*}{\left(\dot{Q}_W^*\right)^{1/3}} \tag{6.51}$$

The flame tilt formulation using the above two dimensionless numbers, U_W^* and $\dot{Q}_W^*$, has been attempted by Oka et al. According to the results of their experiment using a 0.1 m × 0.1 m propane gas burner, they proposed the following expression for the flame tilt angle, $\tan\theta$ (Oka et al., 2000):

$$\tan\theta = 2.72 U_W^* \left(\dot{Q}_W^*\right)^{-1/2} \quad (\theta \geq 40°) \tag{6.52}$$

Note that the power index of $\dot{Q}_W^*$ in Eq. (6.52) is not identical to that of Eq. (6.51). As Eq. (6.51) only shows a variant form of U_F^*, it should be viewed as an implication that θ is dominated by the two dimensionless numbers, U_W^* and $\dot{Q}_W^*$.

Another model proposed by Hu et al. takes a somewhat different form (Hu et al., 2013) from Eq. (6.49). The tilt angle of the flame was determined by the ratio of wind velocity to a characteristic uprising velocity of the flame supported by the buoyancy strength of the fire. The results of a wind tunnel experiment sourced by ethanol and heptane square pool fires were correlated by the dimensionless expression as follows:

$$\tan\theta = 9.1\left[\frac{\rho_\infty c_P \Delta T_{fl} U_\infty^5}{\dot{m}_B'' W^2 \Delta H_C g^2} \cdot \left(\frac{T_\infty}{\Delta T_{fl}}\right)^2\right]^{1/5} \tag{6.53}$$

where ΔT_{fl} is the flame temperature above the ambient, and ΔH_C is the heat of combustion. However, if we recall that $\dot{m}_B'' W^2 \Delta H_C$ in this equation corresponds to the HRR, $\dot{Q}$, Eq. (6.53) can be transformed into:

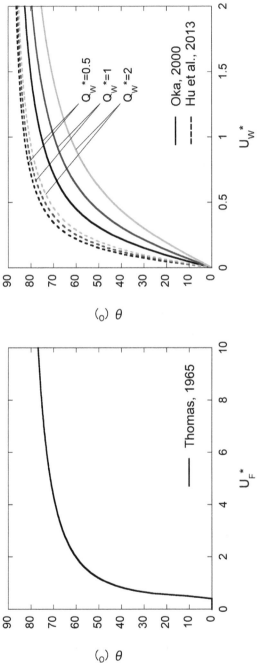

Figure 6.16 Tilt angle of a flame in a crosswind under different dimensionless wind velocities.

$$\tan\theta = 9.1 U_W^* \left(\dot{Q}_W^* \right)^{-1/5} \left(\frac{\Delta T_{fl}}{T_\infty} \right)^{-1/5} \tag{6.54}$$

The expression takes the form of a composite function of the two dimensionless numbers, U_W^* and $\dot{Q}_W^*$, with an additional term of the normalized temperature. It should be noted that ΔT_{fl} varies depending on various conditions, including the type and geometry of the fuel. Thus, it is not that straightforward to accurately evaluate ΔT_{fl} in Eq. (6.54). However, given that $\Delta T_{fl} / T_\infty$ can be regarded as approximately constant in the continuous flame regime of a turbulent diffusion flame as discussed in Chapter 5, Eq. (6.54) eventually follows the form similar to Eq. (6.49).

Figure 6.16 compares θ predicted by the three different models given by Eqs. (6.49), (6.52), and (6.53). However, the panels are separated as the definitions of dimensionless wind velocities differ between the Thomas model and the other two models. In the Hu model, we had to specify the normalized temperature rise, $\Delta T_{fl} / T_\infty$, additionally. Following the discussion in Chapter 5, we assumed that $\Delta T_{fl} = 880$ K in reference to the value of turbulent diffusion flames in a quiescent environment. In all the models, the change in θ was steep when θ was below the range from 50 to 60°, while the change in θ was gentle when θ was over this range. Comparing the Oka and Hu models, the latter predicted the larger θ. This could be attributed to several differences in the experimental conditions, including the fuel type, inflow, and floor boundary conditions.

WORKED EXAMPLE 6.5

A wood crib with a base dimension of 3 m × 3 m is burning in a crosswind with a velocity of $U_\infty = 5$ m/s. The MLR per unit area is $\dot{m}_B'' = 0.05$ kg·m⁻²·s⁻¹. Evaluate the geometry of the flame formed above this wood crib.

SUGGESTED SOLUTION

The flame length, L_{fl}, above a rectangular wood crib can be calculated using Eq. (6.41). Substituting the given values to the variables of this equation, we obtain L_{fl} as follows:

$$L_{fl} = 2 \times 70 \times \left(\frac{0.05}{1.2 \times \sqrt{9.8 \times 3}} \right)^{0.86} \left(\frac{5}{\sqrt{9.8 \times 3}} \right)^{-0.22} = 3.25 \text{ m}$$

In this calculation, we assumed that the ambient gas density is $\rho_\infty = 1.2$ kg·m⁻³.

Next, we calculate the tilt angle, $\cos\theta$, using Eq. (6.49). The dimensionless wind velocity, U_F^*, can be obtained from Eq. (6.47) as follows:

$$U_F^* = \frac{5}{\left(0.05 \times 9.8 \times 3/1.2\right)^{1/3}} = 4.67 \geq 1$$

By substituting U_F^* into Eq. (6.49), we obtain $\cos\theta$ as follows:

$$\cos\theta = 0.7 \times \left(4.67\right)^{-0.49} = 0.329$$

Chapter 7

Ignition and fire spread processes

A large amount of thermal energy is released during the combustion of fuels in outdoor fires. It is partially transferred to surrounding fuels by thermal radiation and convection. The heated fuels raise their temperature, thermally decompose, and release flammable gases into the gas phase. Ignition of the flammable gases creates new fire sources that may cause further ignition of surrounding fuels. The successive occurrence of such an ignition of fuel is the basic mechanism of fire spread in outdoor fires.

The ignition process of fuels varies depending on their state: gas, liquid, or solid. Ignition of gaseous fuels occurs when sufficient thermal energy is supplied to the mixture of the fuel and oxygen (flammable gas mixture), provided the concentration is within its flammability limit. Ignition of liquid fuels does not occur in its original state, but the phase transition from the liquid to the gas is required. The liquid fuel evaporates due to external heating and mixes with oxygen to form a flammable gas mixture in the gas phase. Once gasified, the subsequent ignition process is similar to that of gaseous fuels. Although there are some exceptions, the ignition process of solid fuels generally follows that of liquid fuels. It is thermally decomposed due to external heating and releases the decomposed gases into the gas phase to form a flammable gas mixture. However, the thermal decomposition (pyrolysis) of solid fuels generally accompanies the formation of a layer of char residue on the material surface. Due to a difference in the thermo-physical property between the char and virgin material, char residue formation causes a discontinuous change in the heat conduction within the material subject to heating. In addition, oxygen diffusion causes char oxidation, accompanying heat generation near the material surface.

Solid vegetative materials, including wildland fuels (horizontally layered fallen leaves, twigs, bark and branches, and shrubs and tree canopies) and structural fuels (wooden structural members and materials), are the major types of fuels in outdoor fires. Thus, in this chapter, we focus on the ignition process of solid vegetative materials. We discuss how external heating progresses the ignition process of solid fuels: temperature elevation and

DOI: 10.1201/9781003096689-7

thermal decomposition of the material, and the release of decomposed flammable gases into the gas phase followed by ignition.

The type of fire spread in outdoor fires can be categorized into two in terms of the mode of heat transfer between fuels. They are the fire spread between continuously and discretely distributed fuels. The difference between the two is that the fire spread in the former continues without interruption as long as the fuel distribution continues, whereas the fire spread in the latter may be interrupted depending on the separation distance between fuels. We discuss that the rate of fire spread through continuously distributed fuels can be formulated by extending the ignition model of an individual solid combustible. We also discuss the critical separation distance of discretely distributed fuels through two different approaches, deterministic and probabilistic.

7.1 IGNITION PROCESS OF A SOLID

Vegetative materials, the most common fuels in outdoor fires, are organic compounds forming a char layer above the virgin layer during pyrolysis. In comparison with non-charring materials, the discontinuous change in the heat conduction due to the char layer formation complicates the ignition process of charring materials. Figure 7.1 schematically illustrates the transition of a typical ignition process of charring materials, which comprises six subprocesses from (a) to (f), in association with the change of distributions of temperature and concentrations of chemical species (oxygen and flammable gases) in and around the material (Quintiere, 2006; Torero et al., 2010).

(a) The surface temperature of the material is raised due to external heating. The temperature difference between the surface and the interior causes thermal energy transfer by conduction.

(b) If the material temperature is elevated, water contained in the material starts to evaporate. As the latent heat required for the evaporation of water is large, the internal temperature stagnates at around 100°C until the evaporation is completed. The duration of the temperature stagnation depends on the intensity of heating and the amount of water contained within the material. If the temperature inside the material elevates even higher, pyrolysis of the material starts. Pyrolysis produces gaseous pyrolyzate that contains flammable components and solid char that remains as a residue on the material surface. As the thermo-physical property of char is different from that of virgin materials due to its high porosity, the behavior of heat conduction changes at the interface of the two materials.

(c) Diffusion of oxygen into the pore structure of the char layer causes its oxidation (surface combustion). Surface combustion is an exothermic reaction that raises the temperatures of the char layer and the virgin layer beneath it. The start of surface combustion can be

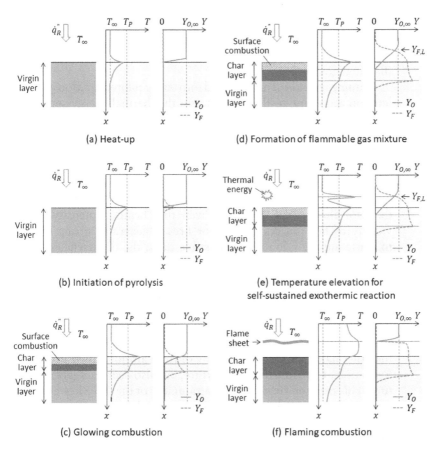

Figure 7.1 Schematic representation of a typical sequence of events leading to flaming ignition of a charring material (modified from Torero et al., 2010).

considered as the onset of glowing or smoldering ignition. Glowing ignition is a direct surface oxidation process, but differs from smoldering ignition in that smoldering ignition is a self-sustained process, whereas glowing ignition is supported by external heating (Babrauskas, 2002b).

(d) In addition to external heating, the transfer of heat generated in the surface combustion enhances the pyrolysis of the virgin layer. This increases the supply rate of flammable gases into the gas phase. The released flammable gases mix with oxygen in the air and form a flammable gas mixture.

(e) An exothermic reaction occurs in the gas phase if the concentration of the flammable gas exceeds its lean flammability limit, $Y_{F,L}$, and sufficient thermal energy is supplied to the flammable gas mixture. The chemical reaction generally proceeds to 'thermal runaway', in which a rapid temperature elevation is attained due to positive feedback of the heat generation, inducing further heat generation (flaming ignition).

Flaming ignition can be divided into either piloted ignition or auto-ignition according to the difference in the thermal energy supply. In the pilot ignition, a pilot (such as a small flame, surface glowing of material, sparks) in the flammable gas mixture is the direct cause of flaming ignition. Whereas in the auto-ignition, an increase of the bulk temperature of the flammable gas mixture without the presence of any pilot proceeds to the onset of the chemical reaction. Thus, a higher intensity of external heating is required for the auto-ignition than for the piloted ignition. Given that heating intensity in the vicinity of fire sources in outdoor fires is generally strong enough to cause auto-ignition of combustible fuels, the occurrence of both types of ignition are mixed in outdoor fires (Frankman et al., 2012).

The fresh air supply has two opposing effects on combustion: promotion and suppression. A moderate supply of fresh air to combustible fuels subject to external heating promotes the mixing of flammable gas and oxygen necessary for the onset of flaming ignition. Surface oxidation of the char layer may also be promoted. On the other hand, an excessive supply of fresh air not only prevents the formation of a flammable gas mixture, but also decreases its temperature. This may interfere with the attainment of the three key elements of combustion (oxygen, flammable gas, and thermal energy)

(f) In the flaming combustion phase, the material receives heat from the flame, promoting pyrolysis. The formation of a chain reaction between the heat generation in the gas phase and the pyrolysis in the solid phase establishes sustained combustion of the material. On the other hand, the flame formation above the material inhibits the diffusion of oxygen from the ambient into the char layer. Thus, the surface combustion of the char layer is suppressed when enveloped by the flame. The heat generated in the flaming combustion is transferred to adjacent materials and may cause another ignition.

In Figure 7.1, the ignition process comprises a series of sub-processes. However, not all fuels subject to external heating follow these sub-processes in sequence, but some fail to reach the flaming ignition. For example, in a weak external heating condition, the material temperature may not reach a level sufficient to start pyrolysis, or the supply rate of flammable gas may not reach a level sufficient to attain a flammable gas mixture within the flammability limit. On the other hand, in a strong external heating condition, flaming ignition may start without an apparent occurrence of glowing combustion as the flammable gas mixture attains the flammability requirement within a short time after pyrolysis begins. In actual fire environments, the ignition mode of a solid vegetative material can be categorized as follows (Babrauskas, 2002b):

(I) Glowing ignition.
(II) Ignition that involves a two-stage process – transition from glowing ignition to flaming ignition.
(III) Flaming ignition.

Figure 7.2 shows the experimentally observed times to ignition of wood pieces placed in a heated furnace, indicating that there are regimes of different ignition modes (Akita, 1959). For all ignition modes, the time to ignition, t_{ig}, becomes shorter as the furnace temperature, T_∞, increases. In ascending order, the onset temperatures of the ignition modes are glowing ignition, piloted ignition, then auto-ignition. As glowing combustion is a relatively slow oxidation reaction compared with flaming combustion, it generally lasts longer. Glowing combustion may shift to flaming combustion if the material condition attains the onset requirements. Examples of the onset requirements in outdoor fires include the formation of a flammable gas mixture due to a change in wind condition, or the supply of thermal energy necessary for ignition due to a change in external heating condition.

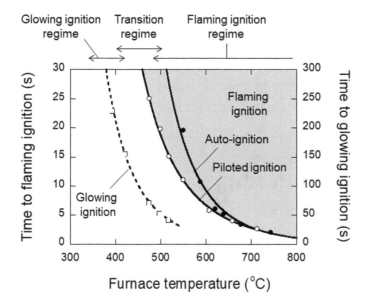

Figure 7.2 Experimentally observed regimes of different ignition modes of wood (Akita, 1959).

In assessing the hazard of outdoor fires, the time to flaming ignition of externally heated combustible materials is often critical. Time to flaming ignition of solid material, t_{ig}, can be considered as the sum of the pyrolysis time, t_P (the time for the internal temperature of the heated material to reach the pyrolysis temperature, T_P), the mixing time, t_M (time for the pyrolyzate gas to mix with fresh air to form a flammable gas mixture within a flammability limit), and the chemical reaction time, t_C (the time for the flammable gas mixture to start the chemical reaction on the external supply of thermal energy) (Fernandez-Pello, 1995; Quintiere, 2006):

$$t_{ig} = t_P + t_M + t_C \tag{7.1}$$

According to an estimate by Quintiere, t_M and t_C are generally sufficiently short compared to t_P (Quintiere, 2006). In such a case, Eq. (7.1) can be simplified as follows:

$$t_{ig} \cong t_P \tag{7.2}$$

Figure 7.3 shows a comparison between t_{ig} and t_P of black PMMA (polymethyl methacrylate) subject to radiant heating, $\dot{q}_e''$ (Dakka et al., 2002). Although there are some discrepancies between t_{ig} and t_P, especially in the small range of $\dot{q}_e''$, it demonstrates that the relationship given by Eq. (7.2) provides a reasonable approximation for t_{ig}.

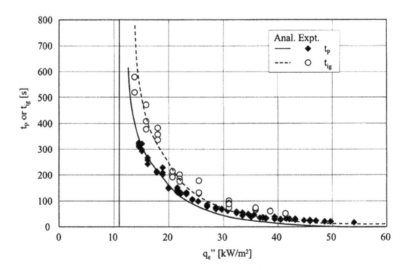

Figure 7.3 Comparison of times to ignition and pyrolysis for black PMMA using the LIFT apparatus (ASTM 1321) (Dakka et al., 2002).

Thermal decomposition of solid combustible materials is often idealized in the following Arrhenius-type reaction rate equation:

$$\dot{m}_F''' = A_S \exp\left(-\frac{E_S}{RT}\right) \tag{7.3}$$

where A_S is the pre-exponential factor, E_S is the activation energy for the pyrolysis specific to the solid, R is the universal gas constant, and T is the absolute temperature. Among these parameters, A_S and E_S are constants unique to each combustible material. Eq. (7.3) indicates that the supply rate of flammable gas is a nonlinear function of T in which a critical temperature might exist that allows a significant flammable gas concentration to make ignition possible (Quintiere, 2006). In other words, the evaluation of t_P can be reduced to a problem of internal temperature elevation of an externally heated solid combustible material (Thomson et al., 1988). This is a desirable ignition property of solid combustible materials in evaluating the hazard of fire spread between fuel objects in outdoor fires.

7.2 TIME TO IGNITION

In the previous section, we demonstrated that the pyrolysis time of a solid combustible material subject to external heating, t_P, can be approximated to the time to ignition, t_{ig}. In addition, t_P can be obtained as the time required for the internal temperature of the solid combustible material to reach the pyrolysis temperature, T_P. Major combustible materials involved in outdoor fires are vegetative fuels, including wildland fuels (horizontally layered fallen leaves, twigs, bark and branches, and shrubs and tree canopies) and structural fuels (wooden structural members and materials). In this section, the process of internal temperature rise of a solid combustible material under external heating is described, based on the discussion of heat conduction in Chapter 3.

7.2.1 Thermal thickness of a solid

The response behavior of a solid combustible material subject to external heating is significantly affected by its thermal thickness. Thermal thickness is different from physical thickness. When a thermally thin material is subject to external heating, the temperature inside the material immediately becomes uniform. In contrast, when a thermally thick material is subject to external heating, a temperature distribution is formed inside the material as it takes time for the thermal energy to diffuse into the material. In this subsection, we discuss the criteria for determining whether a given material is thermally thick or thin.

As stated in Chapter 3, the 1D heat conduction equation takes the form:

$$\frac{\partial T}{\partial t} = \frac{k}{\rho c}\left(\frac{\partial^2 T}{\partial x^2}\right) + \frac{\dot{q}'''}{\rho c} \tag{7.4}$$

This is a partial differential equation of first order in time and second order in space. Therefore, the initial and boundary conditions, which generally take the following forms, are required to obtain its solution:

Initial condition: $T\big|_{t=0} = f(x)$ (7.5a)

Boundary condition: $-k\dfrac{\partial T}{\partial x}\bigg|_{x=0} = \varepsilon\dot{q}_R'' - \varepsilon\sigma(T_S^4 - T_\infty^4) - h(T_S - T_\infty)$ (7.5b)

where $\dot{q}_R''$ is the external heat flux incident on the material surface, ε is the absorptivity, h is the heat transfer coefficient, T_∞ is the ambient temperature, and T_S is the surface temperature. Eq. (7.5b) is nothing but a heat balance equation at the solid surface exposed to the ambient gas at the temperature T_∞ and external heating with the heat flux $\dot{q}_R''$.

Non-dimensionalization is a partial or full removal of physical dimensions from an equation involving physical quantities by substituting variables. Making the heat conduction equation dimensionless allows us to clarify the relationships between variables, thus elucidating the governing dimensionless parameters of the phenomena. For non-dimensionalization, we consider the following dimensionless parameters:

$$T^* = \frac{T - T_\infty}{T_0 - T_\infty}, \; T_S^* = \frac{T_S - T_\infty}{T_0 - T_\infty}, \; Q^* = \frac{\dot{q}'''L^2}{k(T_0 - T_\infty)}, \; Q_S^* = \frac{\dot{q}_S''L}{k(T_0 - T_\infty)}, \; X^* = \frac{x}{L}$$

$$\tag{7.6}$$

In this equation, the net radiative heat flux term is represented by $\dot{q}_S'' = \varepsilon\dot{q}_R'' - \varepsilon\sigma(T_S^4 - T_\infty^4)$. The superscript '*' denotes that the parameter is dimensionless. By substituting these dimensionless parameters into Eqs. (7.4) and (7.5b), we obtain:

$$\frac{\partial T^*}{\partial Fo^*} = \frac{\partial^2 T^*}{\partial X^2} + Q^* \tag{7.7a}$$

$$\frac{\partial T^*}{\partial X^*}\bigg|_{X^*=0} = Bi^*T_S^* - Q_S^* \tag{7.7b}$$

where Fo^* is a dimensionless number called the Fourier number representing the ratio of the thermal energy transfer to the amount retained in the solid material. Bi^* is another dimensionless number called the Biot number representing the ratio of the thermal energy transfer by convection to conduction:

$$Fo^* = \frac{kt}{\rho c L^2} = \frac{\alpha t}{L^2} \tag{7.8a}$$

$$Bi^* = \frac{hL}{k} \tag{7.8b}$$

These are the dominant dimensionless numbers in heat conduction. Note that the definition of the Nusselt number, which we discussed in Chapter 3, takes the same expression as that of the Biot number. However, the difference between the two lies in k ; it is the thermal conductivity of the solid in the Biot number, whereas it is the fluid for the Nusselt number.

Let us consider a plane wall made of a single material with both sides exposed to the ambient gas of uniform temperature, T_∞, as shown in Figure 7.4. As the convective heat flux at the surface should correspond to the internal heat flux by conduction:

$$\frac{k}{d}(T_S - T_0) = h(T_\infty - T_S) \tag{7.9}$$

where T_0 is the temperature at the center, and $T_S (> T_0)$ is the surface temperature. If the internal temperature profile is linearly approximated for the heat conduction calculation in Eq. (7.9) for simplicity, we obtain:

$$\frac{T_S - T_0}{T_\infty - T_S} = \frac{hd}{k} = Bi^* \tag{7.10}$$

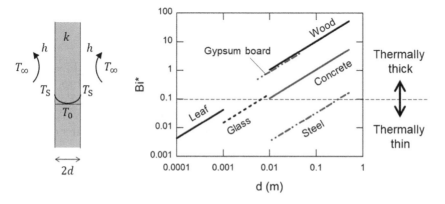

Figure 7.4 Estimated Biot numbers of plane walls with different thermal conductivities.

The plane wall can be considered thermally thin if the numerator on the left-hand side of Eq. (7.10) is sufficiently small compared to the denominator; in other words, when the Biot number, Bi^*, is small, or $T_S \cong T_0$. Although it is difficult to specify a general threshold for a thermally thin object, $Bi^* < 0.1$ can be considered as a reference (Incropera et al., 2006).

The estimated B^* for some representative materials are also included in Figure 7.4, in which $h = 0.015\,\mathrm{kW \cdot m^{-1} \cdot K^{-1}}$ was assumed. d was varied within a general range for each material. In addition to the values of k given in Table 3.1 in Chapter 3, that of leaves was taken from measurements by Hays (Hays, 1975). Invoking $Bi^* < 0.1$ as the threshold, leaf, glass, and steel are thermally thin materials, whereas concrete, gypsum board, and wood are thermally thick materials.

It should be noted that the effect of water content is not reflected in the above discussion. As water content delays the temperature rise of an object subject to external heating, there are often cases when the thermally thin assumption is not suitable for temperature-rise calculations. For example, live foliage, one of the common combustible materials in crown fires, has a generally high water content (green woods can have a moisture content of more than 100%) (USDA, 2011). It has been reported that live forest fuels subject to external heating were still actively releasing water at ignition, implying that there are steep temperature gradients within these physically thin fuels (McAllister et al., 2014). For such a case, Eq. (7.9) can be reformulated by assuming that uniform water evaporation occurs within the object:

$$\frac{k}{d}(T_S - T_0) - \dot{q}_V''' d = h(T_\infty - T_S) \tag{7.11}$$

where $\dot{q}_V'''$ is the rate of water evaporation per unit volume. The expression equivalent to Eq. (7.10) can be obtained by transforming this equation:

$$\frac{T_S - T_0}{T_\infty - T_S} = \frac{hd}{k} + \frac{\dot{q}_V''' d^2}{k(T_\infty - T_S)} = Bi^* + \frac{\dot{q}_V''' d^2}{k(T_\infty - T_S)} \tag{7.12}$$

Compared to Eq. (7.10), we have an additional term, $\dot{q}_V''' d^2 / k(T_\infty - T_S)$, on the right-hand side of Eq. (7.12). The condition for a material to be thermally thin is narrowed because the whole right-hand side of Eq. (7.12), including the additional term, needs to be less than 0.1. Thus, leaves could be categorized into a thermally thick material depending on their water content.

7.2.2 Ignition of a thermally thin solid under constant exposures

In advance of considering thermally thick materials, the ignition condition of thermally thin materials is discussed. Time to ignition of a thermally thin solid subject to external heating, t_{ig}, can be obtained as the time for the temperature of the entire body, T, to reach the ignition temperature, T_{ig}.

(1) Convection boundary condition

Assume a thermally thin material with a thickness, d, and a temperature, T, is exposed to hot ambient gas with a temperature, T_∞, as shown in Figure 7.5. If only convective heat transfer is considered, the conservation of the thermal energy of the material can be expressed as follows:

$$\rho c d \frac{dT}{dt} = h S (T_\infty - T) \tag{7.13}$$

Solving this differential equation under the initial condition, $T = T_0$ at $t = 0$, we obtain the following equation expressing the time development of T:

$$T - T_\infty = (T_0 - T_\infty) \exp\left(-\frac{h}{\rho c d} t\right) \tag{7.14}$$

The time to ignition of the material, t_{ig}, can be calculated as the time for the plane wall to reach its ignition temperature, T_{ig}. By letting $t = t_{ig}$ and $T = T_{ig}$ in Eq. (7.14), t_{ig} takes the form:

$$t_{ig} = \frac{\rho c d}{h} \ln\left(\frac{T_\infty - T_0}{T_\infty - T_{ig}}\right) (T_\infty > T_{ig}) \tag{7.15}$$

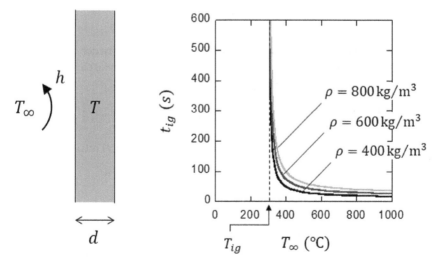

Figure 7.5 Time to ignition of a thermally thin material with the convection boundary condition.

However, note that this expression is applicable only when $T_\infty > T_{ig}$. We invoke the Taylor series expansion, which is:

$$\ln(1+\beta) = \beta - \frac{\beta^2}{2} + \frac{\beta^3}{3} - \frac{\beta^4}{4} + \cdots \quad (-1 \le \beta \le 1) \tag{7.16}$$

This equation can be approximated as $\ln(1+\beta) \cong \beta$ when $|\beta| \ll 1$. Thus, Eq. (7.15) can be simplified as follows:

$$t_{ig} = \frac{\rho c d}{h} \ln\left(1 + \frac{T_{ig} - T_0}{T_\infty - T_{ig}}\right) \cong \frac{\rho c d (T_{ig} - T_0)}{h(T_\infty - T_{ig})} \quad (T_\infty > T_{ig}) \tag{7.17}$$

Some calculation results for the t_{ig} of a thermally thin leaf using Eq.(7.17) are also shown in Figure 7.5. In this calculation, we assumed $h = 0.015\,\mathrm{kW \cdot m^{-1} \cdot K^{-1}}$, $d = 0.3 \times 10^{-3}$ m, $T_{ig} = 573$ K, and $c = 1.62\,\mathrm{kJ \cdot kg^{-1} \cdot K^{-1}}$. The density, ρ, was varied in three steps, 400, 600, and 800 kg·m^{-3}. Although a long t_{ig} was required the T_∞ only slightly exceeded T_{ig}, t_{ig} steeply decreased as T_∞ increased.

(2) External radiation boundary condition

t_{ig} in Eq. (7.15) only considered convective heat transfer at the material surface. To additionally consider the effect of radiative heat flux $\dot{q}''_R$ from an external fire source, the conservation of thermal energy equivalent to Eq. (7.13) takes the form (Figure 7.6):

$$\rho c d \frac{dT}{dt} = \varepsilon \dot{q}''_R - \varepsilon\sigma(T^4 - T_\infty^4) - h(T - T_\infty) \tag{7.18}$$

As the right-hand side of Eq. (7.18) contains the fourth-order term of T, an analytical solution is not readily available. Thus, we linearize the second term of Eq. (7.18) as follows (Quintiere, 1981):

$$\varepsilon\sigma(T_S^4 - T_\infty^4) = \left[\varepsilon\sigma(T^2 + T_\infty^2)(T + T_\infty)\right](T - T_\infty) = h_r(T - T_\infty) \tag{7.19}$$

where h_r is the effective coefficient. By substituting Eq. (7.19) into Eq. (7.18), we obtain:

$$\rho c d \frac{dT}{dt} = \varepsilon \dot{q}''_R - (h + h_r)(T - T_\infty) = \varepsilon \dot{q}''_R - h_T(T - T_\infty) \tag{7.20}$$

where h_T is the total heat transfer coefficient. Solving this differential equation under the initial condition, $T = T_\infty$ at $t = 0$, and the boundary condition, $\dot{q}_R'' = constant$, we obtain the following equation for the time development of the T (Quintiere, 1981):

$$T - T_\infty = \frac{\varepsilon \dot{q}_R''}{h_T}\left[1 - \exp\left(-\frac{h_T t}{\rho c d}\right)\right]$$ (7.21)

By letting $t = t_{ig}$ and $T = T_{ig}$ in Eq. (7.21), t_{ig} can be calculated as follows:

$$t_{ig} = -\frac{\rho c d}{h_T}\ln\left[1 - \frac{h_T}{\varepsilon \dot{q}_R''}\left(T_{ig} - T_\infty\right)\right]\ \left(\varepsilon \dot{q}_R'' > h_T\left(T_{ig} - T_\infty\right)\right)$$ (7.22)

Note that this expression is applicable only when $\varepsilon \dot{q}_R'' > h_T\left(T_{ig} - T_\infty\right)$. Similar to the case of Eq. (7.17), t_{ig} can be simplified by applying the Taylor series expansion as follows:

$$t_{ig} \cong \frac{\rho c d \left(T_{ig} - T_\infty\right)}{\varepsilon \dot{q}_R''}$$ (7.23)

In linearizing the radiative term in Eq. (7.20), we introduced a new variable, h_T. h_T is a cubic function of T, which still remains as an impediment to obtaining an analytical solution. However, the h_T has been eliminated from Eq. (7.23) through the derivation process of t_{ig}.

Figure 7.6 also shows some calculation results for the t_{ig} of a thermally thin leaf when exposed to external radiative heating. The calculation conditions

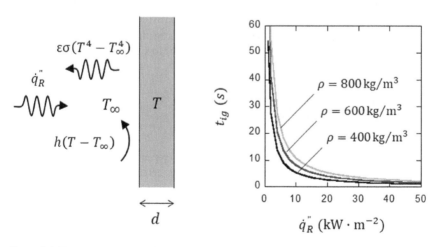

Figure 7.6 Time to ignition of a thermally thin material with the external radiation boundary condition.

follow those assumed in the previous subsection (Figure 7.5). In addition, the absorptivity was assumed to be unity, $\varepsilon = 1$. The result was similar to that of the convection boundary condition in Figure 7.5, in which t_{ig} steeply decreased as $\dot{q}_R''$ increased. However, the heat loss effect, which was originally included in Eq. (7.18), has been eliminated from Eq. (7.23). As a consequence, Eq. (7.23) yields a prediction that ignition occurs under external heating even smaller than its critical value, $\dot{q}_{R,cr}''$. Thus, the applicable range of Eq. (7.23) should be regarded as $\varepsilon\dot{q}_R'' \gg h_T\left(T_{ig} - T_\infty\right)$.

7.2.3 Ignition of a thermally thick solid under constant exposures

Combustible materials involved in fires are of a finite thickness. If combustible materials can be regarded as 1D in the direction of thickness, the temperature distribution of the material can be determined by specifying the boundary conditions at the two surfaces. However, we generally need to resort to numerical approaches to obtain the temperature distribution. To avoid this, we assume that the material is a semi-infinite plane wall. Analytical solutions for the 1D heat conduction equation of semi-infinite solids are available under selected boundary conditions, as discussed in Chapter 3. We can use the analytical solutions to evaluate the time to ignition of heated materials, t_{ig}, as the time for the surface temperature, T_S, to reach the ignition temperature, T_{ig}.

(1) Convection boundary condition

Let us consider the case where the solid surface is exposed to a hot gas of a constant temperature, T_∞. Thermal energy transfer between the solid and the hot gas occurs by convection. By substituting the ignition temperature, T_{ig}, and the time to ignition, t_{ig}, into Eq. (3.25b) in Chapter 3, we obtain the following equation:

$$\frac{T_{ig} - T_0}{T_\infty - T_0} = 1 - \exp\left(\frac{h\sqrt{\alpha t_{ig}}}{k}\right)^2 \mathrm{erfc}\left(\frac{h\sqrt{\alpha t_{ig}}}{k}\right) \tag{7.24}$$

Recall that the Taylor series expansion of the exponential function and the asymptotic expansion of the complementary error function respectively take forms:

$$\exp\left(\beta^2\right) = 1 + \beta^2 + \frac{\beta^4}{2!} + \cdots \tag{7.25a}$$

$$\mathrm{erf}\left(\beta\right) = \frac{\exp\left(-\beta^2\right)}{\sqrt{\pi}}\left(\frac{1}{\beta} - \frac{1}{2\beta^3} + \frac{3}{4\beta^5} - \cdots\right) \tag{7.25b}$$

As the higher-order terms can be neglected in Eqs. (7.25) when $|\beta| \ll 1$, Eq. (7.24) can be approximated to the following form by introducing $\beta = h_S \sqrt{\alpha t} / k$:

$$\frac{T_{ig} - T_0}{T_\infty - T_0} \cong 1 - \frac{1}{\beta\sqrt{\pi}} = 1 - \frac{k}{h_S\sqrt{\alpha t_{ig}}} \frac{1}{\sqrt{\pi}} \tag{7.26}$$

Thus, t_{ig} can be expressed in an approximate form as follows (Lawson et al., 1952; Akita, 1959; Tanaka, 1993).

$$t_{ig} \cong \frac{k\rho c}{\pi h^2} \left(\frac{T_\infty - T_0}{T_\infty - T_{ig}} \right)^2 \tag{7.27}$$

t_{ig} is proportional to the thermal inertia, $k\rho c$, and inversely proportional to the temperature difference between the ambient and ignition, $T_\infty - T_{ig}$. The form of the equation is similar to Eq. (7.17) for the t_{ig} of thermally thin material under the convection boundary condition.

The result of the calculated t_{ig} of semi-infinite wood as a function of the gas temperature, T_∞, is shown in Figure 7.7. The thermal conductivity of wood, k, was obtained by introducing the density, ρ, and moisture content, $\phi(= 0.12)$, into Eq. (3.5) in Chapter 3. The other conditions follow those assumed in the previous subsection (Figures 7.5 and 7.6). Compared to the case of thermally thin material (Figure 7.5), t_{ig} was significantly longer, and the effect of ρ was notable.

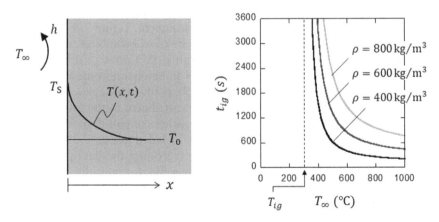

Figure 7.7 Time to ignition of a thermally thick material under the convection boundary condition.

(2) Specified heat flux boundary condition

We consider a case where the incident heat flux, $\varepsilon \dot{q}_R''$, is given at the surface boundary of a semi-infinite plane wall. By substituting the ignition temperature, T_{ig}, and the time to ignition, t_{ig}, into Eq. (3.28b), the following relationship between T_{ig} and t_{ig} can be obtained:

$$T_{ig} - T_0 = 2\varepsilon \dot{q}_R'' \sqrt{\frac{t_{ig}}{\pi k \rho c}} \tag{7.28}$$

This can be transformed to yield the equation for t_{ig} as follows (Carslaw et al., 1959):

$$t_{ig} = \frac{\pi k \rho c}{4} \left(\frac{T_{ig} - T_0}{\varepsilon \dot{q}_R''} \right)^2 \tag{7.29}$$

The obtained equation indicates that t_{ig} is proportional to $k\rho c$ and inversely proportional to $\dot{q}_R''^2$. Similar to Eq. (7.27) for the convection boundary condition, t_{ig} is proportional to $k\rho c$ in Eq. (7.29).

A drawback of the analytical solution given by Eq. (3.28b) is that the heat flux at the surface is given as a constant value, and the change in the rate of heat loss from the surface due to its temperature rise is not accountable. Ignoring the heat loss from the surface should preferably be avoided as it may lead to underestimation of t_{ig}. However, this equation is still useful for providing important insights into the relationships of involved physical parameters. For example, t_{ig} is proportional to $\dot{q}_R''^{-2}$ for a thermally thick material (Eq. (7.29)), whereas it is proportional to $\dot{q}_R''^{-1}$ for a thermally thin material (Eq. (7.23)). In engineering applications, Eq. (7.29) is often used in analyzing t_{ig} under exposure to a fire environment. Figure 7.8 shows the result of ignition tests for three different species of wood subject to radiative heating using the Cone Calorimeter apparatus (ISO 5660). Although the gradients change with different wood species and moisture content, the relationship, $t_{ig}^{-1/2} \propto \varepsilon \dot{q}_R''$ as predicted in Eq. (7.29), holds in all cases. According to Eq. (7.29), the difference in the gradient is attributed to the properties of the heated material, such as $k\rho c$, and T_{ig}. Moisture content is another factor that affects the gradient, but is not explicitly included in Eq. (7.29).

In Figure 7.8, the intersection of the line regressing the data points and the abscissa ($t_{ig}^{-1/2} = 0$) is called the critical heat flux for ignition, $\dot{q}_{cr}''$. This is an estimated threshold of $\varepsilon \dot{q}_R''$ below which the ignition of the material does not occur. It should be noted that $\dot{q}_{cr}''$ does not necessarily represent the lower ignition limit of the incident heat flux under actual heating conditions, $\dot{q}_{min}''$. However, it is virtually impossible to experimentally measure $\dot{q}_{min}''$ as the material needs to be exposed to heating for an extremely long time. In general,

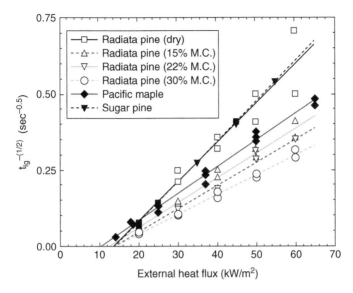

Figure 7.8 Relationships between the external heat flux and the time to ignition for radiata pine, Pacific maple, and sugar pine (Moghtaderi et al., 1997).

$\dot{q}''_{cr}$ is somewhat lower than $\dot{q}''_{min}$ and therefore can be regarded as a conservative estimate of $\dot{q}''_{min}$ (Janssens, 1991b). Thus, for engineering purposes, $\dot{q}''_{cr}$ has been widely used as the nominal value of the critical heat flux.

(3) External radiation with heat loss from the surface

The convection boundary and specified heat flux boundary conditions, which have been discussed above, are hypothetical boundary conditions under which analytical solutions are available. They are incomplete as they only partially consider modes of heat transfer present in actual fire situations. The boundary condition of material exposed to a fire environment can be generalized as follows:

$$-k \frac{\partial T}{\partial x}\bigg|_{x=0} = \dot{q}''_{net} = \varepsilon \dot{q}''_R - \varepsilon\sigma\left(T_S^4 - T_\infty^4\right) - h\left(T_S - T_\infty\right) \qquad (7.30)$$

where $\dot{q}''_R$ is the incident heat flux from an external heat source, and ε is the absorptivity. Involvement of a non-linear term of T_S on the right-hand side of this equation inhibits derivation of its analytical solution. Thus, by expressing the radiative and convective terms in an integrated manner using the total heat transfer coefficient, h_T, as given by Eq. (7.19), Eq. (7.30) can be transformed as (Tanaka, 1993):

$$-k\frac{\partial T}{\partial x}\bigg|_{x=0} = \varepsilon\dot{q}_R'' - h_T\left(T_{ig} - T_\infty\right) = h_T\left[\left(T_\infty + \frac{\varepsilon\dot{q}_R''}{h_T}\right) - T_{ig}\right] \qquad (7.31)$$

In other words, the boundary condition in Eq. (7.30) can be replaced by the convection boundary condition for a semi-infinite solid hypothetically exposed to a hot gas at a temperature, $T_\infty + \varepsilon\dot{q}_R'' / h_T$. Thus, by substituting h with h_T, T_∞ with $T_\infty + \varepsilon\dot{q}_R'' / h_T$, and T_0 with T_∞ in Eq. (7.27), the relationship between $\varepsilon\dot{q}_R''$ and t_{ig} can be obtained as follows (Tanaka, 1993):

$$t_{ig} \cong \frac{k\rho c}{\pi h_T^2}\left[\frac{\varepsilon\dot{q}_{\bar{R}}''}{\varepsilon\dot{q}_{\bar{R}}'' - h_T\left(T_{ig} - T_\infty\right)}\right] = \frac{k\rho c}{\pi h_T^2}\left[1 - \frac{h_T\left(T_{ig} - T_\infty\right)}{\varepsilon\dot{q}_{\bar{R}}'' - h_T\left(T_{ig} - T_\infty\right)}\right]^2 \qquad (7.32)$$

This equation is expected to yield a more realistic t_{ig} compared to Eq. (7.27) under the convection boundary condition or Eq. (7.29) under the specified heat flux boundary condition.

Some calculation results of t_{ig} of semi-infinite wood under different $\dot{q}_R''$ are shown in Figure 7.9. In the calculation, we assumed that $h_T\left(T_{ig} - T_\infty\right) = \dot{q}_{cr}'' = 15$ kW $\cdot$ m^{-2}. The other conditions follow those assumed in the previous examples (Figures 7.5, 7.6, and 7.7). In contrast to the t_{ig} for a thermally thin solid under the external radiation boundary condition where there was no lower ignition limit (refer to Figure 7.6), $\dot{q}_{cr}''$ is introduced as the lower critical value of ignition in Eq. (7.32).

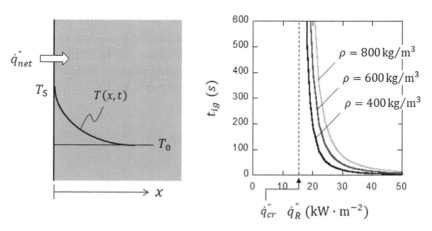

Figure 7.9 Time to ignition of a thermally thick material under external radiation with heat loss from the surface.

(4) Engineering correlations

The t_{ig} expressed by Eqs. (7.27), (7.29), and (7.32) are based on the temperature rise analytically obtained for an inert semi-infinite solid under specific boundary conditions. They do not explicitly consider factors that may affect the ignition behavior, such as the change in thermo-physical properties of thermally degraded material, the heat generation due to surface oxidation, and the temperature stagnation due to water evaporation and pyrolysis. As analytical solutions for the heat conduction equation are available only for limited cases of ideal boundary conditions, it should be noted that the theoretical models in Eqs. (7.27), (7.29), and (7.32) do not fully represent actual ignition conditions. Thus, they have not necessarily achieved accurate predictions for actual ignition problems. Engineering correlations have been obtained by adjusting the theoretical models to better predict the t_{ig} observed in experiments.

Based on the theoretical solutions for a semi-infinite solid as discussed above, Janssens derived a thermal model of piloted ignition. The model was embedded with a power-law correlation:

$$\left(\dot{q}_R'' - \dot{q}_{cr}''\right)t_{ig}^n = \text{constant} \tag{7.33}$$

which follows the form of Eq. (7.32). In Eq. (7.33), the power index, n, is determined experimentally. By using piloted ignition data obtained from the cone calorimeter and LIFT (ASTM) apparatuses, Janssens derived the following correlation for the dimensionless irradiance, φ, and the dimensionless time to ignition, τ_{ig} (Janssens, 1991a):

$$\varphi = \frac{\dot{q}_R''}{\dot{q}_{cr}''} = 1 + 0.73\left(\frac{1}{\tau_{ig}}\right)^{0.547} \tag{7.34}$$

in which τ_{ig} takes the form:

$$\tau_{ig} = \frac{h_{ig}^2 t_{ig}}{k\rho c} \tag{7.35a}$$

$$h_{ig} = \frac{\dot{q}_{cr}''}{T_{ig} - T_0} = 0.015 + \frac{\sigma\left(T_{ig} - T_0\right)^4}{T_{ig} - T_0} \tag{7.35b}$$

The ignition data points with the regression result are plotted in Figure 7.10 (Janssens, 1991a). The result suggests that $\left(t_{ig}\right)^{-0.547}$ is proportional to $\dot{q}_R''$. The obtained power index for t_{ig}, which is −0.547, is similar to the theoretically predicted value of −0.5 in Eq. (7.32). Based on this correlation, the time to ignition of various solid materials is often correlated with the power index of −0.55 (Babrauskas, 2002b).

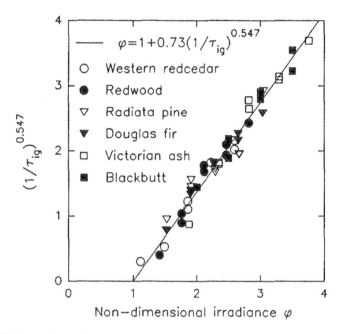

Figure 7.10 Correlation of ignition data with the Janssens model (Janssens, 1991a).

(5) Critical conditions for ignition

The use of the thermally thick solid models discussed above requires some fuel-specific parameters, including the critical ignition temperature, T_{ig}, critical ignition heat flux, $\dot{q}''_R$, and thermal inertia, $k\rho c$. Table 7.1 summarizes these parameters of woods obtained by regressing experimental data by several investigators (Janssens, 1991a; Tran et al., 1992; Spearpoint et al., 2001). The parameters ranged from 0.073 to 1.01 $kJ^2 \cdot s^{-1} \cdot m^{-4} \cdot K^{-2}$ for $k\rho c$, from 571 to 657 K for T_{ig}, and from 9.7 to 16.0 $kW \cdot m^{-2}$ for $\dot{q}''_{cr}$. The large variations are largely due to differences in the attributes of the specimens, such as the species, moisture content, and grain orientations. In comparison to softwoods, hardwoods generally have higher $k\rho c$, T_{ig}, and $\dot{q}''_{cr}$, indicating relatively low ignitability of hardwoods under external heating. Although the effect of moisture content is not apparent from the values of the parameters in Table 7.1, Janssens reported that T_{ig} rises by 2 K for each percent of moisture content increase (Janssens, 1991b).

WORKED EXAMPLE 7.1

Assume that a board of maple is subject to constant external heating of $\dot{q}''_R = 25$ kW $\cdot$ m^{-2}. Calculate t_{ig} when the initial temperature of the board is $T_0 = 20°C$. Use the Stefan-Boltzmann constant $\sigma = 5.67 \times 10^{-11}$ kW $\cdot$ m^{-2} $\cdot$ K^{-4},

Table 7.1 Piloted ignition properties of selected wood species (Janssens, 1991a; Tran et al., 1992; Spearpoint et al., 2001).

Species	Type	$k\rho c$ $(kJ^2 \cdot s^{-1} \cdot m^{-4} \cdot K^{-2})$	T_{ig} (K)	$\dot{q}_{cr}''$ $(kW \cdot m^{-2})$
Woods (oven-dried, perpendicular to grain orientation), Janssens, 1991a.				
Douglas fir	Softwood	0.158	623	13.0
Radiata pine	Softwood	0.156	622	12.9
Redwood	Softwood	0.141	637	14.0
Western redcedar	Softwood	0.087	627	13.3
Victorian ash	Hardwood	0.260	584	10.4
Blackbutt	Hardwood	0.393	573	9.7
Woods (moisture content between 8.06 and 9.71%, perpendicular to grain orientation), Tran et al., 1992.				
Redwood	Softwood	0.073	636	12.42
Southern pine	Softwood	0.183	593	10.68
Basswood	Hardwood	0.141	571	10.00
Red oak	Hardwood	0.360	588	10.53
Woods (moisture content between 4.8 and 8.6%, perpendicular to grain orientation), Spearpoint et al., 2001.				
Douglas fir	Softwood	0.25	657	16.0
Redwood	Softwood	0.22	577	15.5
Maple	Hardwood	0.67	627	13.9
Red oak	Hardwood	1.01	648	10.8

the surface absorptivity $\varepsilon = 1$, and the heat transfer coefficient $h = 0.015 \text{ kW} \cdot \text{m}^{-1} \cdot \text{K}^{-1}$.

SUGGESTED SOLUTION

Let us compare the t_{ig} calculated by three different models in Eqs. (7.29) (specified heat flux boundary condition), (7.32) (external radiation with heat loss from the surface), and (7.34) (engineering correlation by Janssens) are used.

First, Eq. (7.29) yields:

$$t_{ig} = \frac{\pi k \rho c}{4}\left(\frac{T_{ig} - T_0}{\varepsilon \dot{q}_R''}\right)^2 \quad t_{ig} = \frac{\pi \times 0.67}{4} \times \left(\frac{627 - 293}{1 \times 25}\right)^2 = 93.9 \text{ s}$$

Second, the t_{ig} can be calculated by assuming that $h_T\left(T_{ig} - T_\infty\right) = \dot{q}_{cr}''$ in Eq. (7.32):

$$t_{ig} \cong \frac{0.67}{\pi\left(13.9/(627 - 293)\right)^2}\left[\frac{1 \times 25}{1 \times 25 - 13.9}\right]^2 = 624.6 \text{ s}$$

Third, the dimensionless time to ignition, τ_{ig}, and the coefficient, h_{ig}, can be obtained from Eq. (7.35) as follows:

$$\tau_{ig} = \left(\frac{\dot{q}_R'' / \dot{q}_{cr}'' - 1}{0.73}\right)^{-1/0.547} = \left(\frac{25/13.9 - 1}{0.73}\right)^{-1/0.547} = 0.849$$

$$h_{ig} = \frac{\dot{q}_{cr}''}{T_{ig} - T_0} = \frac{13.9}{627 - 293} = 0.0416$$

By substituting τ_{ig} and h_{ig} into Eq. (7.34), t_{ig} can be obtained as follows:

$$t_{ig} = \tau_{ig} \cdot \frac{k\rho c}{h_{ig}^2} = 0.849 \times \frac{0.67}{0.0416^2} = 328.3\text{s}$$

Figure 7.11 shows the changes in t_{ig} under different $\dot{q}_R''$ conditions. The calculation results varied among the three models. Eq. (7.29) showed a different pattern from the others. Ignition was predicted even below the critical heat flux, $\dot{q}_{cr}''$, as the heat loss from the material surface was ignored. As for Eqs. (7.32) and (7.34), the calculation results showed a similar $t_{ig} - \dot{q}_R''$ curve. However, Eq. (7.32) yielded a relatively longer t_{ig} in high $\dot{q}_R''$ ranges.

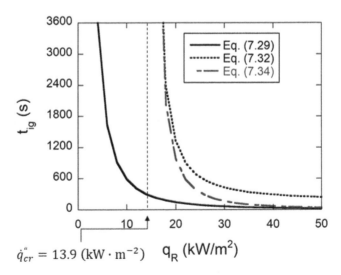

Figure 7.11 Comparison of time to ignition of a slab of wood calculated by three different models.

7.2.4 Ignition of a thermally thick solid under time-varying exposures

In this section, we have discussed the time to ignition of combustible materials subject to constant external heating. However, in actual fires, combustible materials are subject to heating that varies with time. Thus, the expressions derived in the above discussion may not be applicable to certain fire problems in their original forms. If the procedural complexity is not an issue, one can evaluate the time to ignition of a material subject to time-varying heating by numerically calculating the internal temperature, but this is not the subject of this section.

An alternative approach to considering the time dependency of thermal exposure conditions is to focus on the cumulative amount of incident heat flux on the combustible material. The expression for t_{ig} shown in Eq. (7.29) infers that the following relationship between $\dot{q}_R''$ and t_{ig} holds even when the heating changes with time, provided the physical properties of the combustible material do not change significantly:

$$\dot{q}_R''^2 t_{ig} = \frac{\pi k \rho c}{4} \left(T_{ig} - T_0 \right)^2 \tag{7.36}$$

Assuming a negligible change in the physical properties of the combustible material, the right-hand side of Eq. (7.36) can be regarded as a material-specific constant. In the original form of Eq. (7.36), which is Eq. (7.29), the $\dot{q}_R''$ was assumed to be constant. However, if we allow its variation over time, Eq. (7.36) can be modified as follows:

$$\int_0^{t_{ig}} \dot{q}_R''^2 dt = C_{ig} \tag{7.37}$$

where C_{ig} is the material-specific constant. Moreover, considering that there is a lower limit of incident heat flux required for ignition, $\dot{q}_{cr}''$, the following relationship can be derived by further modifying Eq. (7.37):

$$\int_0^{t_{ig}} \left(\dot{q}_R'' - \dot{q}_{cr}'' \right)^2 dt = C_{ig} \tag{7.38}$$

The constant, C_{ig}, should be determined experimentally. If the value of C_{ig} is unknown, a reference value can be obtained by substituting the thermo-physical properties in Tables 7.1 and 7.2 into the right-hand side of Eq. (7.36).

Based on a similar concept, Shields and Silcock introduced a method to predict t_{ig} according to the *FTP* (flux time product) defined below (Shields et al., 1994; Silcock et al., 1995; DiDomizio et al., 2016):

$$\int_0^{t_{ig}} \left(\dot{q}_R'' - \dot{q}_{cr}'' \right)^N dt = FTP \qquad (7.39)$$

Eq. (7.39) can be viewed as a generalized form of Eq. (7.38) with the power index substituted with the flux time product index, N, which needs to be determined experimentally. In their reports, a protocol based on the *FTP* was used to analyze ignition data obtained from the Cone Calorimeter under the heat flux in the range $20 - 70$ kW $\cdot$ m^{-2} for different orientations and ignition modes for three cellulosic materials, i.e., softwood, chipboard, and plywood. The values for $\dot{q}_{cr}''$, N, and *FTP* are summarized in Table 7.2 by material type, specimen orientation, and ignition mode (Shields et al., 1994). The experimentally determined values of N were close to the theoretically inferred value of 2 for all the tested materials. However, the value of the *FTP* on the right-hand side of Eq. (7.39) varied with different orientations and ignition modes, even among the same material type. Note that $\dot{q}_{cr}''$ shown in Table 7.2 are consistently larger for the vertically mounted specimens than the horizontally mounted specimens. Shields et al. concluded that this was due to a difference in the convective heat transfer coefficients at the material surface (Shields et al., 1994).

Table 7.2 FTPs using mean ignition time measured by the Cone Calorimeter test data (Shields et al., 1994).

Material	Orientation	Ignition mode	$\dot{q}_{cr}''$ $(kW \cdot m^{-2})$	N	FTP $((kW \cdot m^{-2})^N \cdot s)$
Softwood (20 mm)	Horizontal	Gas flame	10.5	2.2	37960
		Spark	10.0	2.2	44079
	Vertical	Gas flame	12.0	1.9	11794
		Spark	12.0	1.9	16502
Chipboard (15 mm)	Horizontal	Gas flame	9.0	1.7	8634
		Spark	9.0	1.7	9921
	Vertical	Gas flame	10.5	1.7	8823
		Spark	10.0	1.7	11071
Plywood (12 mm)	Horizontal	Gas flame	8.0	1.7	9921
		Spark	8.5	1.5	5409
	Vertical	Gas flame	10.0	1.8	15265
		Spark	10.0	2.0	42025

WORKED EXAMPLE 7.2

Assume a board of radiata pine with a thickness of 20 mm is placed vertically. The board receives radiative heating from an approaching flame. Calculate the time to ignition of the board when the radiative heating changes over time at a rate of $+0.1\,\mathrm{kW\cdot m^{-2}\cdot s^{-1}}$, i.e., the heat flux transferred to the board is given by $\dot{q}'' = 0.2t\,(\mathrm{kW\cdot m^{-2}})$.

SUGGESTED SOLUTION

Eq. (7.39) can be used to calculate the time to ignition of the board. In Table 7.2, the material-specific conditions of radiata pine, which is softwood, are the critical heat flux for ignition $\dot{q}''_{cr} = 12.0\,\mathrm{kW\cdot m^{-2}}$, power index $N = 1.9$, and $FTP = 11794\,(\mathrm{kW\cdot m^{-2}})^{N}\cdot \mathrm{s}$. The radiative heating incident on the board increases linearly with time. However, its contribution to the ignition of the board is minimal in the initial phase. Thus, we consider the cumulative amount of $\dot{q}''$ only when it exceeds $\dot{q}''_{cr}$. According to the given assumption, $\dot{q}''$ exceeds $\dot{q}''_{cr}$ when $t = 60\,\mathrm{s}$. Thus, Eq. (7.39) can be expressed as:

$$\int_{60}^{t_{ig}} (0.2t - 12.0)^{1.9}\, dt = 11794$$

This equation can be solved using the partial integration method, which yields $t_{ig} = 165\,\mathrm{s}$.

7.3 FIRE SPREAD IN A CONTINUOUS FUEL BED

Fire spread can be viewed as a process of successive ignition of combustible materials. In this section, we discuss the mechanism of fire spread using the ignition models described above. Fire spread follows different processes depending on how combustible materials are distributed: continuous or discrete. In this section, we discuss fire spread in a continuous fuel bed.

7.3.1 Rate of fire spread based on surface temperature

The behavior of fire spread (flame propagation) in a continuous fuel bed is dependent on various conditions, including the thermo-physical properties (specific heat, thermal conductivity, density, and pyrolysis temperature), the morphology (porosity, material thickness, and inclination angle), and the external conditions (the wind, and the external heating). A common approach to examining the effect of these conditions on flame propagation is to divide the combustible material into two segments, burnt and unburnt (Williams, 1976). If we define the interface between the two segments as the flame front, the changing rate of the interface position can be viewed as the flame propagation rate. By introducing the positional coordinate of the interface, x_{ig}, the rate of flame propagation, v_{fl}, can be expressed as follows:

$$v_{fl} = \frac{dx_{ig}}{dt} \tag{7.40}$$

The interface of the two segments can also be regarded as the surface where the pyrolysis of the combustible material starts.

To quantify this relationship, we focus on the energy conservation of a control volume in the combustible material that increases in temperature due to heating from the steadily propagating flame. This is schematically described in Figure 7.12, with the control volume enclosed by the dotted line. The left and right panels of Figure 7.12 show the conservational relationships of thermal energy in thermally thin and thermally thick materials, respectively. The temperature becomes uniform instantaneously upon heating in thermally thin material, whereas a temperature profile is formed in thermally thick material.

We first consider the flame propagation in thermally thin material. To simplify the discussion, we assume that the material properties are homogeneous and unchanged due to combustion. If the net heat flux transferred from the burning segment to the unburnt segment is $\dot{q}''$, the energy conservation relationship for the control volume can be expressed as follows:

$$\rho c d \left(T_P - T_S\right) v_{fl} = \dot{q}'' d_H \cong \dot{q}''_{fl} d_H \tag{7.41}$$

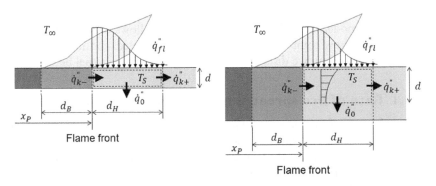

(a) Thermally thin case   (b) Thermally thick case

Figure 7.12 Schematic of surface flame spread in a continuous fuel bed (modified from Quintiere, 2006).

where ρ is the density, c is the specific heat, T_P is the pyrolysis temperature, and T_S is the material temperature. $\dot{q}''$ consists of heat fluxes from different sides of the control volume, $\dot{q}''_{fl}$, $\dot{q}''_{k-}$, $\dot{q}''_{k+}$, and $\dot{q}''_0$. In Eq. (7.41), we considered that $\dot{q}''_{fl}$ is dominant among the others. By transforming Eq. (7.41), the v_{fl} can take the form (Fernandez-Pello, 1995; Thomas PH, 1995; Quintiere, 2006):

$$v_{fl} \cong \frac{d_H}{\rho cd\left(T_P - T_\infty\right) / \dot{q}''_{fl}} = \frac{flame\ heated\ length}{time\ to\ ignition} \tag{7.42}$$

According to this equation, the smaller the density, ρ, specific heat, c, and thickness, d, the larger the v_{fl}.

The underlying implications of Eq. (7.42) can be better understood by separating it into the numerator and denominator (Quintiere, 2006). The numerator represents the length of the control volume heated by the flame, whereas the denominator represents the time to ignition of the material, t_{ig}, subject to external heating, $\dot{q}''_{fl}$. Thus, Eq. (7.42) gives a value with the dimension of velocity by dividing 'flame heated length' by 'time to ignition'. The relationship is also applicable when analyzing the flame propagation velocity through thermally thick materials. According to Eq. (7.29), the t_{ig} of a thermally thick combustible material subject to $\dot{q}''_{fl}$ can be expressed as follows (Carslaw et al., 1959):

$$t_{ig} \cong \frac{\pi k \rho c}{4}\left(\frac{T_P - T_\infty}{\dot{q}''_{fl}}\right)^2 \tag{7.43}$$

where T_∞ is the initial temperature of the combustible material. Assuming that the flame heated length is unchanged between thermally thin and thermally thick materials, v_{fl} can be calculated by substituting T_∞ by the surface temperature, T_S, in Eq. (7.43) (Fernandez-Pello, 1995; Thomas PH, 1995; Quintiere, 2006), which is:

$$v_{fl} \cong \frac{\dot{q}''^2_{fl} d_H}{\frac{\pi}{4} k \rho c\left(T_P - T_S\right)^2} \tag{7.44}$$

Note that the variables in Eq. (7.44), such as $\dot{q}''_{fl}$ and d_H are dependent on the combustion characteristics of the material. Thus, they are generally determined experimentally (Quintiere et al., 1984).

7.3.2 Rate of fire spread based on incident heat flux

The above discussion focused on the material temperature for determining the flame front position. As an alternative to this, the following equation represents the rate of flame propagation as the rate of pyrolysis proportional to the incident heat flux, $\dot{q}''$ (Frandsen, 1971; Rothermel, 1972):

$$v_{fl} = \frac{\dot{q}''}{\rho \Delta H_{ig}} = \frac{heat\ received}{heat\ required\ for\ ignition} \qquad (7.45a)$$

$$\Delta H_{ig} = c_D\left(T_P - T_\infty\right) + m\left\{c_V\left(T_V - T_\infty\right) + L_V\right\} \qquad (7.45b)$$

where ΔH_{ig} is the heat required for ignition, c_D and c_V are the specific heat of dry fuel and water, respectively, m is the fuel moisture content, T_V is the vaporization temperature, and L_V is the latent heat of vaporization. In contrast to Eqs. (7.42) and (7.44), Eq. (7.45) explicitly incorporates the effect of moisture content into its formulation. This is a favorable structure of equation for the analysis of wildland fires, where moisture content plays an important role in the rate of flame spread.

The concept expressed in Eq. (7.45) is the theoretical basis for one of the most widely used models for predicting flame spread in wildland fires (Rothermel, 1972). The conservational relationship of thermal energy for the control volume described in Figure 7.12 also applies to this case, i.e., $\dot{q}''$ can be divided into four component heat fluxes at the boundary of the control volume, $\dot{q}''_{fl}$, $\dot{q}''_{k-}$, $\dot{q}''_{k+}$, and $\dot{q}''_0$. Among these, Rothermel assumed that the horizontal heat flux, $\dot{q}''_{k-}$ (originally, heat flux at a specific depth in the fuel), is the dominant component (Rothermel, 1972):

$$v_{fl} = \frac{\dot{q}''_{k-}\left(1 + \phi_W + \phi_S\right)}{\varepsilon \rho \Delta H_{ig}} \qquad (7.46)$$

where ε is the porosity coefficient, and ϕ_W and ϕ_S are the additional terms representing the effect of wind and slope, respectively. The terms, ϕ_W and ϕ_S, in Eq. (7.46) are the empirical terms introduced to improve the prediction accuracy in exchange for the theoretical rigor of the model.

WORKED EXAMPLE 7.3

Calculate the rate of flame propagation, v_{fl}, over a thermally thick Douglas-fir board on the ground. Use the following thermo-physical properties for the calculation: thermal conductivity $k = 0.14 \times 10^{-3}$ kW $\cdot$ m^{-1} $\cdot$ K^{-1}, density $\rho = 510$ kg $\cdot$ m^{-3}, specific heat $c = 1.62$ kJ $\cdot$ kg^{-1} $\cdot$ K^{-1}, and pyrolysis temperature $T_P = 623$ K. Additionally, the incident heat flux $\dot{q}''_{fl} = 50$ kW $\cdot$ m^{-2}, the flame heated length $d_H = 0.001$ m, and the ambient temperature $T_\infty = 300$ K, are all assumed constant.

SUGGESTED SOLUTION

The flame propagation rate can be calculated using Eq. (7.44) for thermally thick materials. First, calculate the thermal inertia:

$$k\rho c = \left(1.4 \times 10^{-3}\right) \times 510 \times 1.62 = 0.116 \text{ kJ}^2 \cdot \text{s}^{-1} \cdot \text{K}^{-2} \cdot \text{m}^{-4}$$

By introducing this into Eq. (7.44), we obtain the flame propagation rate as follows:

$$v_{fl} \cong \frac{50^2 \times 0.001}{\frac{\pi}{4} \times 0.116 \times \left(623 - 300\right)^2} = 2.64 \times 10^{-4} \text{ m / s} = 1.58 \times 10^{-2} \text{ m / min}$$

Let us now assume that the thickness of the Douglas-fir board is $d = 0.0005\,\text{m} = 0.5\,\text{mm}$ and that it is a thermally thin material. In this case, the flame propagation rate can be calculated by Eq. (7.42). Thus, by substituting the given parameters:

$$v_{fl} \cong \frac{0.001}{510 \times 1.62 \times 0.0005 \times \left(623 - 300\right) / 50} = 3.75 \times 10^{-4} \text{ m / s}$$
$$= 2.25 \times 10^{-2} \text{ m / min}$$

The obtained v_{fl} was larger for the thermally thin case than for the thick case.

7.4 FIRE SPREAD BETWEEN DISCRETE FUEL OBJECTS

As in the case of a continuous fuel bed, the mechanism of fire spread between discretely distributed fuel objects can be analyzed using the ignition models described above. The difference in the mode of fire spread between discretely distributed fuel objects and a continuous fuel bed is that each fuel object needs to be considered as an independent fire source in the former case. We need to put more emphasis on the geometrical relationships between the fuel objects when discussing the mechanism of fire spread.

7.4.1 Rate of fire spread

Figure 7.13 schematically shows the process of fire spread between discrete fuel objects aligned in a straight line. Let us focus on the fuel object enclosed by dotted lines, which receives external heating from an adjacent burning fuel object. The heated fuel object elevates its temperature and gets ignited when the temperature exceeds its critical condition. As the flame develops following ignition, the ignited fuel object becomes a new fire source for the surrounding fuel objects. The fire spread between discretely distributed fuel

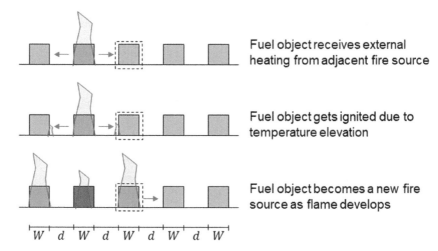

Fuel object receives external
heating from adjacent fire source

Fuel object gets ignited due to
temperature elevation

Fuel object becomes a new fire
source as flame develops

Figure 7.13 Schematic process of fire spread between discrete fuel objects.

objects can be represented by the successive occurrence of such an ignition process. If all the fuel objects are identical (both thermo-physically and geometrically) and aligned at an identical separation distance, the steady rate of fire spread, v_{fl}, can be expressed as follows:

$$v_{fl} = \frac{d + W}{t_{ig} + t_{gr}} \tag{7.47}$$

where d is the separation distance between the fuel objects, W is the side length of the fuel object, t_{ig} is the time to ignition of a fuel object, and t_{gr} is the time for a fuel object to become a new fire source following ignition. Theoretical formulation of t_{gr} generally involves difficulty as it is the time for flame development over a 3D fuel object. Thus, an empirical formulation is generally employed for t_{gr}.

Only a limited number of expressions are available for quantifying the rate of fire spread between independent fuel objects. An example is an algebraic formula for the rate of fire spread in an urban area proposed by Hamada (Hamada, 1951). Historically, urban areas in Japan suffered from numerous conflagrations as they generally comprised of wooden buildings with low fire safety performances. By analyzing the record of urban conflagrations in Japan, Hamada empirically derived an expression for the $t_{ig} + t_{gr}$ for the steady fire spread downwind (Hamada, 1951):

$$t_{ig} + t_{gr} = \frac{3 + 3d/8 + 1.4W/(5 + 0.5U_\infty)}{1.6(1 + 0.1U_\infty + 0.007U_\infty^2)} \tag{7.48}$$

where U_∞ is the velocity of the ambient wind. The inclusion of the U_∞^2 term in the denominator indicates that $t_{ig} + t_{gr}$ rapidly decreases as U_∞ increases. The expression also indicates an increase in $t_{ig} + t_{gr}$ with an increase in d and W in the numerator.

7.4.2 Critical separation distance for fire spread

Maintaining sufficient spacing between fuel objects is an effective way to prevent fire spread in outdoor fires. In Section 7.2, we discussed that there is a lower critical limit of incident heat flux required for ignition of combustible materials, $\dot{q}_{cr}''$. For example, $\dot{q}_{cr}''$ of woods ranges from 10 to 15 kW·m^{-2}, according to the data summarized in Table 7.1. To prevent the occurrence of fire spread, a certain distance d should be maintained between fuel objects so that the incident heat flux on the surface of the heat-receiving fuel object, $\dot{q}''$, is below $\dot{q}_{cr}''$ in a supposed fire environment:

$$\dot{q}'' < \dot{q}_{cr}'' \tag{7.49}$$

In Chapter 3, we demonstrated that the fraction of the radiative energy emitted from a fire source i that is incident on a heat-receiving point j can be represented by the view factor, F_{ji}. F_{ji} is defined solely by the geometrical relationship between the fire source and the heat-receiving point. Let us consider one general situation: a rectangular-shaped radiation surface i (with sides of length $2a$ and $2b$) parallel and directly opposite to a heat-receiving point j. The fire source i is assumed to be a black body, and its temperature is constant at T_i. Using the formula for F_{ji} given in Chapter 3, Eq. (7.49) can be transformed as follows:

$$\dot{q}'' = \left(\sigma T_i^4\right) F_{ji} = \left(\sigma T_i^4\right) \times \frac{4}{2\pi}\left(\frac{X}{\sqrt{1+X^2}}\tan^{-1}\frac{Y}{\sqrt{1+X^2}} + \frac{Y}{\sqrt{1+Y^2}}\tan^{-1}\frac{X}{\sqrt{1+Y^2}}\right) < \dot{q}_{cr}'' \tag{7.50}$$

where σ is the Stefan-Boltzmann constant, and $X = \dfrac{a}{d}$ and $Y = b/d$ represent the geometrical configuration between i and j. In Eq. (7.50), $\dot{q}''$ is expressed as a function of d. However, it is not straightforward to analytically obtain d from this equation. Thus, the minimum requirement for d is generally obtained graphically from the relationship between $\dot{q}''$ and d (ref. Worked example 7.4).

If the geometry of the fire source i can be approximated by a circle of radius r, we can derive an expression in a more tractable form. Using the equation for F_{ji} described in Chapter 3, the relationship equivalent to Eq. (7.50) can be expressed as follows:

$$\dot{q}'' = \left(\sigma T_i^4\right) F_{ji} = \left(\sigma T_i^4\right) \times \frac{r^2}{d^2 + r^2} < \dot{q}''_{cr} \tag{7.51}$$

By transforming this equation, the following requirement for preventing ignition can be obtained:

$$d > r \sqrt{\frac{\sigma T_i^4}{\dot{q}''_{cr}} - 1} \tag{7.52}$$

The critical value of d is in proportion to r and the square root of σT_i^4.

WORKED EXAMPLE 7.4

Calculate the critical distance for preventing the ignition of a fuel object next to a heat source, d. Assume that the heat-receiving point j is directly opposite the square radiation surface i. However, consider three different geometries for the heat source i : A ($a = b = 1$ m), B ($a = b = 2$ m), and C ($a = b = 3$ m). The heat source can be regarded as a black body, and its temperature is $T_i = T_\infty + \Delta T = 300 + 880 = 1180$ K, referring to the discussion in Chapter 5. Critical heat flux for ignition is $\dot{q}''_{cr} = 15$ kW $\cdot$ m^{-2} according to the data in Table 7.1.

SUGGESTED SOLUTION

Figure 7.14 shows the calculation results of the incident heat flux, $\dot{q}''$, with varying d using Eq. (7.50). To find d under a critical condition of $\dot{q}''_{cr} = 15$ kW $\cdot$ m^{-2}, we draw a horizontal line from 15 kW $\cdot$ m^{-2} on the vertical

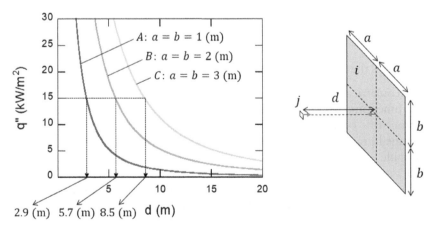

Figure 7.14 Estimation of minimum separation distance for avoiding ignition of a fuel object.

axis extending to each curve. Draw a perpendicular line down the abscissa from the intersection of the horizontal line and the curve. The foot of the perpendicular line is the d that needs to be maintained for each heat source. The results were 2.9 m, 5.7 m, and 8.5 m for the heat sources A, B, and C, respectively.

For reference, d were also calculated using Eq. (7.52) by approximating the square radiant surface i by a circle of the same area. The results were 2.8 m, 5.7 m, and 8.5 m for the heat sources A, B, and C, respectively. The agreement was reasonable. Although there are limited cases where the shape of actual fire sources can be approximated by a circle, Eq. (7.52) would still be useful in obtaining a quick estimate for the separation requirement.

7.4.3 Probability of fire spread

We now focus on the probabilistic aspect of building safety from external fire sources. In general, the probability of fire spread of a building subject to external heating can be evaluated by comparing the intensity of the external heating versus the fire resistance performance of the building. The building is less likely to be ignited if its fire resistance performance is higher than the intensity of external heating. Conversely, the building is more likely to be ignited if its fire resistance performance is lower than the intensity of external heating. However, it is often impractical to fully specify factors affecting the intensity of external heating in outdoor fires, as it depends on various uncertain factors, including the fuel and weather conditions. It is also impractical to fully specify factors affecting the fire resistance performance of buildings, as it also depends on various uncertain factors, including the property variation in construction materials. An effective approach to quantify such uncertainties is to regard the intensity of external heating and the fire resistance performance as probabilistic quantities.

The intensity of external heating and the fire resistance performance of a building are represented by the random variables, H (heating) and R (resistance), respectively. The probability density functions of H and R are denoted by $f_H(h)$ and $f_R(r)$, respectively. H and R are expressed in the same units so that they can be directly compared. In this case, an event $H \geq R$ indicates the occurrence of fire spread, whereas $H < R$ indicates the non-occurrence or failure of fire spread. In other words, a building is safe either if the probability of $H \geq R$ is sufficiently small or if the probability of $H < R$ is sufficiently large. The probability of $H \geq R$ is called the probability of fire spread or exceedance probability of fire spread, p_{fs}. The probability of $H < R$ is called the probability of survival, p_s. They can be respectively expressed as follows:

$$p_{fs} = P[H \geq R] \tag{7.53a}$$

$$p_s = P[H < R] = 1 - p_{fs} \tag{7.53b}$$

When H and R take positive values and are probabilistically independent, the relationship between the two probability density functions, $f_H(h)$ and $f_R(r)$, can be schematically described, as shown in Figure 7.15. In the left panel of Figure 7.15, the probability for H being within the range from h to $h + dh$ can be expressed as:

$$P[h < H \leq h + dh] = f_H(h) \tag{7.54}$$

The probability of fire spread due to external heating of intensity h can be regarded as the probability of R being lower than h, which can be expressed as follows:

$$P[R \leq h] = \int_0^h f_R(r)dr = F_R(h) \tag{7.55}$$

where F_R is the cumulative distribution function of R. Given that H and R are independent, the probability that the fire spread to occur and that the H to be within the range from h to $h + dh$ can be expressed as:

$$P[(h < H \leq h + dh) \cap (R \leq h)] = f_S(s)ds \int_0^s f_R(r)dr = f_S(s)F_R(s)ds \tag{7.56}$$

As the external heating, H, can take values between 0 and infinity, p_{fs} can be expressed as:

$$p_{fs} = \int_0^\infty f_S(s)\left[\int_0^s f_R(r)dr\right]ds = \int_0^\infty f_S(s)F_R(s)ds \tag{7.57}$$

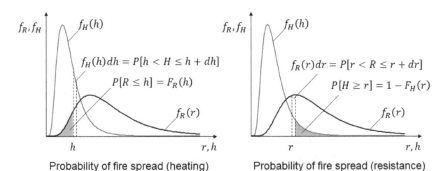

Probability of fire spread (heating)   Probability of fire spread (resistance)

Figure 7.15 Fire spread probability based on the relationship between external heating and fire resistance performance.

On the other hand, if we focus on the fire resistance performance, R, the probability for R to be within the range from r to $r + dr$ is $f_R(r)$. The probability for fire spread to occur to a building with fire resistance performance r is the probability for H being greater than r, which is $1 - F_H(r)$. Therefore, p_{fs} can be expressed as follows:

$$p_{fs} = \int_0^\infty f_R(r) \left[\int_r^\infty f_H(h)dh \right] dr = \int_0^\infty f_R(r)\left[1 - F_H(r) \right] dr \qquad (7.58)$$

Either Eq. (7.57) or Eq. (7.58) is the basic equation for p_{fs}. In general, it is difficult to solve these equations analytically. However, if $f_S(s)$ and $f_R(r)$ follow either a normal or log-normal distribution, p_{fs} can be solved analytically.

(1) When both H and R follow a normal distribution

Let us assume that the external heating, H, and the fire resistance performance, R, are normally distributed. If their means and variances are represented by μ and σ^2, respectively, they can be expressed as:

$$H = Normal(\mu_H, \sigma_H^2), \text{ and } R = Normal(\mu_R, \sigma_R^2) \qquad (7.59)$$

The normal distribution has a range from $-\infty$ to $+\infty$. Actual values of H and R are positive, but the effect of the range from $-\infty$ to 0 in the normal distribution is assumed to be negligible. The variable Z obtained by subtracting H from R can be regarded as the safety margin of the building. Z is also normally distributed, and its probability distribution can be expressed as:

$$Z = Normal(\mu_Z, \sigma_Z^2) = Normal(\mu_R - \mu_S, \sigma_R^2 + \sigma_S^2) \qquad (7.60)$$

Using the variable Z, the probability of fire spread, p_{fs}, can be expressed as follows:

$$p_{fs} = P[S \ge R] = \int_{-\infty}^z f_Z(z)dz = \int_{-\infty}^z \frac{1}{\sqrt{2\pi}\sigma_Z} \exp\left[-\frac{(z - \mu_Z)^2}{2\sigma_Z^2} \right] dz \qquad (7.61)$$

Further, by substituting $\zeta = (z - \mu_Z) / \sigma_Z$ into Eq. (7.61), we obtain:

$$p_{fs} = \frac{1}{\sqrt{2\pi}} \int_{-\infty}^{-\mu_Z/\sigma_Z} \exp\left(-\frac{\zeta^2}{2} \right) d\zeta = \Phi\left(-\frac{\mu_Z}{\sigma_Z} \right) = 1 - \Phi\left(\frac{\mu_Z}{\sigma_Z} \right) \qquad (7.62)$$

where Φ is the cumulative distribution function of the standard normal distribution, which is a normal distribution with the mean at zero and the variance at unity, which takes the form:

$$\Phi(x) = \frac{1}{\sqrt{2\pi}} \int_{-\infty}^{x} \exp\left(-\frac{t^2}{2}\right) dt \tag{7.63}$$

Note that $\beta = \dfrac{\mu_Z}{\sigma_Z}$ is an important variable called the safety index or reliability index, which is often used in the reliability analysis of systems. As evident from Eq. (7.62), β is a quantity that is coupled pairwise with p_{fs} via the standard normal distribution. Thus, when either of p_{fs} or β is given, the other can be determined.

(2) When both H and R follow a log-normal distribution

Let us assume that the external heating, H, and the fire resistance performance, R, follow a log-normal distribution. In such a case, $\ln H$ and $\ln R$ follow a normal distribution:

$$\ln H = Normal\left(\lambda_H, \xi_H^2\right), \text{ and } \ln R = Normal\left(\lambda_R, \xi_R^2\right) \tag{7.64}$$

where λ_H and ξ_H^2 are the mean and variance of $\ln H$, respectively, and λ_R and ξ_R^2 are the mean and variance of $\ln R$, respectively. The mean μ and variance σ^2 of a quantity x, and the mean λ and variance ξ^2 of $\ln x$ can be mutually converted. μ and σ^2 can be expressed in terms of λ and variance ξ^2 as follows:

$$\mu = \exp\left(\lambda + \frac{\xi^2}{2}\right), \text{ and } \sigma^2 = \exp\left(2\lambda + \xi^2\right)\left[\exp\left(\xi^2\right) - 1\right] \tag{7.65}$$

Conversely, λ and ξ^2 can be expressed in terms of μ and σ^2 as follows:

$$\lambda = \ln\left(\frac{\mu}{\sqrt{1 + \sigma^2/\mu^2}}\right), \text{ and } \xi^2 = \ln\left(1 + \frac{\sigma^2}{\mu^2}\right) \tag{7.66}$$

When both H and R were normally distributed in the previous discussion, the safety margin, Z, was expressed as a difference between the two variables, $Z = R - H$. In contrast, in the case of a log-normal distribution, the safety margin, Z, is expressed as a ratio of the two variables, $Z = R / H$. Given that the logarithm of this variable is $\ln Z = \ln R - \ln H$, the $\ln Z$ is also normally distributed:

$$\ln Z = Normal\left(\lambda_Z, \xi_Z^2\right) = Normal\left(\lambda_R - \lambda_H, \xi_R^2 + \xi_H^2\right) \tag{7.67}$$

Thus the Z also follows a log-normal distribution. As p_{fs} is equivalent to the probability for $Z \le 1$, p_{fs} can be expressed as follows:

$$p_{fs} = P[Z \le 1] = \int_0^1 f_Z(z)dz = \int_0^1 \frac{1}{\sqrt{2\pi}\xi_Z z} \exp\left[-\frac{\left(\ln z - \lambda_Z\right)^2}{2\xi_Z^2}\right]dz \tag{7.68}$$

Further, substituting $\zeta = \left(\ln z - \lambda_Z\right)/\xi_Z$ into Eq. (7.68), we obtain:

$$p_{fs} = \frac{1}{\sqrt{2\pi}} \int_{-\infty}^{-\lambda_Z/\xi_Z} \exp\left(-\frac{\zeta^2}{2}\right)d\zeta = \Phi\left(-\frac{\lambda_Z}{\xi_Z}\right) = 1 - \Phi\left(\frac{\lambda_Z}{\xi_Z}\right) \tag{7.69}$$

An equation similar to Eq. (7.62) is obtained for the log-normal distribution. In Eq. (7.69), the safety index is $\beta = \lambda_Z / \xi_Z$.

7.4.4 Vulnerability curves for fire spread

Various vulnerability curves have been developed for evaluating the resistance performance of buildings against natural disasters such as earthquakes and tsunamis (Shinozuka et al., 2000; Rossetto et al., 2003; Porter et al., 2007; Koshimura et al., 2009). Vulnerability curves can be represented by the exceedance probability of a given damage state of a building as a function of a demand parameter. Using the concepts presented in the previous subsection, we can also derive a vulnerability curve for a building subject to external heating.

In deriving a vulnerability curve, we need to determine a hazard metric representative of the heating from external fire sources. In actual fires, there are multiple causes of ignition, including radiative heat transfer from flames, convective heat transfer by wind-blown fire plumes, localized heating due to deposited firebrands, and their combination. In the present consideration, we assume a fire environment where radiative heat transfer is dominant. As per this idea, the probability of fire spread, p_{fs}, can be represented by a function comparing the radiative heat flux from a fire source, $\dot{q}_R''$, as H and the critical value of fire spread, $\dot{q}_{R,cr}''$, as R.

In the formulation, we invoke the framework discussed in the previous subsection, in which both H and R follow a log-normal distribution. The final form of p_{fs} is given by Eq. (7.69) where the statistics of the safety margin, Z, which are λ_Z and ξ_Z, need to be specified. However, a vulnerability curve generally aims to determine the ignition probability of a building under a specific radiative exposure, $\lambda_H = \ln(\dot{q}_R'')$. Therefore, we denote $\ln Z$ in Eq. (7.67) as an explicit function of $\ln(\dot{q}_R'')$ as follows:

$$\ln Z = Normal\left(\lambda_Z, \xi_Z^2\right) = Normal\left(\lambda_R - \ln\left(\dot{q}_R''\right), \xi_Z^2\right) \tag{7.70}$$

By substituting this into Eq. (7.69), p_{fs} takes the form:

$$p_{fs} = prob\left(\dot{q}_R'' > \dot{q}_{R,cr}''\right) = \Phi\left(\frac{\ln\left(\dot{q}_R''\right) - \lambda_R}{\xi_Z}\right) \tag{7.71}$$

The vulnerability curve given by Eq. (7.71) has two model parameters, λ_R and ξ_Z. The parameters for different structural types were regressed using the record of fire spread in the 2016 Itoigawa Fire as shown in Figure 7.16 (Tajima, 2020). Note that the considered structural types were traditional wooden buildings with bare wood on their exterior (TW), wooden buildings with fire-preventive construction for the exterior walls and soffits (FP), and fire-resistant and quasi fire-resistant buildings (FR/QFR). Among these, FR/QFR and FP were structural types defined by the Building Standard Law (BSL) of Japan, with fire spread resistance performance ranked according to the given order. In particular, FR/QFR are mostly non-wooden buildings such as reinforced concrete and steel constructions. FP are generally referred to as wooden buildings with non-flammable sidings or mortar-plastered walls on their exterior. However, TW were built before BSL enforcement and are presently not allowed to be constructed as new buildings. The fire spread resistance performance of TW buildings is considered the least among all the structural types as it includes exposed bare wood on the exterior.

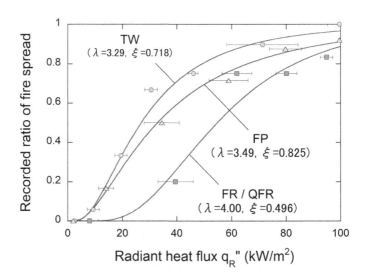

Figure 7.16 Examples of radiation-based vulnerability functions for fire spread in urban fires (modified from Tajima, 2020).

The results show that vulnerability to fire spread was, in ascending order: TW, FP, and FR/QFR at any $\dot{q}_R''$. It should be noted that, in specific conditions, a building did not ignite even under a considerably high $\dot{q}_R''$. This could be attributed to the involvement of an error in the estimated $\dot{q}_R''$ because it ignores the shielding effect of objects, such as trees and walls, between buildings or the fire resistance performances of buildings varied even when they were categorized into the same structural type.

Chapter 8

Firebrands

In outdoor fires, convective and radiative heat transfers from a fire source are the major causes of fire spread to its surroundings. However, as the intensity of convective and radiative heat transfer attenuates with distance from the fire source, the spatial extent of their influence is restricted to an area relatively close to the fire source. On the other hand, firebrands or embers released from a fire source may cause spot ignition far downwind as they are dispersed by the ambient wind while sustaining combustion. The ignition of combustibles far downwind of a fire source causes an increase in the rate of fire spread, which disrupts efforts to prevent fire spread by firefighters.

In contrast to fire spread caused by convective and radiative heat transfer, fire spread caused by spot ignition can be characterized by its highly probabilistic nature. Let us assume that two firebrands are simultaneously released into an airflow from the same position above a fire source. They are lofted in the buoyancy-induced updraft, dispersed downwind by the ambient wind, and finally fall onto the ground when the buoyancy effect diminishes with distance from the fire source. Will they land in the same position on the ground as they are released from the same position simultaneously? Almost certainly no. They will probably trace fairly distinct dispersion trajectories due to the high sensitivity of dispersion behavior to the firebrand properties and the airflow condition.

For the firebrand properties, the shape, mass, and thermo-chemical state are unique to individual firebrands as they are generated through thermal degradation of natural materials. Even if the properties may be similar, they are never completely identical. On the other hand, even minor property differences can cause a considerable difference in the aerodynamic forces applied to firebrands.

For the airflow condition, eddies of various scales are formed by buoyant flows above the fire source and ambient wind over the ground surface, forming strong turbulent flows involving random fluctuations. Even a minor shift in position can cause a dramatic change in the aerodynamic

DOI: 10.1201/9781003096689-8

forces applied to firebrands. The cumulative effect of such differences during the dispersion process leads to a variation in the final landing positions of firebrands.

The hazard of fire spread by spot ignition cannot be evaluated adequately without incorporating a probabilistic perspective. Thus, in this chapter, a probabilistic framework for evaluating the fire spread due to spot ignition is first presented. We then discuss the occurrence of three successive sub-processes of fire spread due to spot ignition: generation, dispersion, and ignition. However, as the spot ignition behavior involves a wide range of topics that cannot be fully explained in this chapter, interested readers are referred to some review articles for further reading (Koo et al., 2010; Fernandez-Pello, 2017; Caton et al., 2017; Manzello et al., 2020).

8.1 PROCESS OF FIRE SPREAD

The process of fire spread due to spot ignition in outdoor fires can be divided into the following sub-processes as schematically described in Figure 8.1:

1) Generation: Thermal decomposition of solid combustibles in a fire often involves loss of material integrity and detachment of fragmented pieces from the main body. Fragmentation of thermally degraded materials can even be promoted by being impacted by mechanical forces on the collapse of the main body. At the same time, solid combustibles continuously receive buoyancy-induced aerodynamic forces as they are engulfed in the flame. A portion of such fragmented pieces of combustibles is lofted and released into the ambient airflow as firebrands.

2) Dispersion: Lofted firebrands are dispersed downwind as they receive aerodynamic forces based on the relative motion of the airflow to the firebrand. They eventually land on the ground as the gravity effect

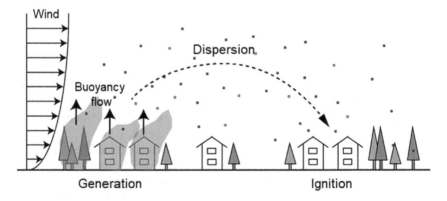

Figure 8.1 Process of fire spread due to spot ignition.

overcomes the buoyancy with distance from the fire source. While being transported, the firebrands sustain combustion, reducing their size and mass. Some firebrands are burnt out during dispersion, while others sustain combustion even after landing on the ground.

3) Ignition: Landing of firebrands is only a sufficient condition for the occurrence of spot ignition, but several other conditions need to be met to reach ignition. The required conditions include the firebrand(s) to sustain combustion after landing; the firebrand(s) to be deposited on a combustible (recipient combustible); and the recipient combustible to get ignited due to heating from the deposited firebrand(s). In some cases, smoldering combustion precedes diffusion combustion when the recipient combustible is ignited. However, the combustion may cease in a smoldering state without shifting to diffusion combustion, unless the decomposition of the recipient combustible accompanies a sufficient amount of flammable gas release into the gas phase.

Large probabilistic variations or uncertainties involved in the three sub-processes are critically important in modeling the spot ignition behavior. As discussed above, the dispersion trajectory of firebrands is a good example of such variations; even if two identical firebrands were released into the airflow from the same location simultaneously, they would land in different locations on the ground. Such probabilistic variations hamper the application of deterministic frameworks for evaluating the hazard of fire spread. Major uncertain factors that can lead to the probabilistic variability of each sub-process of spot ignition are listed below.

1) Generation: Species, shape, mass, thermo-chemical state, initial velocity, and quantity of firebrands released from fire sources.

2) Dispersion: Continuous change in the shape, mass, and thermo-chemical state of lofted firebrands due to combustion. Random fluctuation of the airflow around firebrands.

3) Ignition: Species, shape, mass, and thermo-chemical state of deposited firebrands. Species, shape, and thermo-physical state of recipient combustibles. Contact condition between the deposited firebrands and recipient combustibles. Local wind around the recipient combustibles.

If all these uncertain factors identified, fire spread by spot ignitions could be evaluated deterministically. However, given a large number of uncertain factors involved, a probabilistic/statistical approach would be a realistic option to evaluate the process of spot ignitions rather than a fully deterministic approach. Various existing experiments have demonstrated its applicability to sub-processes of spot ignitions individually.

8.1.1 Process of fire spread (single fire source)

Let us start our discussion with a simple situation by focusing on the relationship between a single fire source generating firebrands (fire source- j) and a combustible as the potential target of spot ignition in the downwind (combustible- i), as shown in Figure 8.2.

Wildland and structural fuels generate many firebrands when burnt in outdoor fires. The shape, mass, and thermo-chemical state imparted in the generation sub-process affect the behavior of individual firebrands in the subsequent dispersion and ignition sub-processes. Thus, the characteristics of generated firebrands, not only the quantity, are important for the hazard evaluation of fire spread due to spot ignitions. Let us classify the characteristics of firebrands into categories $(k = 1,2,3,\cdots)$, so that the firebrands in the same category behave similarly in each sub-processes. As the total number of firebrands generated from a fire source, N_{all}, is the sum of the number of firebrands in each category, N_k:

$$N_{all} = \sum_k N_k \left(k = 1,2,3,\cdots \right) \qquad (8.1)$$

We now focus on combustible-i downwind of fire source-j. As a matter of course, not all the released firebrands will be deposited on the combustible-i.

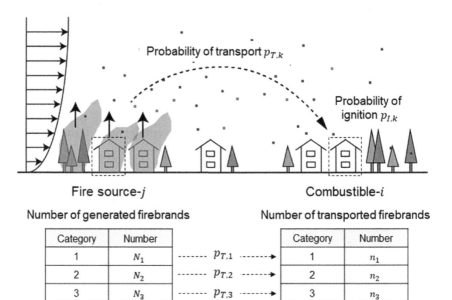

Figure 8.2 Process of fire spread by firebrand dispersion (single fire source).

Rather, a few or even all the released firebrands may land in other locations. Denoting the probability of category-k firebrands released from fire source-j to be transported to the combustible-i by $p_{T,k}$, the number of category-k firebrands being deposited on the combustible-i, n_k, can be expressed as follows:

$$n_k = p_{T,k} N_k \tag{8.2}$$

We further define $p_{I,k}$ as the ignition probability of combustible-i by a single category-k firebrand deposited on it. Assuming that multiple category-k firebrands may be deposited on combustible-i, the ignition probability by any category-k firebrands, $p_{FS(i),k}$, can be expressed as the probability of ignition by at least one of the deposited category-k firebrands. Using n_k in Eq. (8.2), $p_{FS(i),k}$ can be expressed as follows:

$$p_{FS(i),k} = 1 - \left(1 - p_{I,k}\right)^{n_k} \tag{8.3}$$

The above discussion on the ignition of the combustible-i by deposited category-k firebrands can be extended to the ignition of the combustible-i by deposited firebrands of any category. The probability, $p_{FS(i)}$, can be expressed as follows:

$$p_{FS(i)} = 1 - \left(1 - p_{I,1}\right)^{n_1} \left(1 - p_{I,2}\right)^{n_2} \left(1 - p_{I,3}\right)^{n_3} \cdots \tag{8.4}$$

Considering that the ignition probability of a recipient combustible by an individual firebrand is sufficiently small compared to unity ($p_{I,k} \ll 1$), Eq. (8.4) can be transformed into:

$$p_{FS(i)} \cong 1 - \left(1 - n_1 p_{I,1}\right)\left(1 - n_2 p_{I,2}\right)\left(1 - n_3 p_{I,3}\right)\cdots$$
$$= 1 - \prod_k \left(1 - n_k p_{I,k}\right) \tag{8.5}$$

Substituting the value of n_k in Eq. (8.2), the ignition probability of the combustible-i due to firebrands released from the fire source-j, $p_{FS(i)}$, takes the form:

$$p_{FS(i)} \cong 1 - \prod_k \left(1 - N_k p_{T,k} p_{I,k}\right) \tag{8.6}$$

In this equation, the number of firebrands generated from the fire source, N_k, the probability for a category-k firebrand to be transported and deposited on the recipient combustible-i, $p_{T,k}$, and the ignition probability of the recipient combustible-i, $p_{I,k}$, are the variables representing the behavior of each sub-processes of generation, dispersion, and ignition, respectively.

Table 8.1 Calculation conditions for the probability of fire spread (single fire source).

Category	N_k	$p_{T,k}$	$p_{I,k}$
1	100,000	1.0×10^{-3}	1.0×10^{-3}
2	100	1.0×10^{-4}	1.0×10^{-2}
3	1	1.0×10^{-5}	1.0×10^{-1}

* Values are hypothetical and have no physical or statistical basis.

WORKED EXAMPLE 8.1

Consider that firebrands generated from a single fire source-j have a potential impact on a combustible-i in the downwind, as shown in Figure 8.2. Calculate the probability of fire spread, $p_{FS(i)}$, assuming that firebrands generated from the fire source-j are classified into three categories as listed in Table 8.1. The firebrands in the younger numbered categories are smaller in size; they are greater in number and can be scattered farther, but their hazards are smaller.

SUGGESTED SOLUTION

The ignition probabilities of the combustible-i by firebrands of each category are respectively calculated as follows:

$$N_1 p_{T,1} p_{I,1} = 10,000 \times (1.0 \times 10^{-3}) \times (1.0 \times 10^{-3}) = 1.0 \times 10^{-2}$$

$$N_2 p_{T,2} p_{I,2} = 100 \times (1.0 \times 10^{-4}) \times (1.0 \times 10^{-2}) = 1.0 \times 10^{-4}$$

$$N_3 p_{T,3} p_{I,3} = 1 \times (1.0 \times 10^{-5}) \times (1.0 \times 10^{-1}) = 1.0 \times 10^{-6}$$

By substituting them into Eq. (8.6), the probability of fire spread, $p_{FS(i)}$, can be obtained as follows:

$$p_{FS(i)} \cong 1 - (1 - 1.0 \times 10^{-2}) \times (1 - 1.0 \times 10^{-4}) \times (1 - 1.0 \times 10^{-6}) = 0.0101$$

In this hypothetical case, the effect of the category-1 firebrands was dominant.

8.1.2 Process of fire spread (multiple fire sources)

The expression given by Eq. (8.6) assumes that there is only one fire source generating firebrands that potentially impact the concerned combustible.

However, we should consider multiple fire sources for evaluating the hazard of fire spread in actual outdoor fires, as described in Figure 8.3. Denoting the number of fire sources by M, the probability of fire spread to a combustible-i by firebrands generated from any of the upwind fire sources, $p_{FS(i)}$, can be expressed as the probability of fire spread by firebrands generated from at least one of these fire sources, which is as follows:

$$p_{FS(i)} \cong 1 - \prod_{j=1}^{M}\left(1 - p_{FS(j \to i)}\right) \tag{8.7}$$

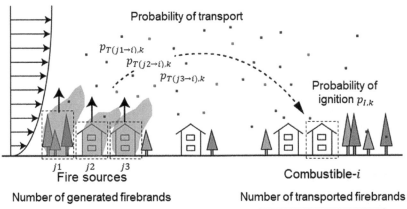

Figure 8.3 Process of fire spread by firebrand dispersion (multiple fire sources).

where $p_{FS(j \to i)}$ is the ignition probability of the recipient combustible-i by firebrands generated from fire source-j. In Eq. (8.7), $p_{FS(j \to i)}$ is nothing but the ignition probability given by Eq. (8.6). Thus, by substituting Eq. (8.6) into Eq. (8.7), $p_{FS(i)}$ can be restated as:

$$p_{FS(i)} \cong 1 - \prod_{j=1}^{M} \prod_{k} \left(1 - N_{j,k} p_{T(j \to i),k} p_{I,k} \right) \tag{8.8}$$

where $N_{j,k}$ is the number of category-k firebrands generated from fire source-j. In the above equation, the probability of a category-k firebrand being transported and deposited on combustible-i, $p_{T(j \to i),k}$, differs if the fire source is different. However, the ignition probability, $p_{I,k}$, is indifferent to the fire sources, j, provided the firebrands are classified into the same category, k.

WORKED EXAMPLE 8.2

Consider a case where combustible-i is under a potential impact of firebrands generated from three fire sources ($j1$, $j2$, and $j3$), as shown in Figure 8.3. Calculate the probability of fire spread to the combustible-i, which is $p_{FS(i)}$, assuming that firebrands generated from the fire sources are classified into three categories as listed in Table 8.2.

SUGGESTED SOLUTION

The ignition probabilities of the combustible-i by firebrands generated from each fire source and category are respectively calculated as:

For fire source-$j1$: $N_1 p_{T(j1 \to i),1} p_{I,1} = 1.0 \times 10^{-2}$

$$N_2 p_{T(j1 \to i),2} p_{I,2} = 1.0 \times 10^{-4}$$

$$N_3 p_{T(j1 \to i),3} p_{I,3} = 1.0 \times 10^{-6}$$

Table 8.2 Calculation conditions for the probability of fire spread (multiple fire sources).

Fire source	Category	N_k	$p_{T(j \to i),k}$	$p_{I,k}$
j1	1	100,000	1.0×10^{-3}	1.0×10^{-3}
	2	100	1.0×10^{-4}	1.0×10^{-2}
	3	1	1.0×10^{-5}	1.0×10^{-1}
j2	1	100,000	5.0×10^{-3}	1.0×10^{-3}
	2	100	5.0×10^{-4}	1.0×10^{-2}
	3	1	5.0×10^{-5}	1.0×10^{-1}
j3	1	100,000	1.0×10^{-2}	1.0×10^{-3}
	2	100	1.0×10^{-3}	1.0×10^{-2}
	3	1	1.0×10^{-4}	1.0×10^{-1}

* Values are hypothetical and have no physical or statistical basis.

For fire source-$j2$: $N_1 p_{T(j2 \to i),1} p_{I,1} = 5.0 \times 10^{-2}$

$$N_2 p_{T(j2 \to i),2} p_{I,2} = 5.0 \times 10^{-4}$$

$$N_3 p_{T(j2 \to i),3} p_{I,3} = 5.0 \times 10^{-6}$$

For fire source-$j3$: $N_1 p_{T(j3 \to i),1} p_{I,1} = 1.0 \times 10^{-1}$

$$N_2 p_{T(j3 \to i),2} p_{I,2} = 1.0 \times 10^{-3}$$

$$N_3 p_{T(j3 \to i),3} p_{I,3} = 1.0 \times 10^{-5}$$

By substituting them into Eq. (8.8), the probability of fire spread, $p_{FS(i)}$, can be obtained as follows:

$$p_{FS(i)} \cong 0.155$$

As shown in Table 8.2, the breakdowns of the number of generated firebrands, N_k, and the ignition probability, $p_{I,k}$, are the same among the fire sources, $j1$, $j2$, and $j3$. The only difference is in the probability associated with the transport, $p_{T(j \to i),k}$, which is attributed to the difference in the locations of the three fire sources relative to the recipient combustible-i and the behaviors of airflow along the flight paths.

8.2 GENERATION

Firebrands are often generated by the combustion of vegetative fuels, including wildland fuels (horizontally layered fallen leaves, twigs, bark and branches, and shrubs and tree canopies) and structural fuels (wooden structural members and materials). Although there are sources other than vegetative fuels, such as hot metal particles generated by the arc of electric wires (Fernandez-Pello, 2017), they are not considered here. A general procedure of firebrand generation during the combustion of vegetative fuel is described in Figure 8.4. Combustion of a vegetative fuel accompanies thermal decomposition of virgin material, by which flammable gas is released into the gas phase, and a residual char layer remains on the fuel surface. Thermal decomposition of virgin material causes a decrease in its density and structural integrity, but at this stage, with a minor effect on the volume and shape (Barr et al., 2013). Burning of vegetative fuels generally accompanies oxidation (surface combustion) of the char layer as the surface temperature of the char layer is maintained high due to irradiation from the flame, and oxygen is abundantly available outdoors. The release of flammable gas may continue until all the virgin material is consumed. However, oxidation of the char layer continues even after decomposition of the virgin material has ended. Oxidation causes the thinning (i.e., increase of porosity) of the char layer and the recession of its surface, which leads to a further reduction of the structural integrity as a material (Barr et al., 2013). As burning

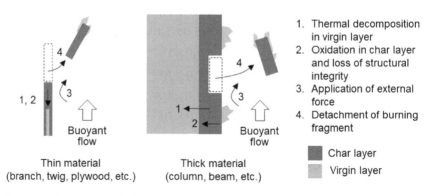

Figure 8.4 Procedure of firebrand generation during combustion of vegetative fuels.

fuels are continuously exposed to aerodynamic forces associated with the fire-induced buoyancy flow, the degraded material may be fragmented into pieces and detached from the main body to be released into the flow field as firebrands. Another type of force that may be applied to burning fuels is the mechanical force accompanied by the collapse of the main body. The mechanical force is considered as a major cause of firebrand generation in structural fires (Hayashi et al., 2014).

The process of firebrand generation is also affected by the condition of the source material, as shown in Figure 8.4. Thermal degradation of thin materials (e.g., branches and twigs in wildlands, and shingles and plywood boards used in structures) generally extends over its entire body. Thus, the size and shape of generated firebrands depend on those of the main body. On the contrary, thermal degradation of thick materials (e.g., bark in wildlands, and columns and beams in structures) generally remains on the surface of the source material. Thus, in general, firebrands from thick materials are generated by delamination of the char layer at the surface. In such a way, the size and shape of generated firebrands are less dependent on those of the main body. Rather, they are affected by the other factors of the source material, such as the porosity and thickness of the charred component.

Figures 8.5 and 8.6 show collected samples of firebrands generated in a prescribed fire in a forest and an urban fire, respectively (El Houssami et al., 2016; Takeya et al., 2017). Although they are charred fragments of vegetative materials in a more specific sense, we simply call them firebrands in this chapter. These two images indicate that firebrand shapes depend on their presumed sources. Firebrands in a wildland fire comprise granular and flaky firebrands from bark, and cylindrical and rod-like firebrands from branches and twigs. On the other hand, firebrands in an urban fire comprise blocky and granular firebrands from relatively large structural members such as columns and beams, and platy firebrands from shingles and plywood boards.

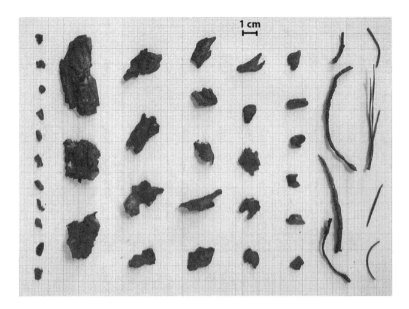

Figure 8.5 Firebrands collected in a prescribed fire in a pine forest (El Houssami et al., 2016).

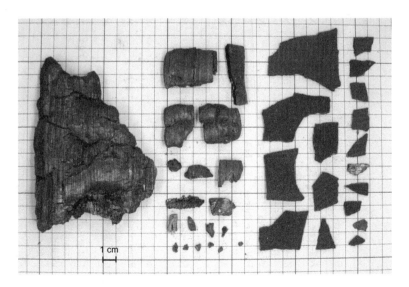

Figure 8.6 Firebrands collected in an urban fire in Itoigawa city in 2016.

Table 8.3 summarizes several characteristics of firebrands collected in different experiments and actual fires. Items include the fire size (MW), the wind velocity (m·s^{-1}), the projected area (cm^2), the weight (g), the density (kg·m^{-3}), and the mass fraction of generated firebrands to the initial source (%).

Table 8.3 Characteristics of firebrands collected in different experiments and actual fires.

Literature	Fire source	Fire size (MW)	Ambient wind velocity (m·s⁻¹)	Projected area of a firebrand (cm²)	Mass of a firebrand (g)	Firebrand density (kg·m⁻³)	Mass fraction of generated firebrands to the initial source (%)
Yoshioka et al., 2004	Full-scale building	~6.3	~4	NA	0.0174–0.0620	NA	NA
Manzello et al., 2008	Korean pine and Douglas-fir	2.1–16.6	~0	NA	Large percentage less than 0.3 g	NA	0.08–0.28
Miura et al., 2011	Wood crib	2–6	3–7	NA	0.0027–0.0314	NA	0.06–1.64
Hayashi et al., 2014	Full-scale three-story building	NA	4.6 (0.9)	2.62–9.25 (2.70–7.74)	0.124–0.820 (0.211–0.963)	76–86	NA
Manzello et al., 2014	Angora fire in 2007	NA	4.5–6.7	More than 85% were smaller than 0.5 cm²	NA	NA	NA
Suzuki et al., 2014	Full-scale building	~0.1 kg/m²s	6	More than 90% were less than 10 cm² and 25% less than 1 cm²	More than 90% less than 1 g and 56% less than 0.1 g	NA	NA
El Houssami et al., 2016	Prescribed fire in a pine forest	0.5–3.2 MW/m	3.0	About 80% in range of 0 to 1 cm²	Majority between 0.005 and 0.02 g	NA	NA
Suzuki et al., 2018	Itoigawa fire in 2016	NA	~9	More than 60% less than 2 cm²	More than 60% less than 0.1 g	NA	NA
Hedayati et al., 2019	Full-scale structural assemblies	NA	5.36–17.88	2.10–4.87 (2.72–7.87)	0.09–0.38 (0.24–1.44)	NA	NA
Himoto et al., 2021	Wood crib in an open-top compartment	0.3–0.7	1.4–3.2 (0.01–0.11)	0.0884–0.178 (0.132–0.350)	0.0095–0.0383 (0.0037–0.0116)	29.7–74.5 (7.74–20.6)	0.00526–0.0176

* Values in parentheses are the standard deviations.
** Range of values represents that of multiple test cases.

Although a comparison between these cases is not necessarily possible, large variations exist in the firebrand characteristics. Quantification of such variations is required for using the evaluation framework presented in Section 8.1.

8.2.1 Shape and mass

The shape and mass characterize the behavior of firebrands in the dispersion and ignition sub-processes that follow the generation sub-process. Of these, the shape of firebrands is inherently 3D. However, their irregularity and fragile nature have been impediments to intact measurement of the 3D structure. Alternatively, the shape of firebrands is often represented by the projected area as viewed from the top. Figure 8.7 shows frequency distributions of the projected area of firebrands collected in a wind tunnel experiment (Himoto et al., 2021). In the experiment, wood cribs of different thicknesses were burnt in an open-top combustion chamber for ejecting firebrands into a flow field. A large variation exists among the measured projected area, with a small number of relatively large firebrands mixed in with the majority of relatively small firebrands. However, note that Figure 8.7 summarizes the size of the individually measurable firebrands. Thus, the frequency distributions do not fully represent the actual size of the firebrands, but there were unmeasurable firebrands of even smaller sizes. Thermal energy maintained by a firebrand generally increases as its size increases, provided the surface combustion is occurring homogeneously on its surface. On the other hand, smaller firebrands could have a larger contact area with recipient combustibles when deposited on them. At the same time, the duration of surface combustion that maintains the hazardous condition

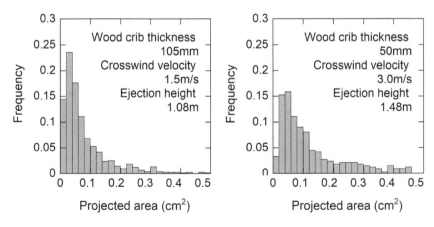

Figure 8.7 Frequency distribution of projected area of firebrands collected in a wind tunnel experiment (Himoto et al., 2021).

of firebrands is relatively short for smaller firebrands. Currently, there is no quantitative level of certainty as to the lower bound of the firebrand size below which the impact on the recipient combustible is negligible.

Many experiments have investigated the shape and mass of firebrands, as summarized in Table 8.3. These variables are dependent on the fire source condition. However, a limited number of models can generalize the relationship between the fire source condition and the firebrand shape and mass.

Barr et al. developed a model for brand breakage for a tree with self-similar branching features (Barr et al., 2013) (Figure 8.8). The model predicts the removal of branches of varying diameters and their release into the ambient flow with a simplified analysis of the moment and strength of individual branches. The breakage event is assumed to occur when the time-increasing stress at the branch junction accompanied by the thinning due to oxidation equals the critical stress of the specimen. The occurrence of a breakage event is determined by comparing the magnitudes of the dimensionless failure criterion, D_{ox} / D_o, and the dimensionless time evolution of the oxidizing branch, D_{cr} / D_o, which are respectively given by:

$$\frac{D_{cr}}{D_o} = \left[\left(\frac{\sigma_w}{\sigma_{cr}} \right)^2 + \left(\frac{\sigma_d}{\sigma_{cr}} \right) \right]^{1/2} - \left(\frac{\sigma_w}{\sigma_{cr}} \right) \tag{8.9a}$$

$$\frac{D_{ox}}{D_o} = \left[1 - \frac{t_{ox}}{\tau} \left(\frac{\sigma_w}{\sigma_{cr}} \right)^{-3/2} \right]^{2/3} \quad \left(D_{ox} > D_{cr} \right) \tag{8.9b}$$

where σ_w is the characteristic stress associated with the weight of the branch, σ_d is the characteristic stress associated with drag on the branch, σ_{cr} is the

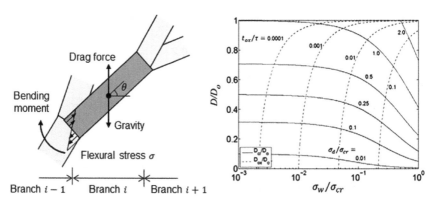

Figure 8.8 Comparison of D_{ox} / D_o and D_{cr} / D_o for determining the breakage of individual branches (Barr et al., 2013).

critical flexural stress for wood specimens, t_{ox} is the oxidation time, and τ is the time constant. A breakage event is determined by the intersection of Eq. (8.9) (Figure 8.8). If D_{ox}/D_o is greater than D_{cr}/D_o, then the branch remains on the plant, but if D_{ox}/D_o decreases as the oxidation progresses, the branch fractures when reaching $D_{ox}/D_o = D_{cr}/D_o$. The breakage model can be coupled with a plume model to predict the size distribution of firebrands lofted from a fractal tree.

Another example of the breakage model was presented by Caton-Kerr et al. (Caton-Kerr et al., 2019). In order to model the breakage of thermally degraded cylindrical wooden dowels, they conducted a set of experiments to test the mechanical response of materials after combustion. Based on the scale analysis of the experimental results, they reported that there were two distinct failure regimes dominating the breakage and ultimately the formation of firebrands as follows:

$$\Pi_2 = \begin{cases} e^{-13.51}\Pi_1^{-0.21} \left(\Pi_1 < 5.14 \times 10^{-13} \right) \\ e^{-131.51}\Pi_1^{-4.38} \left(\Pi_1 \geq 5.14 \times 10^{-13} \right) \end{cases} \tag{8.10}$$

The two dimensionless parameters are the ratio of the average burning rate of the material to its scaled mechanical stiffness, Π_1, and the recoverable plastic strain in the transverse direction of the dowel, Π_2. These are respectively defined as follows:

$$\Pi_1 = \frac{\alpha \dot{m}}{P_\infty D_0^3} \frac{\rho_\infty}{\rho_0} \tag{8.11a}$$

$$\Pi_2 = \frac{F_{max}L_0 v_{RT}}{E_L D_0^3} \frac{\rho_0}{\rho_s} \tag{8.11b}$$

where F_{max} is the critical force of breakage, ρ_0 is the initial density of the material, L_0 is the length of the material, D_0 is the initial dowel diameter, v_{RT} is the Poisson's ratio in the radial plane and in the transverse direction, α is the thermal diffusivity, $\dot{m}$ is the mass loss rate, E_L is the modulus of elasticity in the longitudinal direction, ρ_s is the density of wood cell wall material, P_∞ is the ambient pressure, and ρ_∞ is the ambient gas density. In Eq. (8.10), the first regime demonstrates that the recoverable plastic strain is weakly affected by the burning rate parameter, Π_1. The second regime demonstrates that the material strain becomes strongly affected by changes in the burning rate.

While the above two models adopted a formulation focusing on the breakage mechanism of thermally degraded wood branches, the model proposed by Himoto et al. is an example of statistical analysis of experimental data (Himoto et al., 2022). They approximated the distribution of the projected area, A_P, of firebrands with a log-normal distribution:

$$p\left(A_P|\lambda,\xi\right) = \frac{1}{\sqrt{2\pi}\xi A_P}\exp\left\{-\frac{1}{2}\left(\frac{\ln A_P - \lambda}{\xi}\right)^2\right\} \tag{8.12a}$$

$$\lambda = \ln\left(\frac{\mu_A}{\sqrt{1 + \sigma_A^2/\mu_A^2}}\right) \tag{8.12b}$$

$$\xi = \sqrt{\ln\left(1 + \frac{\sigma_A^2}{\mu_A^2}\right)} \tag{8.12c}$$

where μ_A and σ_A are the mean and standard deviation of A_P, respectively. By regressing the data of a full-scale burn experiment of a three-story wooden building (Hayashi et al., 2014), they obtained the following correlations for μ_A and σ_A:

$$\frac{\mu_A}{d^2} = 0.445W^{*1.07} \tag{8.13a}$$

$$\frac{\sigma_A}{d^2} = 0.0107 \tag{8.13b}$$

where d is the representative width of the fuel component, and W^* is the dimensionless parameter that controls A_P. W^* was derived from the balance between the buoyancy and gravitational force in the vertical direction above the fire source (Himoto et al., 2021):

$$W^* = \frac{\rho_\infty}{\rho_P}\frac{W_0^2}{gd} \tag{8.14}$$

where ρ_∞ and ρ_P are the ambient gas and firebrand densities, respectively, g is the acceleration due to gravity, and W_0 is the initial ejection velocity of the firebrand from the fire source. The effect of the fire size is not explicitly included in W^*, but is represented by W_0. One of the advantages of this model is its probabilistic representation of the variation of firebrand size. However, the data used for regressing the model was collected from a structural fire, and its applicability to wildland fires has still not been verified.

WORKED EXAMPLE 8.3

Assume three fire sources, A, B, and C, with heat release rates (HRR) of 1×10^3, 1×10^4, and 1×10^5 kW, respectively. Estimate the frequency distribution of A_P generated from each fire source. The representative width of the fuel component is $d = 10$ cm, and the ambient air and firebrand densities are $\rho_\infty = 1.2$ kg·m^{-3} and $\rho_P = 70$ kg·m^{-3}, respectively:

SUGGESTED SOLUTION

The theoretical model for turbulent diffusion flames by Baum et al. is used to calculate the vertical velocity of the updraft above the fire source, W (Baum et al., 1989) (refer to Chapter 5). The model divides the flame structure into continuous flame, intermittent flame, and plume regimes. The velocities corresponding to each regime are given by:

$$W = \begin{cases} 6.83\dot{Q}^{1/5}\left(\dfrac{H}{\dot{Q}^{2/5}}\right)^{1/2} & (continuous\ flame:H \le 0.08\dot{Q}^{2/5}) \\ 1.85\dot{Q}^{1/5}\ (intermittent\ flame:0.08\dot{Q}^{2/5} < H \le 0.20\dot{Q}^{2/5})\ (m/s) \\ 1.08\dot{Q}^{.1/5}\left(\dfrac{H}{\dot{Q}^{.2/5}}\right)^{-1/3} & (plume:0.20\dot{Q}^{2/5} < H) \end{cases}$$

where z is the height above the fire source, $D = \left(\dot{Q}/c_P\rho_\infty T_\infty g^{1/2}\right)^{-2/5}$ is the scaling length of the fire, $\dot{Q}$ is the HRR, c_P is the specific heat, ρ_∞ is the ambient gas density, and g is the acceleration due to gravity. Representing the initial ejection velocity, W_0, by W in the intermittent flame regime, which takes the highest value throughout the entire fire plume, we obtain $W_0 = 7.4, 11.7, 18.5$ m·s^{-1} for the fire sources A, B, and C, respectively. By introducing W_0 into Eq. (8.14) and then W^* into Eq. (8.13), we obtain $\mu_A = 0.306, 0.820,$ and 2.20 cm for the fire sources A, B, and C, respectively. Figure 8.9 shows the cumulative distribution function of the A_P by Eq. (8.12).

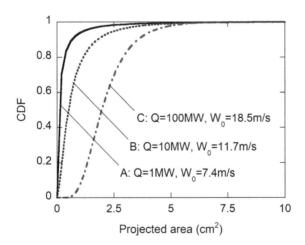

Figure 8.9 Estimated cumulative distribution function (CDF) of the projected area of firebrands.

Mass is another important quantity that characterizes the spot ignition behavior of generated firebrands. Table 8.3 summarizes the ranges of mass of an individual firebrand collected in existing experiments. However, it is the density of firebrands rather than the mass that is primarily used in the analysis of spot ignition behavior. However, they were measured only in limited experiments. The ranges of firebrand density were reportedly 76–86 $kg \cdot m^{-3}$ in the full-scale burn experiment of a three-story wooden building (Hayashi et al., 2014), and 29.7–74.5 $kg \cdot m^{-3}$ in the wind tunnel experiment that burnt wood cribs in an open-top combustion chamber (Himoto et al., 2021). They are the records of firebrands generated from a structural fire or its surrogate representing a fire source in urban fires. The difference between wildland and urban fires concerning the generation environment of firebrands is that combustibles burn in an open environment in wildland fires, while they burn in an enclosure in urban fires. The combustibles in an enclosure are exposed to intense heating for a longer duration as the ejection path of their fragmented pieces to the outdoors is restricted. Examples of possible ejection paths from a burning enclosure are broken windows and fallen roofs. Thus, the density of firebrands in urban fires is expected to be lower than that in wildland fires.

8.2.2 Quantity

In the previous subsection, we focused on the characteristics of individual firebrands, shape, and mass (or density). This is important for understanding how individual firebrands can be dispersed in the flow field and ignite the combustible on which they are deposited. In this subsection, we focus on the quantity of firebrand generation used to evaluate the hazard of spot ignitions in Eqs. (8.6) and (8.8).

Several attempts have been made to experimentally quantify the amount of generated firebrands (Manzello et al., 2008; Miura et al., 2011; Filkov et al., 2017; Adusumilli et al., 2021; Almeida et al., 2021; Hajilou et al. 2021; Thompson et al. 2022). However, due to difficulties in measurement, few examples actually quantified the relationship between the consumption rate of the combustible and the generation rate of firebrands. Miura et al. counted the number of generated firebrands during the burn of wood cribs (Japanese red pine) using video records (Miura et al., 2011). They burnt three types of wood cribs, crib A (1,000 mm × 1,000 mm × 450 mm), crib B (1,500 mm × 1,500 mm × 450 mm), and crib C (2,000 mm × 2,000 mm × 450 mm), in different crosswind conditions. Figure 8.10 compares the total mass loss during the burning of the wood cribs and the cumulative number of generated firebrands. In this figure, the steeper the slope of the curve, the greater the number of generated firebrands. The result indicates that the larger the wind velocity and the smaller the dimension of component

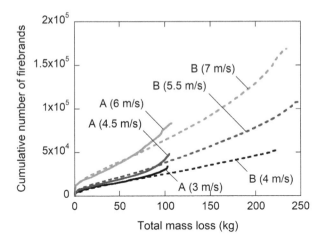

Figure 8.10 The relationship between the total mass loss of wood crib and the total number of generated firebrands (modified from Miura et al., 2011).

wood sticks, the greater the number of generated firebrands. The crosswind not only enhances the combustion intensity of wood cribs due to fresh air supply, but also boosts the detachment of thermally degraded fragments by applying aerodynamic forces

Figure 8.10 also shows an increase in the rate of firebrand generation in the decay phase of combustion. Characteristics of collected firebrands are presumably different between the stages of combustion during which the firebrands were generated. In the early stage, when the progress of thermal degradation remained near the material surface, the structural integrity of wood sticks was only partially lost. Thus, most of the detached wood fragments were fine and had a low thermal energy content. In contrast, in the later stage, the structural integrity of wood sticks was lost due to the progress of thermal degradation. Thus, the detached wood fragments included larger fragments with higher thermal energy content. Figure 8.10 also shows an increase in the rate of firebrand generation in the decay phase of combustion. Such a difference in the characteristics of firebrands may lead to differences in the ignition probability of recipient combustibles in different time stages.

The large quantities of firebrand generation, as exhibited in Figure 8.10, may give an impression as if a large proportion of the burning fuel was fragmented into pieces and released into the flow field as firebrands. However, the proportion of the total mass of the collected firebrands to the initial mass of the wood cribs (conversion ratio) ranged only from 0.06 to 1.64% in the experiment shown in Figure 8.10 (Miura et al., 2011). In another

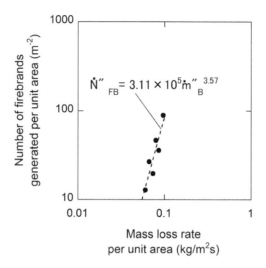

Figure 8.11 Relationship between the MLR per unit base area of fuel and firebrand generation rate per unit base area.

experiment, the conversion ratio ranged from 0.08 to 0.28% when burning Korean pine and Douglas-fir trees (Manzello et al., 2008), and from 0.00526 to 0.0176% when burning wood cribs in an open-top combustion chamber (Himoto et al., 2021). The conversion ratio can be affected by various factors such as the fuel conditions (species, size, and thermo-physical properties) and the burn environment (open or enclosed). In an enclosed space, a considerable amount of detached fragments (i.e., potential firebrands) are consumed before being released outside as the enclosing surfaces block their ejections. This infers a smaller conversion ratio of burning fuel to firebrands in urban fires where fuels burn in enclosed spaces, than in wildland fires where fuels burn in an open environment.

Figure 8.11 shows the relationship between the mass loss rate (MLR) per unit base area, $\dot{m}_B''$, and the firebrand generation rate (number of firebrands generated per unit time) per unit base area, $\dot{N}_{FB}''$, from the experiment by Miura et al. (Miura et al., 2011). An important correlation can be obtained by regressing the data points as follows:

$$\dot{N}_{FB}'' = 3.11 \times 10^5 \, \dot{m}_B''^{3.57} \tag{8.15}$$

The obtained correlation does not explicitly include presumable controlling parameters, such as the crosswind and fuel geometry conditions, as the variables. Although this is attributed to the limited data availability, their effects are partly reflected in the only variable, $\dot{m}_B''$.

8.3 DISPERSION

Figure 8.12 shows the deposition distribution of firebrands collected in a full-scale burn experiment of a three-story wooden building with a total floor area of 2,260 m² (Hayashi et al., 2014). The X-axis is in the wind direction, and the Y-axis is its orthogonal. The average ambient wind velocity during the experiment was 4.6 m·s⁻¹. In the X-axis direction, a relatively large quantity of firebrands was collected near the fire source, but it steeply decreased with distance from the fire source. Although deposited firebrands were collected as far as 700 m downwind of the fire source, dispersion of not a few firebrands was observed further downwind (Hayashi et al., 2014). In the Y-axis, the dispersion distributions of firebrands show a bell-shaped distribution centered at X = 0. In contrast to burn experiments of a structure, those of wildland fuels generally involve simultaneous burn of multiple fuel objects distributed within a certain range. This entails an identification problem of firebrands obscuring from which fuel object the collected firebrands were generated. In other words, firebrands deposited at a specific location may have originated in several different locations, thereby hampering the estimation of their dispersal processes. However, given the phenomenological similarity of the dispersion behavior, the basic characteristics of the deposition distribution when burning a wildland fuel object are expected to be similar to those of a structure.

The dispersion distribution of firebrands has been a relatively common subject of study as it represents the range of spot ignition hazards in outdoor

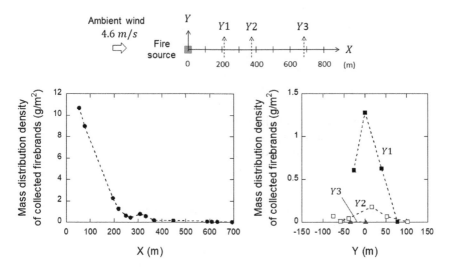

Figure 8.12 Deposition distribution of firebrands collected in a full-scale burn experiment of a three-story wooden building (Hayashi et al., 2014).

fires. Analysis approaches can be divided into experimental, numerical, and investigative, each with its own advantages and disadvantages.

The advantage of the experimental approach over the other approaches is, above all, the capability to obtain reliable data under realistic fire conditions (Manzello et al., 2008; Hayashi et al., 2014; Song et al., 2017; Himoto et al., 2021). In particular, in wind tunnel experiments where cross-wind conditions can be controlled in a reproducible manner, dispersion distributions under different fire conditions can be analyzed comparatively. On the other hand, the experimental approach has not been the primary means of investigation for dispersion studies, as it requires a large-scale experimental facility.

Experimental investigations include the measurement of combustion and aerodynamic characteristics of individual firebrands (Tarifa et al., 1965; Woycheese et al., 1999; Ellis, 2000). The measurement results have been incorporated into the computation of the transport of individual firebrands, which is currently the most common approach to examining the dispersion behavior (Lee et al., 1969; Lee et al., 1970; Mraszew et al., 1977; Tse et al., 1998; Himoto et al., 2005; Anthenien et al., 2006; Sardoy et al., 2007; Kortas et al., 2009; Koo et al., 2012; Tohidi et al., 2017; Anand et al., 2018). The method that does not require experimental facilities has attracted the interest of many researchers. On the other hand, various uncertain factors are involved in the dispersal sub-process of firebrands, which cannot be fully incorporated into the computational procedure. Thus, experimental and investigative methods generally have an advantage in terms of reliability over numerical methods.

Ranges of firebrand dispersion have also been investigated in actual large outdoor fires (Fire Research Institute, 1976; Takeya et al., 2017; Suzuki et al., 2018; Storey et al., 2020). There is no source better than data from reconnaissance surveys for understanding the actual dispersion behavior of firebrands. Meanwhile, the collected firebrands in actual fires are not necessarily suitable materials for quantitative analysis. As combustion of firebrands generally continues even after deposition, there is no guarantee that the shape and mass of the collected firebrands are those at the time of deposition. Furthermore, as actual fires generally involve multiple fire sources, it is virtually impossible to identify from which fire source the collected firebrands were generated.

8.3.1 Motion of a lofted firebrand (translation and rotation)

In the earliest attempts to model the motion of a lofted firebrand, its shape has often been assumed to be a sphere. Such an assumption was introduced for the convenience of analysis rather than the representation of actual firebrand shapes. As demonstrated in Figures 8.5 and 8.6, the shapes of actual

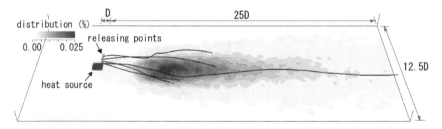

Figure 8.13 An example of dispersal simulation of disk-shaped firebrands in a cross-wind. Trajectories of several lofted firebrands are indicated with curved lines (Himoto et al., 2005).

firebrands are not that ideal. They are mostly non-spherical whose motion exhibits a complex behavior in a turbulent flow field due to a frequent change in its projected area versus relative flow and a pronounced lift effect exerted orthogonally to the relative flow. Figure 8.13 shows a simulated dispersion of disk-shaped firebrands of identical shape and mass from a fire source in a crosswind (Himoto et al., 2005). A certain fraction of firebrands were dispersed for a long distance, tracing complex trajectories. Large variation in the deposition distribution, despite the identical shape and mass, is an important implication of the uncertain nature of firebrand dispersion. In this subsection, we discuss the modeling of the dispersion behavior of non-spherical particles in a flow field

Figure 8.14 shows a schematic representation of the motion of a non-spherical firebrand. We consider a rigid-body particle that does not deform under aerodynamic forces. Its motion in 3D space can be divided into the translation and rotation around the center of gravity of the particle, which can be expressed in terms of the velocity, $\mathbf{u}_P$, and the angular velocity, $\mathbf{\omega}_P$, respectively. These are both defined by the coordinate system with the origin at the center of gravity of the particle, $x_P y_P z_P$ (particle coordinate system). Additionally, the position, $\mathbf{X}_P$, and the attitude angle, $\mathbf{Q}_P$, are defined by the coordinate system on the ground, XYZ (spatial coordinate system). The origin of the spatial coordinate system can be at an arbitrary location. However, it would be most straightforward to set it at the fire source location, as shown in Figure 8.14. In this subsection, uppercase and lowercase characters are used to represent the quantities defined by the spatial coordinate system, XYZ, and the particle coordinate system, $x_P y_P z_P$, respectively. Bold characters represent vector quantities.

Figure 8.15 shows the computational procedure of the firebrand dispersion. First, the airflow motion (velocity, $\mathbf{U}$) around the particle is calculated in the spatial coordinate system, XYZ. This is converted into the quantity in the particle coordinate system, $x_P y_P z_P$, which is $\mathbf{u}$. The angular velocity, $\mathbf{\omega}$,

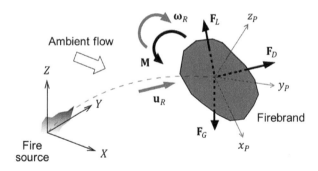

Figure 8.14 Schematic representation of the motion of a non-spherical firebrand.

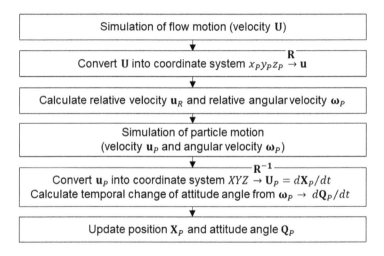

Figure 8.15 Computational procedure of the firebrand dispersion in 3D space.

can be calculated using $\mathbf{u}$. The airflow motion ($\mathbf{u}$ and $\boldsymbol{\omega}$) relative to the particle motion ($\mathbf{u}_p$ and $\boldsymbol{\omega}_p$) can be used to calculate the aerodynamic force and moment ($\mathbf{f}$ and $\mathbf{m}$) applied to the particle. The $\mathbf{f}$ and $\mathbf{m}$ are used to update $\mathbf{u}_p$ and $\boldsymbol{\omega}_p$. The updated $\mathbf{u}_p$ and $\boldsymbol{\omega}_p$ are further converted into $\mathbf{U}_p$ and $\boldsymbol{\Omega}_p$ in the spatial coordinate system, XYZ, respectively. They are used to update the position and the attitude angle ($\mathbf{X}_p$ and $\mathbf{Q}_p$) of the particle. In this computational procedure, $\mathbf{U}$ can be obtained by solving the governing equations of the airflow motion using the computational fluid dynamics (CFD) simulations (Fletcher, 1991; Oran et al., 2001; Ferziger et al., 2002; Poinsot et al., 2005; Yeoh et al., 2009). CFD simulations can provide detailed information necessary for the computation of particle dispersion. However, we focus on the particle motion in a prescribed flow field and do not discuss it further in this subsection.

(1) Equation of motion

The following momentum and angular momentum equations need to be solved simultaneously to describe the motion of a lofted firebrand in a flow field (Goldstein et al., 2013; Crowe et al., 2011):

$$\frac{d}{dt}(\rho_P V_P \mathbf{u}_P) = \mathbf{f} \tag{8.16a}$$

$$\frac{d}{dt}(\mathbf{i}_P \cdot \boldsymbol{\omega}_P) = \mathbf{m} \tag{8.16b}$$

where ρ_P is the particle density, V_P is the particle volume, $\mathbf{f}$ is the aerodynamic, $\mathbf{i}_P$ is the inertia tensor, $\mathbf{m}$ is the moment caused by the rotational motion of the surrounding fluid relative to the particle. The drag, $\mathbf{f}_D$, lift, $\mathbf{f}_L$, and gravity, $\mathbf{f}_G$ are the basic forces that comprise the $\mathbf{f}$.

The time derivatives of $\mathbf{u}_P$ and $\boldsymbol{\omega}_P$ can be obtained by transforming the above equations. In the transformation, we should consider the rotation of the particle coordinate system, $x_P y_P z_P$, which is expressed by the angular velocity, $\boldsymbol{\omega}_P = (\xi_P, \eta_P, \zeta_P)$, with respect to the spatial coordinate system, XYZ, which is the inertial system. As a general example, the time derivatives of an arbitrary vector, $\boldsymbol{\theta} = (\theta_x, \theta_y, \theta_z)$, defined by the particle coordinate system, $x_P y_P z_P$, can be obtained in the following manner. Letting $\mathbf{i}$, $\mathbf{j}$, and $\mathbf{k}$ be the unit vectors of each axis, $\boldsymbol{\theta}$ can be expressed as:

$$\boldsymbol{\theta} = \theta_x \mathbf{i} + \theta_y \mathbf{j} + \theta_z \mathbf{k} \tag{8.17}$$

Differentiating both sides of this equation with respect to time, we obtain:

$$\frac{d\boldsymbol{\theta}}{dt} = \frac{d\theta_x}{dt}\mathbf{i} + \frac{d\theta_y}{dt}\mathbf{j} + \frac{d\theta_z}{dt}\mathbf{k} + \theta_x\frac{d\mathbf{i}}{dt} + \theta_y\frac{d\mathbf{j}}{dt} + \theta_z\frac{d\mathbf{k}}{dt} \tag{8.18}$$

The first three terms on the right-hand side of Eq. (8.18) are the time derivatives in the absence of the axis rotation. So, they are collectively denoted as $d^*\boldsymbol{\theta} / dt$. By referring to Figure 8.16, the time derivatives of the axis vectors in the last three terms of Eq. (8.18) can be expressed as follows:

$$\frac{d\mathbf{i}}{dt} = \lim_{\Delta t \to 0} \frac{\{\mathbf{i} + (\zeta_P \Delta t)\mathbf{j} - (\eta_P \Delta t)\mathbf{k}\} - \mathbf{i}}{\Delta t} = \zeta_P \mathbf{j} - \eta_P \mathbf{k} \tag{8.19a}$$

$$\frac{d\mathbf{j}}{dt} = \lim_{\Delta t \to 0} \frac{\{\mathbf{i} + (\xi_P \Delta t)\mathbf{k} - (\zeta_P \Delta t)\mathbf{i}\} - \mathbf{j}}{\Delta t} = \xi_P \mathbf{k} - \zeta_P \mathbf{i} \tag{8.19b}$$

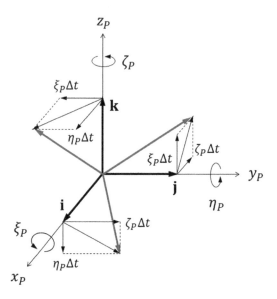

Figure 8.16 Change in the axis vectors accompanied by the particle rotation around the origin (center of gravity of the particle).

$$\frac{d\mathbf{k}}{dt} = \lim_{\Delta t \to 0} \frac{\left\{\mathbf{i} + \left(\eta_P \Delta t\right)\mathbf{i} - \left(\xi_P \Delta t\right)\mathbf{j}\right\} - \mathbf{k}}{\Delta t} = \eta_P \mathbf{i} - \xi_P \mathbf{j} \qquad (8.19c)$$

They are nothing but the external product the $\boldsymbol{\omega}_P$ and $\boldsymbol{\theta}$, which is $\boldsymbol{\omega}_P \times \boldsymbol{\theta}$. To summarize, the time derivative of $\boldsymbol{\theta}$ in Eq. (8.18) takes the form:

$$\frac{d\boldsymbol{\theta}}{dt} = \frac{d^*\boldsymbol{\theta}}{dt} + \boldsymbol{\omega}_P \times \boldsymbol{\theta} \qquad (8.20)$$

The first term represents the time derivative in the absence of the axis rotation, and the second term represents the effect of axis rotation.

The force and moment terms in Eq. (8.16) are dependent on the motion of the surrounding airflow relative to the particle, $\mathbf{u}_R$ and $\boldsymbol{\omega}_R$. As an example, Figure 8.17 describes the direction of the forces ($\mathbf{f}_D$, $\mathbf{f}_L$, and $\mathbf{f}_G$) and the moment, $\mathbf{m}$, applied to a disk-shaped particle. Solid lines indicate the directions of the forces and moment application, and the dotted lines indicate the motion of the surrounding airflow relative to the particle. This relationship for a disk-shaped particle can be similarly applicable to the other shapes.

$\mathbf{f}_D$ is exerted in the direction of the airflow velocity relative to the particle velocity, $\mathbf{u}_R \left(= \mathbf{u} - \mathbf{u}_P\right)$. The magnitudes of $\mathbf{f}_D$ are proportional to the momentum of the particle, $\frac{1}{2}\rho_\infty \left|\mathbf{u}_R\right|^2$. Thus, $\mathbf{f}_D$ can be given by:

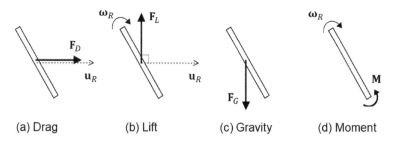

Figure 8.17 Direction of forces and moment applied on a disk-shaped particle.

$$f_D = C_D A_P \cdot \frac{1}{2} \rho_\infty |u_R| u_R \tag{8.21}$$

where C_D is the drag coefficient, and A_P is the projected area of the particle. f_L is exerted in the direction orthogonal to the relative velocity, u_R. The magnitudes of f_L are also proportional to $\frac{1}{2}\rho_\infty |u_R|^2$. Thus, f_L can be expressed as:

$$f_L = C_L A_P \cdot \frac{1}{2} \rho_\infty |u_R|^2 \frac{u_R \times \omega_R}{|u_R \times \omega_R|} \tag{8.22}$$

where C_L is the lift coefficient, and $\omega_R \left(= \omega - \omega_P = \omega - \frac{1}{2}(\nabla \times u) \right)$ is the airflow angular velocity relative to the particle angular velocity.

m is exerted in the direction opposite to ω_R as the rotation attenuates due to the viscous damping effect. In other words, m is exerted in the direction in which the relative angular velocity, ω_P, eventually dissipates. The magnitude of m is also proportional to $\frac{1}{2}\rho_\infty |u_R|^2$. Thus, m can be expressed as:

$$m = -C_M A_P \cdot \frac{1}{2} \rho_\infty |u_R|^2 \frac{\omega_R}{|\omega_R|} \tag{8.23}$$

where C_M is the moment coefficient. Note that the aerodynamic coefficients C_D, C_L, and C_M depend on the shape of the particle, and need to be determined experimentally (Crowe et al., 2011).

In contrast to f_D, f_L, and m, whose direction of action is determined based on the motion of the airflow relative to the particle, f_G is exerted vertically downward:

$$\mathbf{f}_G = (\rho_P - \rho_\infty)V_P \mathbf{g} \tag{8.24}$$

where ρ_∞ is the ambient air density, and $\mathbf{g}$ is the acceleration due to gravity.

(2) Coordinate conversion by Euler angle

A common approach to converting $\mathbf{u}_P$ and $\boldsymbol{\omega}_P$ defined by the particle coordinate system, $x_P y_P z_P$, into the spatial coordinate system, XYZ, is to use the Eulerian angle (Goldstein et al., 2013). The Eulerian angle, $\mathbf{E} = (\alpha, \beta, \gamma)$, is a way of describing the relationship between two rectangular coordinate systems in 3D Euclidean space, defined by the angles of three rotations in the following order (Figure 8.18). First, rotate the spatial coordinate system XYZ by α around the Z -axis and let the resulting coordinate system be $X_1 Y_1 Z_1$. Second, rotate the system $X_1 Y_1 Z_1$ by β around *the* X_1 -axis and let the resulting coordinate system be $X_2 Y_2 Z_2$. Third, rotate the system $X_2 Y_2 Z_2$ by γ around the Z_2 -axis and let the resulting coordinate system be $X_3 Y_3 Z_3$. The rotations of the angles α, β, and γ are represented by the following rotation matrices, respectively:

$$\mathbf{R}_\alpha = \begin{pmatrix} \cos\alpha & \sin\alpha & 0 \\ -\sin\alpha & \cos\alpha & 0 \\ 0 & 0 & 1 \end{pmatrix} \tag{8.25a}$$

$$\mathbf{R}_\beta = \begin{pmatrix} 1 & 0 & 0 \\ 0 & \cos\beta & \sin\beta \\ 0 & -\sin\beta & \cos\beta \end{pmatrix} \tag{8.25b}$$

$$\mathbf{R}_\gamma = \begin{pmatrix} \cos\gamma & \sin\gamma & 0 \\ -\sin\gamma & \cos\gamma & 0 \\ 0 & 0 & 1 \end{pmatrix} \tag{8.25c}$$

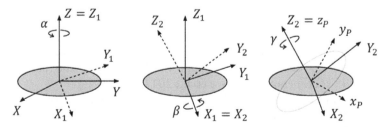

Figure 8.18 Relationship between the particle coordinate system $x_P y_P z_P$ and the spatial coordinate system XYZ defined by the Euler angle.

An arbitrary rotation in 3D Euclidean space can be reproduced by synthesizing these three rotations. Assume that the system $X_3Y_3Z_3$ is equivalent to the particle coordinate system $x_Py_Pz_P$. Then, the conversion matrix from the spatial coordinate system XYZ to the particle coordinate system $x_Py_Pz_P$ can be expressed as:

$$R = \begin{pmatrix} \cos\alpha\cos\gamma - \sin\alpha\cos\beta\sin\gamma & \sin\alpha\cos\gamma + \cos\alpha\cos\beta\sin\gamma & \sin\beta\sin\gamma \\ -\cos\alpha\sin\gamma - \sin\alpha\cos\beta\cos\gamma & -\sin\alpha\sin\gamma + \cos\alpha\cos\beta\cos\gamma & \sin\beta\cos\gamma \\ \sin\alpha\sin\beta & -\cos\alpha\sin\beta & \cos\beta \end{pmatrix}$$

(8.26)

Using R, the airflow velocity defined by the spatial coordinate system XYZ, U, can be converted to that defined by the particle coordinate system $x_Py_Pz_P$, u, as follows:

$$u = RU \qquad (8.27)$$

On the contrary, for the conversion from the particle coordinate system $x_Py_Pz_P$ to the spatial coordinate system XYZ, the inverse matrix, R^{-1}, can be used. This can be obtained by synthesizing the rotation matrices in the opposite order, which takes the form:

$$R^{-1} = \begin{pmatrix} \cos\alpha\cos\gamma - \sin\alpha\cos\beta\sin\gamma & -\cos\alpha\sin\gamma - \sin\alpha\cos\beta\cos\gamma & \sin\alpha\sin\beta \\ \sin\alpha\cos\gamma + \cos\alpha\cos\beta\sin\gamma & -\sin\alpha\sin\gamma + \cos\alpha\cos\beta\cos\gamma & -\cos\alpha\sin\beta \\ \sin\beta\sin\gamma & \sin\beta\cos\gamma & \cos\beta \end{pmatrix}$$

(8.28)

The inverse matrix, R^{-1}, can be used to convert u_R into U_R as follows:

$$U_R = R^{-1}u_R \qquad (8.29)$$

The trajectory of the particle, X_P, can be obtained by integrating the velocity of the particle defined by the spatial coordinate system XYZ, U_R, over time.

By definition, the Euler angle, E, is nothing but the attitude angle, Q_P, of the particle. Thus, in the following discussion, the Euler angle, E, is referred to as the attitude angle, $Q_P = (\alpha, \beta, \gamma)$. Let us consider expressing the attitude angle, $Q_P = (\alpha, \beta, \gamma)$, using the angular velocity of the particle, $\omega_P = (\xi_P, \eta_P, \zeta_P)$. According to the definition in Eqs. (8.25) and (8.26), there is a relationship between ω_P and the axes of each rotation, I, I_1, and K_2, as follows:

$$\omega_P = \dot{\alpha}I + \dot{\beta}I_1 + \dot{\gamma}K_2 \qquad (8.30)$$

where the dot accent $\dot{x}$ denotes the rate of change over time, dx / dt. Note that each term in Eq. (8.30) can be respectively expressed as:

$$\dot{\alpha}\mathbf{I} = \dot{\alpha} \cdot \mathbf{R} \begin{pmatrix} 0 \\ 0 \\ 1 \end{pmatrix} = \dot{\alpha} \begin{pmatrix} \sin\beta\sin\gamma \\ \sin\beta\cos\gamma \\ \cos\beta \end{pmatrix} \tag{8.31a}$$

$$\dot{\beta}\mathbf{I}_1 = \dot{\beta} \cdot \mathbf{R}_\gamma \mathbf{R}_\beta \begin{pmatrix} 1 \\ 0 \\ 0 \end{pmatrix} = \dot{\beta} \begin{pmatrix} \cos\gamma \\ -\sin\gamma \\ 0 \end{pmatrix} \tag{8.31b}$$

$$\dot{\gamma}\mathbf{K}_2 = \dot{\gamma} \begin{pmatrix} 0 \\ 0 \\ 1 \end{pmatrix} \tag{8.31c}$$

By substituting Eq. (8.31) into Eq. (8.30), we obtain:

$$\boldsymbol{\omega}_P = \begin{pmatrix} \xi_P \\ \eta_P \\ \zeta_P \end{pmatrix} = \begin{pmatrix} \dot{\alpha} \cdot \sin\beta\sin\gamma + \dot{\beta} \cdot \cos\gamma \\ \dot{\alpha} \cdot \sin\beta\cos\gamma - \dot{\beta} \cdot \sin\gamma \\ \dot{\alpha} \cdot \cos\beta + \dot{\gamma} \end{pmatrix} \tag{8.32}$$

Eq. (8.32) can be used to derive the time derivative of $\mathbf{Q}_P = (\alpha, \beta, \gamma)$ as follows:

$$\frac{d\mathbf{Q}_P}{dt} = \begin{pmatrix} \dot{\alpha} \\ \dot{\beta} \\ \dot{\gamma} \end{pmatrix} = \begin{pmatrix} \xi_P \dfrac{\sin\gamma}{\sin\beta} + \eta_P \dfrac{\cos\gamma}{\sin\beta} \\ \xi_P\cos\gamma - \eta_P\sin\gamma \\ -\xi_P \dfrac{\sin\gamma}{\tan\beta} - \eta_P \dfrac{\cos\gamma}{\tan\beta} + \zeta_P \end{pmatrix} \tag{8.33}$$

The attitude angle, $\mathbf{Q}_P$, can be obtained by integrating Eq. (8.33) over time.

(3) Coordinate conversion by quaternions

The Euler angle provides an intuitive understanding of the relationship between the spatial coordinate system XYZ and the particle coordinate system $x_P y_P z_P$. However, it does not necessarily provide a stable solution for the attitude angle, $\mathbf{Q}_P$. This is because the right-hand side of Eq. (8.33) contains s inβ or tanβ in the denominators, which may cause divergence of $d\alpha / dt$ and $d\gamma / dt$ when $\beta = 0$ or $\beta = \pi$. In order to avoid such a singularity issue, we can alternatively define $\mathbf{Q}_P$ using the quaternion. The quaternion is a system of numbers that extends the complex numbers, often used to express the rotation of an object in 3D space (Vince, 2011). According to the quaternion concept, the attitude angle, $\mathbf{Q}_P$, can be expressed in terms of the four components, q_0, q_1, q_2, and q_3:

$$\mathbf{Q}_P = \begin{pmatrix} q_0 \\ q_1 \\ q_2 \\ q_3 \end{pmatrix} = \begin{pmatrix} \sin\dfrac{\beta}{2}\sin\dfrac{\gamma-\alpha}{2} \\ \sin\dfrac{\beta}{2}\cos\dfrac{\gamma-\alpha}{2} \\ \cos\dfrac{\beta}{2}\sin\dfrac{\gamma+\alpha}{2} \\ \cos\dfrac{\beta}{2}\cos\dfrac{\gamma+\alpha}{2} \end{pmatrix} \tag{8.34}$$

We now consider the relationship between $\mathbf{Q}_P$ and $\boldsymbol{\omega}_P$. The time derivative of each component of $\mathbf{Q}_P$ can be given by:

$$\frac{d\mathbf{Q}_P}{dt} = \begin{pmatrix} \dot{q}_0 \\ \dot{q}_1 \\ \dot{q}_2 \\ \dot{q}_3 \end{pmatrix} = \begin{pmatrix} \dfrac{\dot{\beta}}{2}\cos\dfrac{\beta}{2}\sin\dfrac{\gamma-\alpha}{2} + \dfrac{\dot{\gamma}-\dot{\alpha}}{2}\sin\dfrac{\beta}{2}\cos\dfrac{\gamma-\alpha}{2} \\ \dfrac{\dot{\beta}}{2}\cos\dfrac{\beta}{2}\cos\dfrac{\gamma-\alpha}{2} - \dfrac{\dot{\gamma}-\dot{\alpha}}{2}\sin\dfrac{\beta}{2}\sin\dfrac{\gamma-\alpha}{2} \\ -\dfrac{\dot{\beta}}{2}\sin\dfrac{\beta}{2}\sin\dfrac{\gamma+\alpha}{2} + \dfrac{\dot{\gamma}+\dot{\alpha}}{2}\cos\dfrac{\beta}{2}\cos\dfrac{\gamma+\alpha}{2} \\ -\dfrac{\dot{\beta}}{2}\sin\dfrac{\beta}{2}\cos\dfrac{\gamma+\alpha}{2} - \dfrac{\dot{\gamma}+\dot{\alpha}}{2}\cos\dfrac{\beta}{2}\sin\dfrac{\gamma+\alpha}{2} \end{pmatrix} \tag{8.35}$$

By substituting Eq. (8.33), Eq. (8.35) can be transformed so that $d\mathbf{Q}_P/dt$ is expressed with the components of $\boldsymbol{\omega}_P = (\xi_P, \eta_P, \zeta_P)$ as follows:

$$\frac{d\mathbf{Q}_P}{dt} = \begin{pmatrix} \dot{q}_0 \\ \dot{q}_1 \\ \dot{q}_2 \\ \dot{q}_3 \end{pmatrix} = \frac{1}{2}\begin{pmatrix} -\dot{q}_2\xi_P - \dot{q}_3\eta_P + \dot{q}_1\zeta_P \\ \dot{q}_3\xi_P - \dot{q}_2\eta_P - \dot{q}_0\zeta_P \\ \dot{q}_0\xi_P + \dot{q}_1\eta_P + \dot{q}_3\zeta_P \\ -\dot{q}_1\xi_P + \dot{q}_0\eta_P + \dot{q}_2\zeta_P \end{pmatrix} \tag{8.36}$$

In contrast to Eq. (8.33), the singularity issue is resolved from Eq. (8.36). This equation can be integrated over time to update the attitude angle, $\mathbf{Q}_P$, maintaining the numerical stability. However, the involvement of rounding errors is generally inevitable when integrating discretized equations. Thus, the following relationship derived from the definition of quaternion in Eq. (8.34) is often used to minimize the accumulation of errors:

$$q_0^2 + q_1^2 + q_2^2 + q_3^3 = 1 \tag{8.37}$$

When using $\mathbf{Q}_P$ in Eq. (8.36), the coordinate conversion matrix, $\mathbf{R}$, and the inverse matrix, $\mathbf{R}^{-1}$ take the following forms, respectively:

$$\mathbf{R} = \begin{pmatrix} -q_0^2 + q_1^2 - q_2^2 + q_3^3 & -2(q_0q_1 - q_2q_3) & 2(q_0q_3 + q_1q_2) \\ -2(q_0q_1 + q_2q_3) & q_0^2 - q_1^2 - q_2^2 + q_3^3 & -2(q_0q_2 - q_1q_3) \\ -2(q_0q_3 - q_1q_2) & -2(q_0q_2 + q_1q_3) & -q_0^2 - q_1^2 + q_2^2 + q_3^3 \end{pmatrix} \tag{8.38a}$$

$$\mathbf{R}^{-1} = \begin{pmatrix} -q_0^2 + q_1^2 - q_2^2 + q_3^3 & -2(q_0q_1 + q_2q_3) & -2(q_0q_3 - q_1q_2) \\ -2(q_0q_1 - q_2q_3) & q_0^2 - q_1^2 - q_2^2 + q_3^3 & -2(q_0q_2 + q_1q_3) \\ 2(q_0q_3 + q_1q_2) & -2(q_0q_2 - q_1q_3) & -q_0^2 - q_1^2 + q_2^2 + q_3^3 \end{pmatrix} \tag{8.38b}$$

8.3.2 Combustion of airborne firebrands

The combustion of firebrands generally is sustained even after being released into the flow field. As lofted firebrands are continuously exposed to airflow, forced convection enhances their combustion, either glowing or nonglowing. The combustion causes a change in their shape and mass, which in turn changes their transport behavior. Due to sustained combustion, some firebrands are burnt up or cease combustion before landing on the ground. However, not a few firebrands sustain combustion even after landing and may cause ignition of combustibles on the ground.

The combustion behavior of a firebrand is affected by its initial characteristics when released into the airflow (shape, mass, and thermochemical state) and the surrounding airflow while being transported (predominantly relative velocity). Several attempts have been made to model the combustion of airborne firebrands (Tarifa et al., 1965; Muraszew et al., 1977; Woycheese et al., 1999; Ellis, 2000; Sardoy et al., 2007). The level of detail varies by model. However, most of the models consider a firebrand as a homogeneous particle with the surface recession due to char oxidation occurring at a uniform rate toward the center.

One of the earliest attempts to experimentally investigate the burning behavior of individual airborne firebrands was conducted by Tarifa et al. (Tarifa et al., 1965). Based on the experimental data, they modeled the rate of density and radius changes of burning wood particles, either spherical or cylindrical, as follows (Tarifa et al., 1965):

$$\frac{\rho_P}{\rho_{P,0}} = \frac{1}{1 + \eta t^2} \tag{8.39a}$$

$$\frac{r_P}{r_{P,0}} = 1 - \left(\frac{\beta + \delta \cdot u_R}{r_{P,0}^2} \right) t \tag{8.39b}$$

where ρ_P is the particle density, r_P is the particle radius, t is the time, and u_R is the airflow velocity relative to the firebrand velocity. The subscripts P and 0 denote the firebrand and initial, respectively. η, β, and δ are the experimentally identified model parameters, which depend on the firebrand conditions such as tree species, and moisture content.

Woycheese et al. incorporated the concept of the droplet combustion model into the combustion modeling of a spherical wood particle in a quiescent atmosphere (Woycheese et al., 1999). The wood is assumed to be a homogeneous solid, and the particle relative velocity is sufficiently high to remove ash from the surface. The regression rate of the spherical diameter, D, is given by (Woycheese et al., 1999):

$$\frac{dD}{dt} = -4\alpha \left(\frac{\rho_\infty}{\rho_P}\right)\left(\frac{\ln(1+B)}{D}\right) \tag{8.40}$$

where α is the thermal diffusivity of air, ρ_∞ is the air density, ρ_P is the particle density, and B is the mass transfer number for wood (=1.2). The surface regression rate, dD/dt, is reportedly an underestimate as the model does not explicitly consider the effects of forced convection (Woycheese et al., 1999).

8.3.3 Range of firebrand dispersal

In Section 8.3.1, we discussed the computational procedure for the particle motion in a prescribed flow field. The range of firebrand dispersal can be evaluated following this computational procedure. However, the flow field that changes over time needs to be determined separately for calculating the aerodynamic forces and moment exerted on flying firebrands. CFD simulations are a viable option for predicting the fire-induced flow field. The effect of fluid motion on particle motion can be evaluated with high precision by maintaining a fine computational grid. However, a drawback of CFD simulations is their high computational cost. As various fire scenarios need to be analyzed in forest or urban disaster management in practice, minimizing the allocation of computational resources assigned to each analysis case is desirable. It is an option to focus the use of coupled simulation of firebrand dispersion and CFD on research purposes. As an alternative, probabilistic distributions can provide a tractable expression for evaluating the range of firebrand dispersal (Muraszew et al., 1976; Albini, 1979; Himoto et al., 2005; Sardoy et al., 2008; Wamg, 2011; Albini et al., 2012; Himoto et al., 2021; Himoto et al., 2022).

Sardoy et al. conducted a numerical analysis on the dispersion of disk-shaped firebrands downwind of a line fire in a crosswind (Sardoy et al., 2008). They reported a bimodal ground-level distribution of firebrands: firebrands deposited in a short-distance range from the fire source were mostly flaming, while they were mostly non-flaming (charring) in a long-distance range. The distributions of the short-distance firebrands were regressed with a log-normal distribution:

$$p(d \mid \mu, \sigma) = \frac{1}{\sqrt{2\pi}\sigma d} \exp\left\{-\frac{(\ln d - \mu)^2}{2\sigma^2}\right\} \qquad (8.41)$$

where μ is the mean and σ is the standard deviation. In generalizing the parameters, they introduced the Froude number, which is defined as:

$$Fr^* = \frac{U_\infty}{\sqrt{gL_C}} \qquad (8.42a)$$

$$L_C = \left(\frac{\dot{Q}'}{c_P \rho_\infty T_\infty g^{1/2}}\right)^{2/3} \qquad (8.42b)$$

where U_∞ is the ambient wind velocity, $\dot{Q}'$ is the heat release rate per unit length $(kW \cdot m^{-1})$, c_P is the specific heat, ρ_∞ is the ambient gas density, T_∞ is the ambient gas temperature, and g is the acceleration due to gravity. They obtained the following expressions for μ and σ by regressing the results of the numerical analysis (Figure 8.19):

$$\mu = \begin{cases} 1.47\left(\dot{Q}' \times 10^{-3}\right)^{0.54} U_\infty^{-0.55} + 1.14 & (Fr^* < 1) \\ 1.32\left(\dot{Q}' \times 10^{-3}\right)^{0.26} U_\infty^{0.11} - 0.02 & (Fr^* > 1) \end{cases} \qquad (8.43a)$$

$$\sigma = \begin{cases} 0.86\left(\dot{Q}' \times 10^{-3}\right)^{-0.21} U_\infty^{0.44} + 0.19 & (Fr^* < 1) \\ 4.95\left(\dot{Q}' \times 10^{-3}\right)^{-0.01} U_\infty^{-0.02} - 3.48 & (Fr^* > 1) \end{cases} \qquad (8.43b)$$

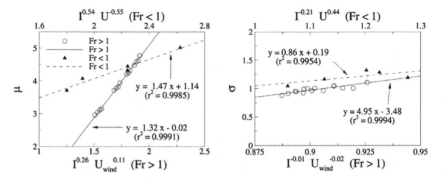

Figure 8.19 Results of the parameter estimation for the log-normal distribution of the dispersion distance of firebrands (Sardoy et al., 2008).

As shown in Figure 8.19, they identified two regimes of fire-induced flow field according to the Froude number, i.e., the buoyancy-driven regime $(Fr^* < 1)$ and the wind-driven regime $(Fr^* > 1)$. The regressions for μ and σ in Eq. (8.43) were obtained separately in each regime.

Another example of probabilistic distribution models also employs a lognormal distribution for representing the variation of the firebrand deposition downwind of a fire source, x_P (Himoto et al., 2022):

$$p(x_P|\lambda,\xi) = \frac{1}{\sqrt{2\pi}\xi x_P}\exp\left\{-\frac{1}{2}\left(\frac{\ln x_P - \lambda}{\xi}\right)^2\right\} \tag{8.44a}$$

$$\lambda = \ln\left(\frac{\mu_x}{\sqrt{1 + \sigma_x^2/\mu_x^2}}\right) \tag{8.44b}$$

$$\xi = \sqrt{\ln\left(1 + \frac{\sigma_x^2}{\mu_x^2}\right)} \tag{8.44c}$$

where μ_x is the mean and σ_x is the standard deviation. By regressing the deposition distribution of firebrands observed in a full-scale burn experiment of a three-story wooden building (Hayashi et al., 2014), they obtained the following relationships for μ_x and σ_x, respectively:

$$\frac{\mu_x}{d} = 41.0B^{*1.06} \tag{8.45a}$$

$$\frac{\sigma_x}{d} = 4.52 \tag{8.45b}$$

where B^* is the dimensionless parameter that controls the dispersion behavior. As schematically described in Figure 8.20, B^* was derived considering the transport of a firebrand (diameter d_P (cm)) ejected from the top of a fire source (height H and vertical velocity W_0) subject to the drag force due to crosswind (velocity U_∞) and gravity (Himoto et al., 2021):

$$B^* = \left(\frac{U_\infty W_0}{g\sqrt{d_P d}}\right)^2 \left(\frac{\rho_\infty}{\rho_P}\right)\left(1 + \sqrt{1 + \frac{2gH}{W_0^2}}\right)^2 \tag{8.46}$$

B^* in Eq. (8.46) can be viewed as the Froude number (the first term) modified by the density ratio (the second term) and the initial effect of ejection (the third term).

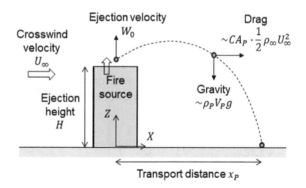

Figure 8.20 Dispersion of a firebrand ejected from the top of a fire source in a crosswind (Himoto et al., 2021).

WORKED EXAMPLE 8.4

Assume a fire source with a height $H = 5$ m and with an HRR $\dot{Q} = 10$ MW in a crosswind with a velocity of $U_\infty = 10$ m·s^{-1}. Calculate the deposition distribution of firebrands of three different diameters ($d_P = 1, 2,$ and 5 cm) downwind of the fire source. The representative width of the fuel component is $d = 10$ cm, and the densities of ambient gas and firebrands are $\rho_\infty = 1.2$ kg·m^{-3} and $\rho_P = 70$ kg·m^{-3}, respectively.

SUGGESTED SOLUTION

The initial ejection velocity of firebrands from the fire source, W_0, can be estimated using the result of Worked example 8.3. The buoyancy-induced vertical velocity in the intermittent regime of the fire plume is $W_0 = 11.9$ m·s^{-1} when $\dot{Q} = 10$ MW. By substituting these values into Eq. (8.46), the dimensionless parameters are $B^* = 1.30, 0.65,$ and 0.26 for firebrands with the diameter $d_P = 1, 2,$ and 5 cm, respectively. Further, by additionally substituting B^* into Eq. (8.45), the averages of the truncated normal distribution are $\mu = 539.2, 255.7,$ and 89.3 m for firebrands with the diameter $d_P = 1, 2,$ and 5 cm, respectively. The estimated cumulative distribution functions (CDF) of the transport distance of firebrands are shown in Figure 8.21.

8.4 IGNITION

We discussed the ignition process of solid fuel subject to external heating in Chapter 7. The ignition mechanism is commonly applicable to the case of firebrand-caused ignition considering the external heating as that of the

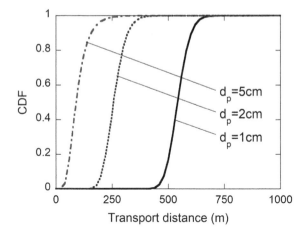

Figure 8.21 Estimated cumulative distribution function (CDF) of transport distance of firebrands with the diameter d_p = 1, 2, and 5 (cm).

deposited firebrands. Flaming combustion starts when the flammable gas mixture formed around the recipient fuel receives sufficient thermal energy either from the deposited firebrands, the other external fire source, or their combination. However, compared to heating from an external fire source such as a flame, heating from a single firebrand is localized, and its intensity depends on the combustion state. Thus, the rate of thermal degradation is generally low at the initial stage of heating, and the concentration of the flammable gas mixture around the recipient fuel may not be maintained within the flammable range. In such a case, smoldering combustion starts, gradually extending its range and increasing the rate of flammable gas release before shifting to flaming combustion.

8.4.1 Factors affecting ignition

The ignition sub-process of solid fuel by deposited firebrands is affected by various conditions (Fernandez-Pello, 2017; Manzello et al., 2020). These can be categorized into four factors as illustrated in Figure 8.22: the conditions of the individual firebrands, the recipient fuel, the deposition of firebrands on fuel bed, and the ambient environment.

(1) Deposited firebrands

The condition of individual firebrands can be represented by the thermo-chemical state, temperature, shape, and mass. These parameters describe the heating capability of firebrands on the recipient fuel.

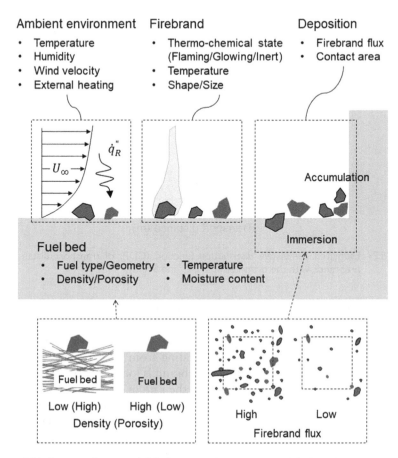

Figure 8.22 Factors affecting solid fuel ignition due to deposited firebrands.

The thermo-chemical state of a firebrand is variable when deposited on fuel. They may be flaming, glowing or smoldering, or inert or burnt out. The heating capability of an individual firebrand is high in this order. A flaming firebrand can heat a wider area of the recipient fuel. However, the flaming combustion of firebrands generally is sustained for a relatively short period of time, given their small content of thermally decomposable components. Flaming combustion shifts to glowing combustion when all the thermally decomposable components are consumed, or the thermal decomposition rate falls below the limit to sustain flaming combustion. Glowing combustion is sustained by the oxidation of the surface char, maintaining the high firebrand temperature. As char oxidation is slow combustion, it generally is sustained for a relatively long period provided sufficient oxygen is supplied. Glowing combustion ceases if any of the temperature, oxygen concentration, or consumable carbon residue falls out of the range required for sustaining combustion.

The shape of firebrands is diverse. The larger the firebrands, the greater the heating capability as a heat source against the recipient fuel, provided the thermo-chemical state and temperature are equivalent. However, firebrand size is not necessarily a sufficient condition for the ignition of recipient fuel. Rather, it is a matter of contact between firebrands and the recipient fuel; the heat released from firebrands can be effectively transferred to the recipient fuel when the contact area is large. Thus, small firebrands with large contact areas, such as those accumulated in the corner of a fuel object, can be more hazardous than large firebrands with small contact areas

(2) Recipient fuel

The condition of recipient fuel can be represented by its species, shape, density (porosity), temperature, and moisture content. These parameters affect the likelihood of the recipient fuel to raise its temperature and release flammable gas when subject to heating.

In wildland fires, litter layers (dead plant materials such as leaves, bark, needles, twigs, and cladodes that have fallen to the ground) are the common recipients of firebrands. Meanwhile, in urban fires, members or parts of the building envelope (eaves, shingles, and vents) and fuel objects placed around buildings are the common recipients of firebrands. These fuels are diverse in thermo-physical properties and shapes, affecting their heat transfer characteristics. If the fuel object is irregularly shaped, firebrands can generally have a large contact area with the fuel object, enhancing the rate of heat transfer to the fuel object. If the surface area of the fuel object relative to the volume is large, the fuel objects raise the temperature at a higher rate than the case with a small surface area.

The fuel moisture content (FMC) varies depending on the weather condition preceding the fire occurrence, resulting in different responses even when subjected to the same heating. The FMC is probably the most important factor affecting the ignition probability. The temperature of a combustible heated by firebrands stagnates near 100°C until the contained water is evaporated. A high FMC delays or even prevents the firebrand-caused ignition as it requires greater thermal energy to evaporate the water content.

(3) Deposition of firebrands on fuel bed

The deposition condition on the fuel bed affects the combustion intensity of the firebrands and the rate of heat transfer to the fuel bed.

The litter layer in forests is a porous or loosely packed fuel bed that contains numerous cavities of different sizes. Firebrands are captured or embedded in these cavities when they land in the fuel bed. The air movement through the cavity provides the deposited firebrands with a moderate oxygen exposure environment even if embedded in the fuel bed. This is an environment suitable for causing ignition of the fuel bed and sustaining the combustion that follows. An effective thermal energy transfer from the burning firebrands to the fuel bed due to the large contact area also increases the ignitability of the fuel bed.

Structures are generally composed of building members and materials with flat surfaces. In contrast to porous fuel beds, the rate of heat transfer to flat members is lower due to a relatively small contact area. However, firebrand-caused ignitions do occur in structures, often around a corner where multiple flat members are assembled. A corner of flat members creates a zone of airflow stagnation, accumulating the firebrands in one place and intensifying the heating. The rate of heat transfer at corners is also increased due to a larger contact area with firebrands.

The heating intensity of firebrands depends not only on the thermochemical state of an individual firebrand, but also on the accumulation density. The number of firebrands per unit area or the intensity of firebrand hazards per unit area, such as the HRR, temperature, or incident heat flux, can be described as the firebrand flux representing the accumulation effect. The accumulation effect is not a simple addition of the individual effects. The firebrands within proximity exchange heat with each other, maintaining a high-temperature condition in the accumulated zone. This activates the thermos-chemical state of individual firebrands, increasing the overall hazard of the accumulated firebrands. As a consequence, an accumulation of multiple firebrands can cause ignition of the fuel bed, even if individual firebrands may not possess enough thermal energy to cause ignition by themselves.

(4) Ambient environment

The ambient environment can be represented by ambient air temperature, humidity, wind velocity, and external heating.

While the scale of temperature rise in a fire environment is an order of $\sim10^3$ K, the change in ambient air temperature is $\sim10^1$ K at most. Thus, its effect on the combustion intensity of firebrands is generally negligible. Rather, the change in ambient temperature is effective in the FMC reduction together with the change in humidity. The moisture content reduction shortens the time to ignition of fuel objects when deposited by firebrands or even prevents ignition. However, it is not the instantaneous values of ambient air temperature and humidity at the moment of firebrand deposition, but the hysteresis effect that controls the FMC. An evaluation for a certain period of time prior to the fire occurrence is required to investigate these effects.

A moderate increase in the ambient wind velocity generally increases the ignition probability as it intensifies the combustion of firebrands. On the other hand, the ignition probability is decreased in a strong wind condition as it prohibits the formation of a flammable gas mixture and its temperature elevation due to an excessive supply of fresh air. However, ignition by firebrands does occur even in a strong wind condition around the corners and backside of terrain and objects where the airflow is stagnant.

External heating refers to radiative and convective heat transfer from the surrounding fire sources to the fuel object, in addition to heating from the deposited firebrands. It reduces the energy required for ignition by the deposited firebrands and intensifies their combustion simultaneously

8.4.2 Probability of ignition

To model the ignition behavior of combustibles under specific external heating, we need to evaluate the 'intensity of heating' and the 'response of combustibles' to it. For example, when we evaluate the time to ignition of combustibles under radiative heating, as discussed in Chapter 7, the radiative heat flux incident on the combustible corresponded to the 'intensity of heating', and the analytical solution of the heat conduction equation corresponded to the 'response of combustibles'. However, physical quantity corresponding to the 'intensity of heating' is still not available when considering firebrand-caused ignitions. This is attributed to the difficulty in generalizing the spatial unevenness of heating associated with the firebrand deposition distribution (Urban et al., 2019a; Hakes et al., 2019; Tao et al., 2021; Bearinger et al., 2021). On the other hand, there is a similar difficulty in identifying the appropriate 'response of combustibles', given that firebrand-caused ignitions often occur in complex firebrand-combustible configurations. There have been several attempts to quantify the time to ignition of combustibles by firebrands (Yin et al., 2014; Wang et al., 2016). However, a general model for firebrand-caused ignition equivalent to the one for radiative-heating-caused ignition is still being sought.

Another issue of firebrand-caused ignitions is that they involve various uncertainties (Hadden et al., 2011; Viegas et al., 2014), as discussed in the previous subsection. It would be more appropriate for practical applications to view firebrand-caused ignitions as probabilistic rather than deterministic events. There are several approaches to modeling probabilistic phenomena. The most basic approach is to abstract the mechanism of the phenomena and theoretically derive the probability distribution that governs the phenomena. For example, binomial distribution can be derived as a discrete probability distribution of the number of successes in a sequence of independent trials. Interestingly, theoretically derived probability distributions are often applicable to cases beyond the originally assumed conditions. For example, the Poisson distribution is derived as a discrete probability distribution that expresses the probability of a given number of events occurring in a fixed time interval. However, the Poisson distribution is widely applied to describe probability distributions with non-negative integers as random variables. In fact, the mechanism of many probabilistic phenomena is often too complex for theoretically deriving the probability distributions, even if they are encountered frequently. In such a case, we need to identify the random variable representing the phenomena and select a known probability distribution suitable for representing the random properties based on observations. The firebrand-caused ignitions fall into this case.

(1) Bernoulli trial

There are two types of probability distributions: discrete and continuous probability distributions. The probability of firebrand-caused ignitions can be regarded as the number of trials in which ignition occurs within a given number of repeated trials. Thus, the random variable of the probability distribution should be an integer greater than or equal to zero. The applicable probability distribution is a discrete probability distribution. Among various discrete probability distributions, the Bernoulli distribution has often been used as the probability distribution for describing the random properties of ignition phenomena. The Bernoulli distribution is a particular form of the binomial distribution, which can be derived from a random trial called the Bernoulli trial.

Let the outcome of a trial be either occurrence or non-occurrence, whose probabilities are θ and $1-\theta$, respectively. Such a trial with only two outcomes, repeated independently, is called the Bernoulli trial. Assuming that n is the number of trials and k is the number of trials in which the event occurs, the probability mass function (PMF) of the binomial distribution can be expressed as follows:

$$\text{Binomial}(k|\theta,n) = {}_nC_k \cdot \theta^k (1-\theta)^{n-k} \quad (k = 0,1,2,\cdots,n) \tag{8.47}$$

where ${}_nC_k$ is the binomial coefficient representing the number of combinations in which an event occurs k times out of n trials, which is given by:

$$_nC_k = \frac{n!}{k!(n-k)!} = \frac{n(n-1)\cdots(n-k+1)}{k!} \tag{8.48}$$

The discrete probability distribution with PMF expressed in the form of Eq. (8.47) is called the binomial distribution. The name 'binomial distribution', comes from the fact that its PMF is the family of positive integers that appear as coefficients in the binomial expansion. Examples of the PMF of the binomial distribution are shown in Figure 8.23. In this figure, histograms are shown for different values of θ under a fixed number of trials of $n = 20$. The PMF is symmetric at $\theta = 0.5$, and becomes asymmetric as the value of θ deviates from this value. The Bernoulli distribution is a particular form of

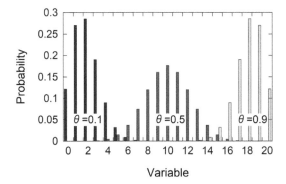

Figure 8.23 Examples of the PMF of the binomial distribution.

the binomial distribution when $n = 1$. The PMF of the Bernoulli distribution can be expressed as follows:

$$\text{Bernoulli}(k|\theta) = \theta^k (1 - \theta)^{1-k} \quad (k = 0,1) \tag{8.49}$$

The occurrence probability of an event, θ, is generally determined either by observations or experiments.

(2) Logistic model

If the occurrence of firebrand-caused ignitions ($k = 1$ for occurrence and $k = 0$ for non-occurrence) follows the Bernoulli distribution, then the model takes the form:

$$k \sim \text{Bernoulli}(\theta) \; (k = 0,1) \tag{8.50}$$

where θ is the ignition ratio (ignition probability). The tilde, ' $\sim$ ', indicates that the left-hand side of the equation follows the probability distribution on the right-hand side. The value of θ is dependent on various factors, such as those listed in the previous subsection. Thus, an additional model is required to express θ in terms of these variables. When using the Bernoulli distribution, the logit function is generally used to formulate θ as a linear function of variables, x_k, which takes the form (Dobson et al., 2008):

$$\text{logit}(\theta) = \log\left(\frac{\theta}{1 - \theta}\right) = \beta_0 + \sum_k (\beta_k x_k) \tag{8.51}$$

where β_0 is the intercept, and β_k are the coefficients. The inverse function of the logit function is called the logistic function, which can be expressed by transforming Eq. (8.51) as follows:

$$\alpha = \log\left(\frac{\theta}{1-\theta}\right) \leftrightarrow \theta = \frac{1}{1+\exp(-\alpha)} = \text{logistic}(\alpha) \qquad (8.52)$$

By using Eq. (8.52), Eq. (8.51) can be transformed as follows:

$$\theta = \text{logistic}\left(\beta_0 + \sum_k (\beta_k x_k)\right) = \frac{1}{1+\exp\left\{-\left[\beta_0 + \sum_k (\beta_k x_k)\right]\right\}} \qquad (8.53)$$

The logistic function is zero when the variable is $-\infty$ and unity when the variable is ∞. This is a feature of the function suitable for representing changes in probability, and is why its inverse function, the logit function, is used to model θ as shown in Eq. (8.51).

Several attempts have been made to formulate θ using the logistic model (Ganteaume et al., 2009; Urban et al., 2019b; Yang et al., 2021). Among these, Urban et al. characterized the capability of single glowing woody firebrands to cause a smoldering ignition in a natural porous fuel bed (coastal redwoods sawdust) in wind tunnel experiments at a constant airflow velocity of $0.5 \text{ m} \cdot \text{s}^{-1}$ (Urban et al., 2019b). Although the ignition phenomenon of natural fuels by firebrands is complex, they focused on investigating the effect of firebrand size, d_{fb}, and fuel moisture content (FMC). Histograms of the measured ignition ratios for different firebrand sizes are shown in Figure 8.24. A multivariate logistic model was fitted to the experimental results. They obtained a smoldering ignition boundary based on the d_{fb} (m) and the FMC (%) as follows:

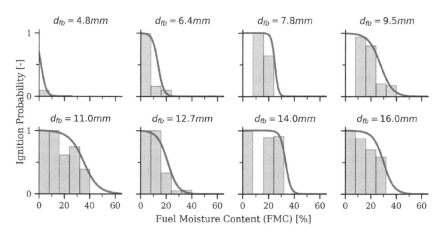

Figure 8.24 Histograms of the observed smoldering ignition ratios for different firebrand sizes (ember diameter, d_{fb}) with the logistic regression curves as a function of FMC (Urban et al., 2019b).

$$p_{SI} = \frac{1}{1 + \exp(-f)} \tag{8.54a}$$

$$f = 5.2591 - 1.186 \times 10^{-4} d_{fb}^{-2} - 1.5528 \times 10^{-1} FMC \tag{8.54b}$$

The regression results show that larger firebrands can ignite sawdust at a relatively high FMC.

Chapter 9

Spatial data modeling

The intensity of heat transferred from a burning fuel object to an unburnt fuel object depends on the positional relationship of the two objects. A small separation distance between the two fuel objects increases the heat transfer rate, resulting in a high probability of fire spread. On the other hand, a large separation distance between the two fuel objects decreases the heat transfer rate, resulting in a low probability of fire spread. Solving such a positional relationship is not that problematic when few fuel objects are involved. For example, if the two fuel objects are directly in front of each other, what we can do is analytically solve the view factor and calculate the radiative heat transfer from the burning fuel object to the unburnt fuel object. However, in large outdoor fires, a considerable number of fuel objects are involved in the fire spread. The shapes of the fuel objects are irregular, and the positional relationships between the fuel objects are complex. It is impractical to obtain all the positional relationships of the involved fuel objects by hand calculations. The use of numerical computations is unavoidable. In this chapter, we discuss the spatial data modeling required for the computational evaluation of fire spread between fuel objects.

The following are some situations in large outdoor fires where the positional relationships between fuel objects are required to evaluate the hazard of fire spread.

- The computation of the radiant heat flux from a burning fuel object (fire source) to another fuel object requires the view factor of the fire source as viewed from the heat-receiving fuel object. The view factor calculation requires the geometrical configuration of the two fuel objects. However, thermal radiation is transferred from the fire source to the heat-receiving object only if the thermal radiation is not obstructed by any other objects between them (Figure 9.1 (a)).
- The range of the convective heating due to a wind-blown fire plume and the spot ignition due to firebrand dispersion depends on the

DOI: 10.1201/9781003096689-9

positional relationships between the fire source and the target fuel objects. However, the positional relationships need to reflect the change in the wind direction and velocity over time (Figure 9.1 (b)).

- Firebrands released from a fire source are transported in a turbulent flow field with random fluctuation. The translation and rotation of a lofted firebrand are determined due to the relative motion of the surrounding airflow to the firebrand. The position and attitude angle of the moving coordinate defined on the flying firebrand need to be tracked to evaluate the trajectory of the firebrand and its final deposition point (Figure 9.1 (c)).

- The geometry of a fuel object and its positional relationships with the other fuel objects can be defined on a coordinate system. The range of firebrand dispersion is often expressed in the form of a probability density distribution with the origin at the location of the fire source. However, the origin of the global coordinate system does not necessarily coincide with the location of the fire source. A new local coordinate system needs to be introduced so that the positions of the fuel objects relative to the fire source can be evaluated (Figure 9.1 (d)).

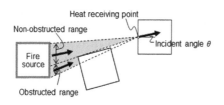

(a) Thermal radiation from a fire source shielded by other objects in the calculation of the view factor.

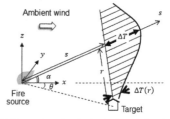

(b) Temperature elevation in the downwind region of a fire plume.

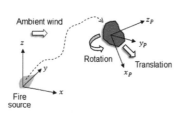

(c) Trajectory of a firebrand released from a fire source in a turbulent flow field.

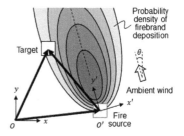

(d) Deposition probability distribution of firebrands released from a fire source.

Figure 9.1 Example situations in outdoor fires that require computation of positional relationships between fuel objects.

9.1 SPATIAL COORDINATE SYSTEM

Fire spread in outdoor fires occurs between fuel objects with miscellaneous geometry and burning characteristics distributed non-homogeneously in a wide range of space. The variation of the positional relationships between fuel objects needs to be expressed using spatial coordinate systems in evaluating the hazard of fire spread between the fuel objects.

9.1.1 3D rectangular coordinate system

A 3D spatial coordinate system is essential for expressing the terrain elevation and the geometry of fuel objects involved in outdoor fires. There are several types of 3D spatial coordinate systems. For generality, it is most straightforward to adopt a rectangular coordinate system, in which the $x-$, $y-$, and $z-$axes are mutually orthogonal. The 3D rectangular coordinate system is also called the 3D Cartesian coordinate system. In the 3D rectangular coordinate system, the $x-$, $y-$, and $z-$axes are in the directions of the thumb, index finger, and middle finger, respectively. The 3D rectangular coordinate system can further be divided into right-handed and left-handed systems. Although there is no technical difference between the two, we need to choose either of the two in advance to ensure consistency in the coordinate computations. In this chapter, we employ the right-handed 3D rectangular coordinate system, as shown in Figure 9.2. Although there is an arbitrariness in defining the direction of each axis, it is conventional to define the $z-$axis in the vertically upward direction and the $x-$axis in the mainstream direction of the ambient airflow, for the analysis of large outdoor fires.

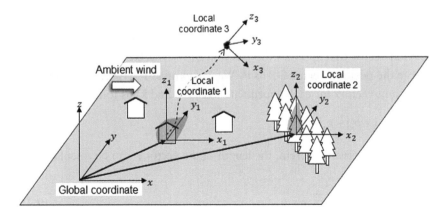

Figure 9.2 3D rectangular coordinate system (right-handed system).

As described in Figure 9.2, the number of coordinate systems introduced into the target space of analysis should not necessarily be one, but multiple coordinate systems can coexist. However, there needs to be a unique coordinate system that can be referenced by other coordinate systems. Such a coordinate system is called the global coordinate system. The local coordinate system is a subordinate coordinate system defined on the global coordinate system. There is no restriction on the number of local coordinate systems introduced into the space. However, the introduction of a local coordinate system is not mandatory. Even if there are multiple fuel objects in the space, all the positional relationships between fuel objects can be expressed solely by the global coordinate system. However, when evaluating the thermal effects from one burning fuel object on another fuel object, the global position of the heat-receiving fuel object needs to be converted relative to the burning fuel object in any case. This is nothing but introducing a local coordinate system with origin defined on the burning fuel object. Examples are shown in Figure 9.2 as local coordinate systems 1 and 2, separately defined on two different fire sources. Another local coordinate system (local coordinate system 3) is defined on one of the firebrands released from the fire source. This is a further subordinate to local coordinate system 1. Such a coordinate system is introduced when the local coordinate system (i.e., a lofted firebrand in this case) involves a translational and rotational motion, while requiring a reference to the fire source. Thus, we can define subordinate local coordinate systems under a local coordinate system.

9.1.2 Vector

A vector is a directed line segment with size and direction. This is a basic means of expressing the positional relationship between objects. Assume a point P located in 3D space, as shown in Figure 9.3. A directed line segment $\overrightarrow{OP}$, which is extended from the origin $O(0,0,0)$ to point $P(p_x, p_y, p_z)$, is a position vector that specifies the position of the P relative to the O. Let us denote the position vector $\overrightarrow{OP}$ as $\mathbf{P}$. We can express $\mathbf{P} = (p_x, p_y, p_z)$ in terms of the unit vectors for each coordinate axis, $\mathbf{i}$, $\mathbf{j}$, and $\mathbf{k}$, as follows:

$$\mathbf{P} = p_x \mathbf{i} + p_y \mathbf{j} + p_z \mathbf{k} \tag{9.1}$$

By the Pythagorean theorem, the length of the vector $\mathbf{P}$ can be calculated as follows:

$$|\mathbf{P}| = \sqrt{p_x^2 + p_y^2 + p_z^2} \tag{9.2}$$

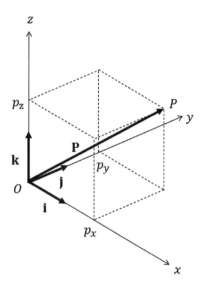

Figure 9.3 Position vector defined in the right-handed 3D rectangular coordinate system.

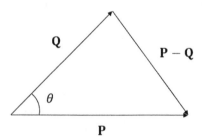

Figure 9.4 Two vectors and the angle between them.

(1) Inner product

The inner product (scalar product) of two vectors is one of the most frequently used operations to evaluate the positional relationship between fuel objects. It can be converted to the angle of the two vectors, θ. The inner product of two vectors, $\mathbf{P} = (p_x, p_y, p_z)$ and $\mathbf{Q} = (q_x, q_y, q_z)$, is defined as the sum of the products of each component as follows (Figure 9.4):

$$\mathbf{P} \cdot \mathbf{Q} = p_x q_x + p_y q_y + p_z q_z \tag{9.3}$$

It is noteworthy that the inner product takes a scalar quantity. The inner product of two vectors can be expressed in an alternative form:

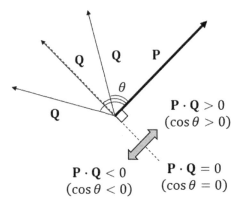

Figure 9.5 Determination of positional relationship of two positional vectors **P** and **Q** based on the sign of the inner product.

$$\mathbf{P}\cdot\mathbf{Q} = |\mathbf{P}||\mathbf{Q}|\cos\theta \tag{9.4}$$

where θ is the angle of the two vectors, **P** and **Q**. By equating Eqs. (9.3) and (9.4), an expression for $\cos\theta$ can be derived as follows:

$$\cos\theta = \frac{\mathbf{P}\cdot\mathbf{Q}}{|\mathbf{P}||\mathbf{Q}|} = \frac{p_x q_x + p_y q_y + p_z q_z}{\sqrt{p_x^2 + p_y^2 + p_z^2}\sqrt{q_x^2 + q_y^2 + q_z^2}} \tag{9.5}$$

One of the features of the inner product can be used to describe another positional relationship between two positional vectors **P** and **Q**, as shown in Figure 9.5. In Figure 9.5, **Q** takes three different angles, θ, relative to **P**. According to the definition of the inner product in Eq. (9.4), the sign of the $\mathbf{P}\cdot\mathbf{Q}$ corresponds solely to the sign of $\cos\theta$. Conversely, the sign of $\cos\theta$ can be determined from the $\mathbf{P}\cdot\mathbf{Q}$ without identifying the value of $\cos\theta$. If $\mathbf{P}\cdot\mathbf{Q} = 0$, then **P** and **Q** are perpendicular to each other ($\cos\theta = 0$). If $\mathbf{P}\cdot\mathbf{Q} > 0$, then **P** and **Q** form an obtuse angle ($\cos\theta > 0$). If $\mathbf{P}\cdot\mathbf{Q} < 0$, then **P** and **Q** form an acute angle ($\cos\theta < 0$). Such a feature of the inner product is useful when computationally judging the visibility of fuel objects from a specific fuel object (refer to Worked example 9.1).

(2) Outer product

The outer product (vector product or cross product) of two vectors is a vector perpendicular to the two vectors, with the length equivalent to the area of the parallelogram bounded by the two vectors. Assuming two positional

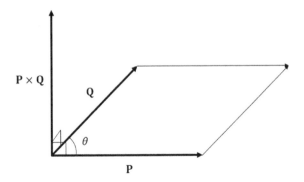

Figure 9.6 Direction and size of the outer product of two vectors.

vectors, $P = \left(p_x, p_y, p_z\right)$ and $Q = \left(q_x, q_y, q_z\right)$, as shown in Figure 9.6, the outer product, $P \times Q$, can be given by:

$$P \times Q = \left(p_y q_z - p_z q_y, p_z q_x - p_x q_z, p_x q_y - p_y q_x\right) \qquad (9.6)$$

In contrast to the inner product that takes a scalar quantity, the outer product takes a vector quantity. Using the unit vectors of each coordinate axis **i, j,** and **k**, the outer product $P \times Q$ can be expressed in an alternative form:

$$P \times Q = \begin{vmatrix} i & j & k \\ p_x & p_y & p_z \\ q_x & q_y & q_z \end{vmatrix} = \left(p_y q_z - p_z q_y\right)i + \left(p_z q_x - p_x q_z\right)j + \left(p_x q_y - p_y q_x\right)k$$

$$(9.7)$$

If the two vectors, **P** and **Q**, are in the direction of the thumb and index finger in the right-handed rectangular coordinate system, respectively, the $P \times Q$ is in the direction of the middle finger. However, note that by definition, $P \times Q$ and $Q \times P$ are the same in length but opposite in direction. Given that the size of the outer product, $P \times Q$, is the area of the parallelogram bounded by the two positional vectors, **P** and **Q**, the absolute value of the outer product $P \times Q$ can be geometrically determined as follows:

$$|P \times Q| = |P||Q|\sin\theta \qquad (9.8)$$

Thus, the unit vector normal to the surface containing the two vectors, **P** and **Q**, can be given by:

$$\mathbf{u}_N = \frac{\mathbf{P} \times \mathbf{Q}}{|\mathbf{P} \times \mathbf{Q}|} \tag{9.9}$$

WORKED EXAMPLE 9.1

Let us assume that four cubic objects are located close together, as shown in Figure 9.7. The cubic object O is at a high-temperature condition and emitting thermal radiation to the other cubic objects A, B, and C. Select cubic objects that receive thermal radiation from one face of the cubic object O (face I).

SUGGESTED SOLUTION

When the normal vector of the face I is given by $\mathbf{P}$, the direction of the relative vector originating from the face I targeting another fuel object, $\mathbf{Q}$, can be determined by the sign of the inner product of the two vectors, $\mathbf{P} \cdot \mathbf{Q}$. If $\mathbf{P} \cdot \mathbf{Q} > 0$, the two vectors $\mathbf{P}$ and $\mathbf{Q}$ are on the same side with respect to face I. On the other hand, if $\mathbf{P} \cdot \mathbf{Q} < 0$, the two vectors $\mathbf{P}$ and $\mathbf{Q}$ are on the opposite side with respect to face I. Assume that $\mathbf{Q}_A$, $\mathbf{Q}_B$, and $\mathbf{Q}_C$ are the positional vectors of the gravitational center of the cubic objects A, B, and C relative to face I, respectively. For the layout of cubic objects in Figure 9.7, the inner products of these relative vectors, $\mathbf{Q}_A$, $\mathbf{Q}_B$, and $\mathbf{Q}_C$, and the normal vector of face I, $\mathbf{P}$, are $\mathbf{P} \cdot \mathbf{Q}_A < 0$, $\mathbf{P} \cdot \mathbf{Q}_B > 0$, and $\mathbf{P} \cdot \mathbf{Q}_C > 0$. Thus, the cubic objects B and C receive radiative heat from face I.

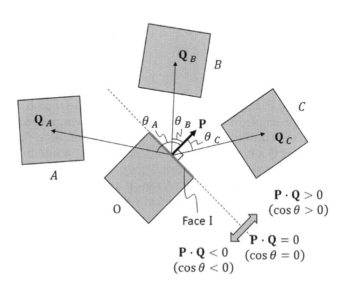

Figure 9.7 Effective range of thermal radiation from a hot cubic object.

9.2 DESCRIPTION OF A 3D OBJECT

In this section, we discuss the description models of a 3D object. The wireframe, surface, and solid models are common description models developed in the field of computer graphics (Shirley et al., 2009). The wireframe model represents a 3D object by ridge lines connecting the vertices on its surfaces possessing no surface information. The amount of data required to describe a 3D object is the lowest among the three models, enabling its use in a computational environment that does not have high image-processing capabilities. The surface model represents a 3D object by adding surface information to the wireframe model. In other words, an object is represented by a polyhedron comprising multiple polygonal surfaces, which is suitable for creating complex surface geometries. As the model only represents the surfaces of the object, it has no information on internal structure. The solid model represents a 3D object with information on the internal structure. Thus, the geometric information of an object, such as the volume, center of gravity, and weight, can be calculated from the solid model data. On the other hand, it is suitable for representing simple geometries such as cubes and cylinders, rather than complex geometries.

In this section, we focus on the surface model to represent 3D objects involved in outdoor fires, given its compatibility with heat transfer computations. We also discuss the multi-point model, which can be considered as a simplification of the wireframe model.

9.2.1 Surface model

In outdoor fires, the heat transfer rate from a burning object to an unburnt object cannot be represented by a single value. It varies from part to part over the surface of the heat-receiving object, depending on the positional relationship between the two objects. The ignition characteristics of the heat receiving object may also vary from part to part over the surface, depending on the material type and shape. If such spatial variations in the external heating and the ignition characteristics over the surface of the heat-receiving object are neglected, the accuracy of fire hazard evaluations is diminished. These spatial variations can effectively be incorporated into the surface model as the boundary condition and the attribute of each constituent polygon.

As a simple example of an object represented by the surface model, Figure 9.8 shows a cube, described as a combination of six square polygons. However, the definition of a polygon is not complete by itself. An arbitrary polygon can be specified by defining the component edges as its subordinate structure. Further, a component edge can be specified by defining the component vertices as its subordinate structure. In other words, a 3D object can be represented by defining the hierarchical structure of the component polygons, edges, and vertices.

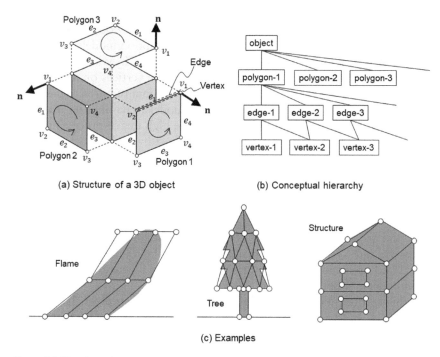

(a) Structure of a 3D object   (b) Conceptual hierarchy

(c) Examples

Figure 9.8 Topological representation and examples of a 3D object by the surface model. The presented configuration of vertices and edges are not necessarily the correct representations of actual objects.

In defining the component surface polygons of a 3D object, we need to identify which side of the object, inside (back side) or outside (front side), each surface polygon faces. The notion of the sides is important for distinguishing objects from surrounding objects and assessing the positional relationship between polygons. Although this may be obvious to our human senses, numerical procedures require a clear statement of the judgment criteria. For example, in the case of a polyhedron receiving radiant heat from a fire source, the affected surface polygons should be limited to those with the fire source on the front side, whereas those with the fire source on the back side should be shielded from radiant heat. In order to systematically determine the front and back sides of surface polygons, we need to define the sequence of vertices that comprise the surface polygons in a consistent manner. In Figure 9.8, the vertices are aligned clockwise when viewed from the outside (front side) and counterclockwise when viewed from the inside (back side).

The consistent sequence of vertices is also used to compute the normal vector of the surface polygons. Suppose there is an N-sided polygon with the vertices represented by the vectors $v_i (i = 1,2,\cdots,N)$. The reference point, v_1, and its adjacent vertices, v_2 and v_N, form two connected edges, $e_1 = v_2 - v_1$ and $e_N = v_N - v_1$ (refer to Figure 9.8). Following Eq. (9.6), the normal vector of the polygon can be obtained as the outer product of the two edges e_1 and e_N, which is $n = e_1 \times e_N$. If the serial number of the vertices i is given clockwise when viewed from the outside of the surface polygon, the normal vector thus obtained, n, always faces outside the object. This feature of polygons can be used to compute the positional relationship between objects numerically.

As shown in Figure 9.8, the surface model can describe various objects that may be involved in outdoor fires. These include flames formed above burning combustibles, plants, and structures. However, a detailed representation of these geometries may accompany an increase in the complexity of the computational procedure. Introducing the surface model into the spatial data modeling is not for rendering objects as if they were real objects, but for effectively representing the positional relationships between objects for assessing the hazard of fire spread. Therefore, the detailedness of the geometry should be accommodated by weighing the computational load against the capability to reasonably assess the hazard of fire spread. For this reason, the configuration of the vertices and edges of the exemplified surface models in Figure 9.8 are not necessarily appropriate representations for each object, but the detailedness can vary depending on its application.

WORKED EXAMPLE 9.2

A cube with a square base of 8 m × 8 m and a height of 8 m is defined in the 3D rectangular coordinate system xyz, as shown in Figure 9.9. One of the eight vertices that comprise the cube is located at the origin of the coordinate system, which is $(0,0,0)$. The 12 edges of the cube are parallel to either of the axes of the coordinate system. Define the cube as the surface model.

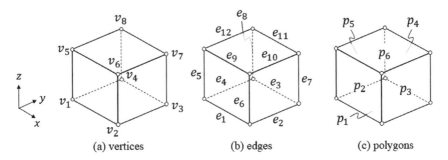

(a) vertices   (b) edges   (c) polygons

Figure 9.9 A cube defined in the right-handed rectangular coordinate system z.

SUGGESTED SOLUTION

The cube consists of six surface polygons, 12 edges, and eight vertices. The surface polygons are defined by the edges and vertices, and the edges are further defined by the vertices. Among this hierarchical data structure, only the vertices at the lowest hierarchical level possess the coordinate values.

Among the eight vertices, let us assume that the vertex v_1 is on the origin $(0,0,0)$. For all the edges to be parallel to either of the axes, the coordinates of the remaining seven vertices can be given as follows:

$$v_2(8,0,0),\ v_3(8,8,0),\ v_4(0,8,0),\ v_5(0,0,8),\ v_6(8,0,8),$$
$$v_7(8,8,8),\ \text{and}\ v_8(0,8,8).$$

Let us denote the vertices comprising the edges and surface polygons with parentheses, $\{\ \}$. With this form of expression, the 12 edges of the cube can be represented as follows:

$$e_1 = \{v_1,v_2\}, e_2 = \{v_2,v_3\}, e_3 = \{v_3,v_4\}, e_4 = \{v_4,v_1\},$$

$$e_5 = \{v_1,v_5\}, e_6 = \{v_2,v_6\}, e_7 = \{v_3,v_7\}, e_8 = \{v_4,v_8\},$$

$$e_9 = \{v_5,v_6\}, e_{10} = \{v_6,v_7\}, e_{11} = \{v_7,v_8\},\ \text{and}\ e_{12} = \{v_8,v_5\}.$$

Each surface polygon of the cube can be defined not only as a set of its component edges, but also as a set of its component vertices. For the latter case, the six polygons of the cube can be specified as follows:

$$p_1 = \{v_4,v_3,v_2,v_1\}, p_2 = \{v_1,v_2,v_3,v_4\}, p_3 = \{v_2,v_3,v_7,v_6\},$$

$$p_4 = \{v_3,v_4,v_8,v_7\}, p_5 = \{v_4,v_1,v_5,v_8\},\ \text{and}\ p_6 = \{v_5,v_6,v_7,v_8\}.$$

Note that the sequence of vertices that comprises each polygon should be clockwise when viewed from the outside (front side), so that the normal vector always points outside the cube.

9.2.2 Multi-point model

The wireframe model is a surface model without surface information, which has the advantage of a simple data structure. There is little reason to use the wireframe model if there is no constraint on the computational capacity, as the surface model has upward compatibility. However, some objects involved in outdoor fires have complex geometries that may not be suitable for representation by the surface model from the viewpoint of computational load. As mentioned earlier, our purpose in representing 3D objects with the description models is not to render objects realistically.

Thus, the use of a simplified description model can be justified as long as it can effectively represent information necessary for evaluating the hazard of fire spread. We introduce a model similar to the wireframe model, which can be described as the multi-point model.

Figure 9.10 shows a topological representation of a 3D object by the multi-point model. The multi-point model represents the object by a sequence of multiple points. Its topological structure is simpler than the surface model described in the previous subsection. The features of the multi-point model can be demonstrated by comparing it with the surface model. In the surface model, the properties and status of an object are represented by each component polygon. The geometry of each component polygon is defined by component edges and vertices. Meanwhile, in the multi-point model, component points do not only determine the geometry of an object, but also possess information on the properties and status of the object as the attributes. In addition, in the surface model, it is required to discern the two sides (front and back) of each component polygon as these polygons are integrated to form a closed volume representing the 3D geometry. Whereas in the multi-point model, such a step of data processing can be omitted. In general, the component points are assumed to have a homogeneous property and status, and thus, do not have directionality.

Flames are a good example of objects that can be represented by the multi-point model. We have described the concept of the multi-point model

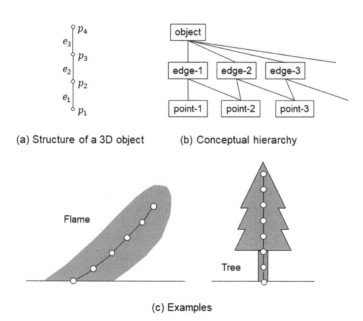

(a) Structure of a 3D object   (b) Conceptual hierarchy

(c) Examples

Figure 9.10 Topological representation and examples of a 3D object by the multi-point model.

in terms of the flame radiative property in Chapter 3. The radiative heat transfer from a flame to the surrounding objects is often calculated by regarding the flame as a sheet or a solid with a homogeneous radiative property (Dayan et al., 1974; Shokri et al., 1989; Shen et al., 2019). The application of such flame models has certain rationality as the temperature within the continuous regime of diffusion flames can be considered constant (refer to Chapter 5). However, this is a simplification of the flame radiative property, which does not account for a variation in the optical length within the flame regime. Representation of the flame radiative property with a serial array of points along the central axis is an effective means to supplement the limitations of the sheet and solid models (Hankinson et al., 2012; Zhou et al., 2017).

The multi-point model can also represent fuel objects that receive external heating. One of its advantages is its capability of easily reflecting the variation of ignition characteristics among different portions of the fuel object. On the other hand, it is not suitable for representing object geometry in as much detail as the surface model. Rather, it is the model for representing object geometry in a simplified manner to reduce the computational complexity. A serial array of points can be used to calculate the radiative heat exchange between the other serial arrays of points in a simple procedure. However, there is a structural difficulty in representing the depth of fuel objects with a serial array of points, which hampers the evaluation of the radiation shielding effect between fuel objects. The applicability of the multi-point model to fuel objects is poorer than to flames, unless fuel objects are sparsely distributed and thus have a minor radiation shielding effect.

9.3 GEOMETRIC COMPUTATIONS IN 3D SPACE

By introducing the description models discussed in the previous section, objects in 3D space can be represented as a set of geometrical features, including points, line segments, and polygons. Even though the configuration of such geometrical features and their positional relationships are obvious to our human senses, they need to be explicitly specified in the description models, which often causes complications in the computational procedure. In this section, we discuss some basic computational procedures for solving geometric problems often encountered in object-to-object heat exchange calculations.

9.3.1 Lines

Assume a line defined by two points, P_1 and P_2. An arbitrary point on this line, P, can be expressed in the following form:

$$P = (1-t)P_1 + tP_2 \qquad (9.10)$$

where t is a real number. If the point $\mathbf{P}$ is on the line segment connecting the two points, $\mathbf{P}_1$ and $\mathbf{P}_2$, t takes the value $0 \le t \le 1$. Eq. (9.10) can be transformed into a different form as:

$$\mathbf{P} = \mathbf{P}_1 + t\left(\mathbf{P}_2 - \mathbf{P}_1\right) \tag{9.11}$$

In this form, $\mathbf{P}$ can be viewed as a point on a straight line starting from $\mathbf{P}_1$ and parallel to the direction vector, $\mathbf{P}_2 - \mathbf{P}_1$. When $t > 0, \mathbf{P}$ is on the same side of $\mathbf{P}_2 - \mathbf{P}_1$ with respect to $\mathbf{P}_1$. When $t < 0$, $\mathbf{P}$ is on the opposite side of $\mathbf{P}_2 - \mathbf{P}_1$ with respect to $\mathbf{P}_1$.

(1) Distance between a point and a line

Consider a line passing through two points, $\mathbf{O}$ and $\mathbf{P}$, and another point separated from the line, $\mathbf{Q}$, as shown in Figure 9.11. The distance from the point to the line, d, can be calculated as the distance from $\mathbf{Q}$ to the foot of the perpendicular to the line, $\mathbf{F}$. To simplify the notation, let $\mathbf{u}$ be the unit vector of the vector, $\mathbf{P} - \mathbf{O} = \mathbf{P}$. As the inner product is zero if the two vectors are orthogonal (cf. Figure 9.5):

$$\left(\mathbf{F} - \mathbf{Q}\right) \cdot \mathbf{u} = 0 \tag{9.12}$$

However, as the foot of the perpendicular is on the line, $\mathbf{F}$ can be expressed with a real number, t, as follows:

$$\mathbf{F} = t\mathbf{u} \tag{9.13}$$

By substituting Eq. (9.13) into Eq. (9.12), t can be obtained as follows:

$$t = \mathbf{Q} \cdot \mathbf{u} \tag{9.14}$$

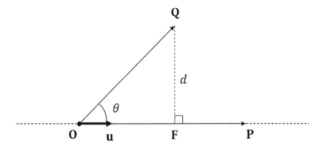

Figure 9.11 Distance between a point and a line.

Referring to the definition of the inner product in Eq. (9.4), the distance, d, can be obtained as follows:

$$d = \sqrt{(F-Q)\cdot(F-Q)} = \sqrt{t^2 - 2t(Q\cdot u)+Q\cdot Q} = \sqrt{Q\cdot Q-(Q\cdot u)^2} \quad (9.15)$$

Note that in this equation, $\mathbf{u}\cdot\mathbf{u} = 1$ as $|\mathbf{u}| = 1$.

(2) Distance between lines

In 2D space, two lines intersect at a point unless they are parallel. In 3D space, there are cases where two lines are neither parallel nor intersect, but are in a twisted position. Assume two lines, $\mathbf{P}(s)$ and $\mathbf{Q}(t)$, as shown in Figure 9.12. They can take the form:

$$\mathbf{P}(s) = \mathbf{P}_0 + s\mathbf{u}_p \tag{9.16a}$$

$$\mathbf{Q}(t) = \mathbf{Q}_0 + t\mathbf{u}_q \tag{9.16b}$$

where $\mathbf{P}_0$ and $\mathbf{Q}_0$ are the points on each line, $\mathbf{u}_p$ and $\mathbf{u}_q$ are the unit vectors of each line, and s and t are real numbers. The distance between the points on each line, $\mathbf{P}(s)$ and $\mathbf{Q}(t)$, can be calculated as:

$$\begin{aligned} d^2 &= |\mathbf{P}(s)-\mathbf{Q}(t)|^2 \\ &= |\mathbf{P}(s)|^2 - 2\mathbf{P}(s)\cdot\mathbf{Q}(t)+|\mathbf{Q}(t)|^2 \\ &= \left(|\mathbf{P}_0|^2 + 2s\mathbf{P}_0\cdot\mathbf{u}_p + s^2\right) - 2\left(\mathbf{P}_0\cdot\mathbf{Q}_0 + s\mathbf{Q}_0\cdot\mathbf{u}_p + t\mathbf{P}_0\cdot\mathbf{u}_q + st\mathbf{u}_p\cdot\mathbf{u}_q\right) \\ &\quad + \left(|\mathbf{Q}_0|^2 + 2t\mathbf{Q}_0\cdot\mathbf{u}_q + t^2\right) \end{aligned} \tag{9.17}$$

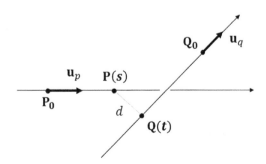

Figure 9.12 Two lines in a twisted position.

In Eq. (9.17), s and t can be regarded as the variables to determine the distance, d. The shortest distance between the two lines can be obtained with the combination of s and t for which d^2 in Eq. (9.17) yields the minimum value. This can be reduced to a problem letting the partial derivatives of d^2 with respect to s and t be zero:

$$\frac{\partial d^2}{\partial s} = 2P_0 \cdot u_p + 2s - 2Q_0 \cdot u_p - 2t u_p \cdot u_q = 0 \tag{9.18a}$$

$$\frac{\partial d^2}{\partial t} = 2Q_0 \cdot u_q + 2t - 2P_0 \cdot u_q - 2s u_p \cdot u_q = 0 \tag{9.18b}$$

Eq. (9.18) can be expressed as a matrix as follows:

$$\begin{bmatrix} 1 & -u_p \cdot u_q \\ -u_p \cdot u_q & 1 \end{bmatrix} \begin{bmatrix} s \\ t \end{bmatrix} = \begin{bmatrix} (Q_0 - P_0) \cdot u_p \\ (P_0 - Q_0) \cdot u_q \end{bmatrix} \tag{9.19}$$

By solving this matrix, the variables s and t can be obtained as:

$$\begin{bmatrix} s \\ t \end{bmatrix} = \begin{bmatrix} 1 & -u_p \cdot u_q \\ -u_p \cdot u_q & 1 \end{bmatrix}^{-1} \begin{bmatrix} (Q_0 - P_0) \cdot u_p \\ (P_0 - Q_0) \cdot u_q \end{bmatrix}$$

$$= \frac{1}{1 - |u_p \cdot u_q|^2} \begin{bmatrix} 1 & u_p \cdot u_q \\ u_p \cdot u_q & 1 \end{bmatrix} \begin{bmatrix} (Q_0 - P_0) \cdot u_p \\ (P_0 - Q_0) \cdot u_q \end{bmatrix} \tag{9.20}$$

The obtained values of s and t can be substituted into Eq. (9.17) to yield d^2.

WORKED EXAMPLE 9.3

Assume a burning object located at the origin $O = (0,0,0)$ of the rectangular coordinate system xyz as shown in Figure 9.13. The fire plume is tilted due to the ambient wind in the direction parallel to the x-axis. The fire plume trajectory is represented by a straight line for simplicity, given by $z = 0.5x$. There is another object at $Q = (20,-10,0)(m)$ exposed to the convective heating of the wind-blown fire plume. If the foot of the perpendicular from this heat-receiving object Q to the fire plume trajectory is given by F, calculate the distances $|O - F|$ and $|Q - F|$.

SUGGESTED SOLUTION

The vector F can be expressed in a coordinate form:

$$F = (x, 0, 0.5x)$$

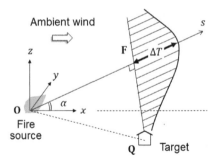

Figure 9.13 Distance from the trajectory of a wind-blown fire plume to a target.

where x is a real variable. With the given coordinates of the target $\mathbf{Q}$, the relative vector $\mathbf{Q} - \mathbf{F}$ can be obtained as follows:

$$\mathbf{Q} - \mathbf{F} = (x - 20, 10, 0.5x)$$

Given that $\mathbf{F} \perp \mathbf{Q} - \mathbf{F}$, the inner product of the two vectors is zero:

$$\mathbf{F} \cdot (\mathbf{Q} - \mathbf{F}) = x(x - 20) + (0.5x)^2 = 0$$

Solving this equation, we obtain $x = 16\,\mathrm{m}$. Thus, the distances $|\mathbf{O} - \mathbf{F}|$ and $|\mathbf{Q} - \mathbf{F}|$ are as follows:

$$|\mathbf{O} - \mathbf{F}| = \sqrt{16^2 + 8^2} = 17.9 \ \mathrm{m}$$

$$|\mathbf{Q} - \mathbf{F}| = \sqrt{4^2 + 10^2 + 8^2} = 13.4 \ \mathrm{m}$$

WORKED EXAMPLE 9.4

Assume a flame tilted in a crosswind, as shown in Figure 9.14. The HRR is $\dot{Q} = 200\,\mathrm{kW}$, the radiative fraction is $\chi_R = 0.3$, the flame length is $L_{fl} = 1.5\,\mathrm{m}$, and the flame tilt angle is $\theta = \pi/4$ rad. Calculate the radiative heat flux incident on a point 2 m downwind from the center of the flame base on the ground. However, use the following equation for calculating the radiative heat flux, $\dot{q}''$, as discussed in Chapter 3:

$$\dot{q}'' = \frac{\chi_R \dot{Q}}{4\pi r^2} \cos\phi$$

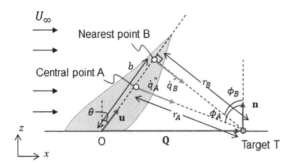

Figure 9.14 Radiative heat transfer from representative point sources.

SUGGESTED SOLUTION

To use the given heat flux equation, we need to identify the representative point of thermal radiation. In this example, we choose two points in the flame for comparison. One is the central point of the flame axis, which is ordinarily used in radiative heat transfer calculations (point A in Figure 9.14). The other is the nearest point on the flame axis from the target point (point B in Figure 9.14).

First, we consider the case with point A as the representative point. Let the center of the flame base be the origin O, the coordinates of the representative point are A $(0.530, 0.530)$, and those of the target are T $(2, 0)$. Thus, we obtain the direct distance between the two points A and T:

$$r_A = \sqrt{(0.530 - 2)^2 + 0.530^2} = 1.56 \text{ m}$$

The incidence angle, $\cos\phi$, can be calculated using Eq. (9.5). By substituting the relative vector of the central point, $\overrightarrow{TA} = \mathbf{A} = (-1.47, 0.530)$, and the normal vector of the ground, $\mathbf{n} = (0,1)$, into Eq. (9.5), we obtain:

$$\cos\phi = \frac{\mathbf{A} \cdot \mathbf{n}}{|\mathbf{A}||\mathbf{n}|} = \frac{-1.47 \times 0 + 0.530 \times 1}{\sqrt{(-1.47)^2 + 0.530^2} \times 1} = 0.339$$

From the given equation, we obtain the incident heat flux, $\dot{q}''$, as follows:

$$\dot{q}'' = \frac{\chi_R \dot{Q}}{4\pi r^2} \cos\phi = \frac{0.3 \times 200}{4 \times 3.14 \times 1.56^2} \times 0.339 = 0.665 \text{ kW} \cdot \text{m}^{-2}$$

Second, we consider the case with point B as the representative point. We use Eq. (9.14) for calculating the distance from the origin to the representative point, b. Substituting the relative vector of the target, $\mathbf{Q} = (2,0)$, and the unit vector of the flame axis, $\mathbf{u} = \dfrac{1}{\sqrt{0.530^2 + 0.530^2}}(0.530, 0.530) = (0.707, 0.707)$, into Eq. (9.14), we obtain:

$$b = \mathbf{Q} \cdot \mathbf{u} = q_x u_x + q_z u_z = 1.41 \text{ m}$$

As $b < L_{fl}$, the representative point is within the flame. Given that the coordinates of the representative point are B $(bu_x, bu_z) = (0.997, 0.997)$, the nearest distance from the flame to the target, r_B, can be calculated using the Pythagorean theorem:

$$r_B = \sqrt{\mathbf{Q} \cdot \mathbf{Q} - (\mathbf{Q} \cdot \mathbf{u})^2} = \sqrt{(0.997 - 2)^2 + (0.997 - 0)^2} = 1.41 \text{ m}$$

As the tilt angle of the flame is $\theta = \pi / 4$ rad, the triangle OBT is an isosceles right triangle. Note that r_B can be alternatively calculated from the following equation:

$$r_B = \sqrt{(q_x^2 + q_z^2) - (q_x u_x + q_z u_z)^2} = \sqrt{(2^2 + 0^2) - (2 \times 0.707 + 0 \times 0.707)^2}$$
$$= 1.41 \text{ m}$$

The incidence angle, $\cos\phi$, can be obtained by substituting the relative vector of the nearest point, $\overrightarrow{TB} = \mathbf{B} = (-1.00, 0.997)$, and the normal vector of the ground, $\mathbf{n} = (0, 1)$, into Eq. (9.5) as follows:

$$\cos\phi = \frac{\mathbf{B} \cdot \mathbf{n}}{|\mathbf{B}||\mathbf{n}|} = \frac{-1.00 \times 0 + 0.997 \times 1}{\sqrt{(-1.00)^2 + 0.997^2} \times 1} = 0.705$$

From the given equation, we obtain the incident heat flux, $\dot{q}''$, as follows:

$$\dot{q}'' = \frac{\chi_R \dot{Q}}{4\pi r^2}\cos\phi = \frac{0.3 \times 200}{4 \times 3.14 \times 1.41^2} \times 0.705 = 1.68 \text{ kW} \cdot \text{m}^{-2}$$

There is a large difference in $\dot{q}''$ between the two cases tested. This is mainly attributed to the difference in $\cos\phi$.

9.3.2 Plane surfaces

A polygon in 3D space can be defined by specifying a point on the plane surface, $\mathbf{P}_0 = (x_0, y_0, z_0)$, and a vector normal to the plane surface, $\mathbf{N} = (a, b, c)$, as shown in Figure 9.15. By definition, the line connecting a point on the plane surface, $\mathbf{P} = (x, y, z)$, and the reference point, $\mathbf{P}_0$, is perpendicular to the normal vector, $\mathbf{N}$. Thus, the inner product of the two vectors, $\mathbf{N}$ and $\mathbf{P} - \mathbf{P}_0$, is zero:

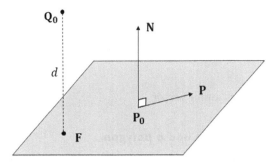

Figure 9.15 A plane surface normal to the vector **N** in 3D space.

$$N \cdot (P - P_0) = 0 \qquad (9.21)$$

Substituting the coordinates into Eq. (9.21), we obtain the following equation describing the plane surface:

$$ax + by + cz - (ax_0 + by_0 + cz_0) = 0 \qquad (9.22)$$

Note that the difference between Eqs. (9.21) and (9.22) is merely in their means of notations, either with vectors or components.

(1) Distance between a point and a plane surface

As shown in Figure 9.15, consider a point Q_0, which is not on the plane surface defined by Eq. (9.21). The distance from Q_0 to the plane surface, d, is nothing but the distance from Q_0 to the foot of the perpendicular on the plane surface, **F**, which can be expressed as:

$$d = \sqrt{|Q_0 - F|^2} \qquad (9.23)$$

The line connecting the two points, Q_0 and **F**, is parallel to the normal vector of the plane surface, **N**. Thus, **F** as viewed from Q_0 can be expressed as:

$$F = Q_0 + tN \qquad (9.24)$$

where t is a real number. As **F** is on the plane surface, we can substitute Eq. (9.24) into Eq. (9.21):

$$N \cdot \{(Q_0 + tN) - P_0\} = 0 \qquad (9.25)$$

By solving this equation, the real number variable, t, can be obtained as follows:

$$t = \frac{\mathbf{N} \cdot (\mathbf{P}_0 - \mathbf{Q}_0)}{|\mathbf{N}|^2} \tag{9.26}$$

Note that $d = t$ if $\mathbf{N}$ is a unit vector.

(2) Intersection of a line and a polygon

Let us assume that a line and a plane surface represent a ray of thermal radiation from a fire source and a surface of a heat receiving fuel object, respectively. A line and an infinite plane surface intersect at some point unless they are parallel. However, as a surface of a heat receiving fuel object has a finite domain, we need to judge whether the intersection point is included in the domain on the plane surface for evaluating the effective range of radiative heat transfer from a fire source. Recall that a line that passes a point, $\mathbf{Q}_0$, and that is in parallel to a unit vector, $\mathbf{u}_q$, can be expressed as follows:

$$\mathbf{Q}(t) = \mathbf{Q}_0 + t\mathbf{u}_q \tag{9.27}$$

where t is a real number. If the plane surface is defined by Eq. (9.21), we can substitute $\mathbf{P}$ in Eq. (9.21) with $\mathbf{Q}(t)$ in Eq. (9.27), which yields:

$$\mathbf{N} \cdot \left\{ (\mathbf{Q}_0 + t\mathbf{u}_q) - \mathbf{P}_0 \right\} = 0 \tag{9.28}$$

By solving this equation, we obtain the real number, t, at the intersection on the plane surface:

$$t = \frac{\mathbf{N} \cdot (\mathbf{P}_0 - \mathbf{Q}_0)}{\mathbf{N} \cdot \mathbf{u}_q} \tag{9.29}$$

Substituting this into Eq. (9.27), the coordinates of the intersection point can be determined. The calculation procedure shown in Eq. (9.27) through Eq. (9.29) is almost the same as the one for calculating the distance between a point and a plane surface shown in Eq. (9.24) through Eq. (9.26), but with a difference in the direction vector of the line, either $\mathbf{u}_q$ or $\mathbf{N}$.

The above calculation procedure determines the intersection point on the plane surface. However, this is not necessarily included in a concerning domain (polygon) of the plane surface. To judge whether the intersection point is included in a polygon, we divide the polygon into multiple triangles and individually judge the inclusion in each triangle. As shown in Figure 9.16,

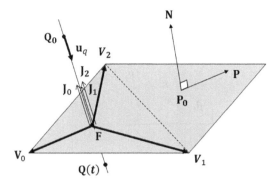

Figure 9.16 Intersection of a line and a polygon on a plane surface.

we can define three outer products of relative vectors (the triangle vertices, V_0, V_1, and V_2, relative to the intersection point, F) for each triangle:

$$J_0 = (V_0 - F) \times (V_1 - F) \tag{9.30a}$$

$$J_1 = (V_1 - F) \times (V_2 - F) \tag{9.30b}$$

$$J_2 = (V_2 - F) \times (V_0 - F) \tag{9.30c}$$

The vector obtained as the outer product is orthogonal to the two vectors being multiplied. However, its direction is determined by the sequence in which the two vectors are multiplied. Given this feature of the outer product, we can judge whether the F is included in the triangle according to the signs of the outer products in Eq. (9.30), i.e., if all the signs are the same, the F is included in the triangle. However, to apply this rule, the vertices need to be aligned in a consistent sequence.

The above approach for judging the intersection of a line and a polygon can be used to calculate the view factor of a radiating surface, A_2, as viewed from a heat-receiving surface, A_1. As discussed in Chapter 3, the view factor can be defined solely by the geometric relationship between the two surfaces as follows (Figure 9.17):

$$F_{12} = \frac{1}{A_1} \int\limits_{A_1} \int\limits_{A_2} \left(\frac{\cos\beta_1 \cos\beta_2}{\pi r^2} \right) dA_2 dA_1 \tag{9.31}$$

where dA is the infinitesimal element of the surfaces, β is the angle between the normal of the infinitesimal element and the line connecting the target infinitesimal element, and r is the distance between the two infinitesimal elements. As the analytical solution of Eq. (9.31) can be obtained only when

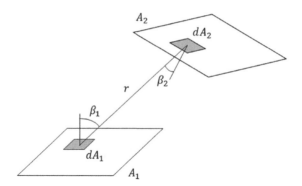

Figure 9.17 Calculation of the view factor between polygons.

the two surfaces are in simple geometrical relationships, view factors between surfaces of fuel objects involved in outdoor fires are often calculated numerically. If we ignore the efficiency of the computational algorithm (Cohen et al., 1993), we can directly solve Eq. (9.31) numerically:

$$F_{12} = \frac{1}{A_1} \sum_{A_1} \sum_{A_2} \left(\frac{\cos\beta_1 \cos\beta_2}{\pi r^2} \Delta A_2 \Delta A_1 \right) \tag{9.32}$$

where ΔA is the area of the infinitesimal element. The values of the variables in Eq. (9.32), $\cos\beta_1$, $\cos\beta_2$, and r, can be obtained relatively easily following the procedures described above. However, there is often a case that the thermal radiation from a fire source to a fuel object is often shielded by another fuel object in between. In other words, the view factor calculation using Eq. (9.32) additionally requires a judgment on whether the line connecting the two infinitesimal elements on A_1 and A_2 has any intersection with other surfaces that could shield the radiation.

(3) Area of a polygon

The area of a polygon can be obtained by dividing the polygon into multiple triangles and summing up the area of each triangle. Given that the size of the outer product of two vectors is the area of the parallelogram bounded by the two vectors, the area of a triangle, ΔA, can be obtained as follows:

$$\Delta A = \frac{1}{2} \left| (\mathbf{V}_1 - \mathbf{V}_0) \times (\mathbf{V}_2 - \mathbf{V}_0) \right| \tag{9.33}$$

where $\mathbf{V}_0$, $\mathbf{V}_1$, and $\mathbf{V}_2$ are the position vectors of the three vertices shown in Figure 9.18. Note that in Eq. (9.33), we need to take the absolute value of

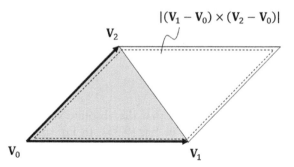

Figure 9.18 Calculation of the area of a triangle using the outer vector.

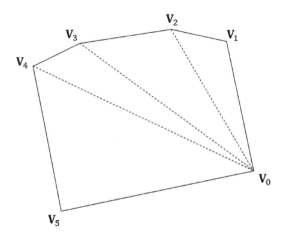

Figure 9.19 Calculation of the area of a convex polygon.

the outer vector as it can be negative depending on the sequence of the three vertices.

Consider a convex polygon with N vertices. The polygon can be divided into $N - 2$ triangles containing a common vertex, V_0, without any overlap. If the polygon is a hexagon, the area can be divided into four triangles, which are $V_0V_1V_2$, $V_0V_2V_3$, $V_0V_3V_4$, and $V_0V_4V_5$ as shown in Figure 9.19. The area of the polygon, A, can be calculated by adding the area of each triangle, ΔA, using Eq. (9.33):

$$A = \frac{1}{2}\sum_{i=1}^{N-2}\left|\left(V_i - V_0\right)\times\left(V_{i+1} - V_0\right)\right| \qquad (9.34)$$

Eq. (9.34) is not applicable to concave polygons, such as the one shown in Figure 9.20. The area of an arbitrary polygon, convex and concave, can be calculated using the following equation:

$$A = \frac{1}{2}\left|\sum_{i=1}^{N-2}\left(\mathbf{V}_i - \mathbf{V}_0\right)\times\left(\mathbf{V}_{i+1} - \mathbf{V}_0\right)\right| \tag{9.35}$$

The difference from Eq. (9.34) is that the absolute value symbol is added outside the summation symbol in Eq. (9.35). In Eq. (9.35), the area of a polygon is calculated by taking the sum of the signed areas of the subdivided triangles obtained by the outer product calculation, then taking its absolute value.

As an example of the area calculation using Eq. (9.35), we use a concave octagon in Figure 9.20. With the common vertex, $\mathbf{V}_0$, the octagon can be divided into six triangles, $\mathbf{V}_0\mathbf{V}_1\mathbf{V}_2$, $\mathbf{V}_0\mathbf{V}_2\mathbf{V}_3$, $\mathbf{V}_0\mathbf{V}_3\mathbf{V}_4$, $\mathbf{V}_0\mathbf{V}_4\mathbf{V}_5$, $\mathbf{V}_0\mathbf{V}_5\mathbf{V}_6$, and $\mathbf{V}_0\mathbf{V}_6\mathbf{V}_7$. As part of the triangles protrudes outside the octagon domain, it is obvious that Eq. (9.34) does not yield a correct area of the octagon. The following equation calculates the signed area of each triangle:

$$\Delta A = \frac{1}{2}\left(\mathbf{V}_i - \mathbf{V}_0\right)\times\left(\mathbf{V}_{i+1} - \mathbf{V}_0\right)\ \left(i = 1,6\right) \tag{9.36}$$

For simplicity, let us assume that the octagon is in the xy plane. In such a case, $\mathbf{V}_0\mathbf{V}_4\mathbf{V}_5$ is the only triangle whose area has a negative value. The triangles with positive and negative values are shown in the bottom-left and bottom-right panels of Figure 9.20, respectively. The circled numbers in the shaded areas indicate the number of times the area is added to the total area. In the bottom-left panel, the domain of the added triangles protrudes outside the octagon domain. There is also a domain where the area of the triangle is added twice. In contrast, the only triangle with a negative value is $\mathbf{V}_0\mathbf{V}_4\mathbf{V}_5$ in the bottom-right panel. Comparing the two, the protruded domain and the domain added twice inside the octagon in the bottom-left panel overlap the triangle in the bottom-right panel. Thus, by adding all the signed areas of the triangles, we obtain the signed area of the polygon, excluding the overlapped area. Thus, the area of the octagon, A, can be obtained as:

$$A = \frac{1}{2}\left|\sum_{i=1}^{6}\left(\mathbf{V}_i - \mathbf{V}_0\right)\times\left(\mathbf{V}_{i+1} - \mathbf{V}_0\right)\right| \tag{9.37}$$

Eq. (9.35) is a generalized form of Eq. (9.37). For convex polygons, the signed areas of all the triangles are either positive or negative. Thus, the equation is also applicable to the computation of the area of convex polygons.

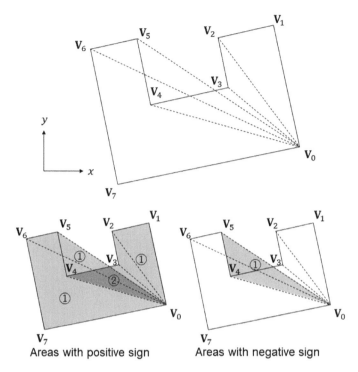

Areas with positive sign   Areas with negative sign

Figure 9.20 Calculation of the area of a concave polygon. Circled numbers are the number when the corresponding shaded area is added to the total area.

WORKED EXAMPLE 9.5

Assume that the coordinates of the octagon vertices in Figure 9.20 are $V_0 = (20,3,0)$, $V_1 = (18,21,0)$, $V_2 = (11,20,0)$, $V_3 = (12,11,0)$, $V_4 = (4,10,0)$, $V_5 = (3,21,0)$, $V_6 = (-2,20,0)$, and $V_7 = (0,0,0)$. The units are all in m. Calculate the area of the octagon, A.

SUGGESTED SOLUTION

Using Eq. (9.36), the signed areas of the six triangles, $V_0V_1V_2$, $V_0V_2V_3$, $V_0V_3V_4$, $V_0V_4V_5$, $V_0V_5V_6$, and, $V_0V_6V_7$, can be obtained as $\Delta A_{012} = 64$ m^2, $\Delta A_{023} = 32$ m^2, $\Delta A_{034} = 36$ m^2, $\Delta A_{045} = -84.5$ m^2, $\Delta A_{056} = 53.5$ m^2, and $\Delta A_{067} = 203$ m^2. Among the six triangles, $V_0V_4V_5$ is the only triangle with a negative value. By taking the absolute value of the sum of the signed areas as expressed in Eq. (9.37), the area of the octagon can be obtained as 304 m^2.

Chapter 10

Fire spread simulation

To assess the hazard of fire spread in outdoor fires, we need to predict the spatial development of the burning area following ignition. This chapter combines the knowledge of various aspects of outdoor fires described in Chapters 2 through 9 to enable predictive calculations of fire spread behavior. However, fire spread in outdoor fires is a complex phenomenon involving various physical and chemical processes. Although we cannot avoid employing mathematical descriptions in formulating the phenomenon, readers might find it insufficient to capture the entire picture of the phenomenon. Thus, we have prepared a sample code to provide readers with an opportunity for hands-on learning of the predictive calculations of fire spread behavior. In the sample code, we employed simple models among several alternatives for each physical and chemical process. If all the details were considered, a more elaborated prediction would be possible. However, this would complicate the computational procedure and might even obscure the nature of the fire spread behavior. We simplified the process of fire spread by considering only the primary factors that affect the behavior. Simplification ensures the clarity of the overall structure of the sample code. In return, there is still room for improvement in terms of simulation accuracy. Readers are encouraged to modify the sample code to suit their interests.

10.1 AN OVERVIEW OF THE MODEL DEVELOPMENT HISTORY

In this section, we briefly overview various approaches that have been proposed to predict the behavior of fire spread in outdoor fires, including wildland, WUI, and urban fires. We discuss how the sample simulation model is positioned within this context. Interested readers are referred to the literature for a more comprehensive review of existing approaches (Weber, 1991; Perry, 1998; Pastor et al., 2003; Lee et al., 2008; Sullivan, 2009).

DOI: 10.1201/9781003096689-10

10.1.1 Development of predictive methods

Among the three major categories of outdoor fires (wildland, WUI, and urban fires), long histories of research exist on wildland and urban fires. In contrast, the research on WUI fires has become active in recent years. The research communities have acknowledged the phenomenological similarities between wildland and urban fires (Williams, 1982). However, prediction methods have been developed independently in each field. Achievements of one have rarely contributed to the development of the other. The growing research interest in WUI fires may be creating an opportunity to fill the research gaps among the related research communities.

(1) Flame propagation in ideal combustible spaces

In wildlands, fuel components can be categorized according to their stratified structure: the ground, surface, and canopy fuel layers. Predictive models of fire spread have been constructed in a way that is consistent with the characteristics of each fuel layer (Pastor et al., 2003). In this subsection, we focus on flame propagation behavior in the surface fuel layer, which is the most fundamental component in evaluating the hazard of wildland fires. The formulation of flame propagation in surface fuels was first attempted by Fons (Fons, 1946). This was a physics-based model deriving the flame propagation rate from the conservational relationship of energy within a combustible material. The model has been a precursor to many other models of a similar concept (Hottel et al., 1965; Albini, 1967; Thomas, 1967; Anderson, 1969; Pagni et al., 1973; Albini, 1985; Weber, 1989; De Mestre et al., 1989). However, an analytical solution to the system of conservation equations, differential equations involving time and spatial derivatives, is generally available only in limited initial and boundary conditions. Thus, the range of applicability of physics-based models is generally limited to idealized combustibles, such as 1D homogeneous surfaces, despite its scientific rigor. Thus, they have not necessarily been successful in ensuring sufficient accuracy in predicting fire spread in actual wildland fires. Against this background, semi-physical or semi-empirical models were proposed, in which the physics-based models were modified based on some empirical considerations (Frandsen, 1971; Rothermel, 1972; Catchpole et al., 1998). One of the most widely used models for predicting fire spread in practice, the Rothermel model, is categorized as the semi-physical model.

Methods employing empirical or statistical approaches have also been developed for predicting fire spread in wildland fires (McArthur, 1966; Noble et al., 1980; Forestry Canada Fire Danger Group, 1992; Cheney et al., 1998; Viegas, 2004). The advantage of such models is, above all, their capability to predict fire spread in a simple procedure. However, empirical models generally encounter challenges in ensuring generality, as their validity is maintained only under similar conditions the to which the models were parameterized. This is one of the reasons why methods employing a physical or semi-empirical approach are preferred in practical forestry management.

In contrast to the case of wildland fires, quantitative methods for predicting the behavior of fire spread in urban areas have started with empirical and statistical formulations (Hamada, 1951; Horiuchi, 1972; Aoki, 1987). This is partly due to the difficulty in expressing the thermal energy transfer in non-homogeneous combustible spaces in a general form of equation (a group of buildings with different shapes and combustion characteristics) and integrating them into a unified system of equations for building-to-building fire spread. The development of technical knowledge on the fire behavior inside buildings had to precede the emergence of physics-based models for predicting fire spread in urban areas.

(2) Fire spread in actual combustible spaces

Mathematical models for the behavior of fire spread simulate only the primary features of the phenomena and do not necessarily simulate every detail of the physical and chemical processes. No matter how detailed the simulation is, some simplifications are inevitable when modeling the phenomena despite the difference in degree. Hence, although we use the term 'actual' combustible spaces, the target combustible spaces are 'ideal' to a certain extent. Earlier prediction methods characterized the 'ideal' combustible spaces as 'homogeneous', 'continuous', and '1D' combustible spaces. In contrast, we characterize the features of 'actual' combustible spaces as 'homogeneous or non-homogeneous', 'continuous or discrete', and '2D or 3D' combustible spaces.

There are two major backgrounds in which the advancement of the prediction methods from the ideal to actual combustible spaces has been achieved. One is the development of computational environments, which enabled fast computing of large and complex systems of equations. The other is the development and popularization of Geographical Information Systems (GIS), which enabled the creation, processing, and integration of spatial data on forests and urban areas. The prediction methods developed against this background include the application of the wave propagation principle to predict flame propagation in a 2D and discontinuous combustible space (Fujita, 1975; Anderson et al., 1982), cell-based simulations or the cellular automaton, in which a large number of cells arranged in a grid change their state over time as they interact with neighboring cells (Kourtz et al., 1971; Frandsen et al., 1979; Clarke et al., 1994; Ohgai et al., 2007; Zhao, 2011; Jiang et al., 2021), and the level-set approach, which numerically solves for a level-set function representing the interface between the non-burning and burning zones (Rehm et al., 2009; Lautenberger, 2013; Bova et al., 2016). Such extensions have improved the prediction accuracy of fire spread behavior and encouraged the practical application in management and planning services. Software packages implementing the above prediction methods are available as routinely used decision support tools (Burgen et al., 1984; Forestry Canada Fire Danger Group, 1992; Scott, 1999; Cruz et al., 2015). In addition, operational tools have been developed that integrate prediction methods into GIS for managing geographic information of forests and urban areas on computers (Vasconcelos et al., 1992; TFD, 1997; Finney, 1998; Lee et al., 2002; Cousins et al., 2002; Thomas et al. 2002; Ren et al. 2004; Dimopoulou, 2004; Ohgai et al. 2007; Kalabokidis et al. 2013).

An important advancement in the prediction of wildland fire spread is the application of computational fluid dynamics (CFD) techniques (Grishin, 1996; Linn, 1997; Morvan et al., 2002; Porterie et al., 2005; Mell et al., 2007; Filippi et al. 2009; Morvan et al. 2009; Mandel et al. 2011; Tang, 2017). This was a leap from the conventional prediction methods based on the thermal energy conservation of a solid fuel bed in a steady crosswind. CFD simulations numerically solve the basic equations of fluid motion, enabling a detailed phenomenological analysis of fire spread coupling the combustion and atmospheric dynamics. On the other hand, CFD simulations accompany several limitations. They require a powerful computational environment sufficient to perform large-scale numerical computations and the users to have a certain level of expertise to ensure reasonable predictions. Thus, its application in practical forestry management has been limited despite its strong predictive capabilities. However, the range of their applications is expected to extend, assuming a steady improvement of the available computational resources in the future.

The characteristics of urban fires that should be considered in the predictions of fire spread are that they are a group of enclosure fires that occur in buildings whose envelopes possess a variety of fire resistance performances. The combustion reaction in an enclosure fire occurs in response to the ventilation-induced fresh air inflow through openings. The amount of generated heat is partially vented from the openings, and the remaining is absorbed by the envelope. We additionally need to consider the complex geometrical configurations of buildings for heat transfer calculations and the difference in the ignitability of building envelopes when subject to heating. The development of engineering methods for predicting fire behavior in buildings has been a basis for the development of physics-based prediction models of fire spread in urban areas (Himoto et al., 2002; Iwami et al., 2004; Himoto et al., 2008; Lee et al., 2010; Lu et al., 2017; Rafi et al., 2018; Sarreshtehdari et al., 2021).

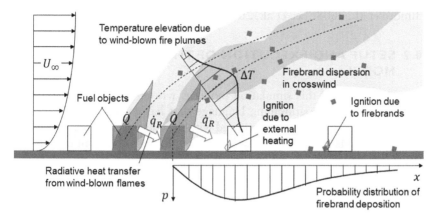

Figure 10.1 Schematic representation of the presented simulation model (reproduction of Figure 1.1).

10.1.2 Presented simulation model

Figure 10.1 shows a schematic diagram of the simulation model presented in this book. In this model, an outdoor fire is regarded as a group of multiple burning fuel objects. Prediction of the burning behavior of individual fuel objects under the thermal influence of the other burning fuel objects leads to prediction of the overall fire behavior. However, the fuel objects are discretely distributed at varying distances and orientations from the other fuel objects. A fire plume formed above a burning fuel object is tilted in a crosswind and heats the other fuel objects in the leeward direction. The heating intensity from a burning fuel object to the other fuel object is determined not only by the burning intensity represented by the heat release rate (HRR) but also by their positional relationships. The closer the target fuel object to the heat source, the stronger the heating intensity and the shorter the time required for ignition, leading to an increase in the fire spread rate. In addition, a burning fuel object releases a large number of firebrands. The firebrands are scattered downwind, causing the ignition of fuel objects located at distant locations. The scattering range of firebrands and the ignition due to deposited firebrands are evaluated probabilistically.

According to the overview of the existing prediction methods discussed in the previous subsection, the present model can be categorized as a 'physics-based' model predicting the behavior of fire spread between 'discretely' distributed '3D' fuel objects. The shape and combustion characteristics of assumed fuel objects are arbitrary. However, we do not consider the burning of combustibles in enclosures to keep the model simple. Readers interested in fire spread involving enclosure fires are referred to related literature (Himoto et al., 2008; Lee et al., 2010).

10.2 SETUP AND EXECUTION OF THE SIMULATION MODEL

The source code of this simulation model is written in Python. Python is a general-purpose programming language that has gained extensive users since its release in 1991. It is permeated with the concept of facilitating users to write readable yet efficient code in the simplest possible manner. Such a feature attracts users, including novices, to participate in the Python communities. This is why we wrote our source code in Python.

Simulated results of the model are saved in KML files, a file format that can be loaded by Google Earth. Google Earth is a computer program that renders a 3D representation of Earth based primarily on satellite imagery developed and provided by Google Inc. The simulated fire spread dynamics are visualized and overlaid on satellite imagery at an arbitrary location on Earth.

Python and Google Earth must be installed in advance to run this simulation model. Both are available free of charge. There are various types of Python distributions, some of which are provided in the form of packages that include libraries (extensions) and an integrated development environment for the convenience of users (e.g., Anaconda package). However, to avoid any problems associated with the version compatibility of the libraries, it is recommended to download and install the installation files from the Python home website:

www.python.org/

10.2.1 Source code

The source code of the simulation model is available on GitHub. GitHub is a platform for software development originally intended to facilitate project management by multiple software developers. It is often used to release source codes to the public, as in the present case. The URL for downloading the source code is as follows:

https://github.com/khimoto298/lofd/

Anyone can download the source code by creating a GitHub account.

This simulation model was tested with Python 3.10.4. As described below, the version dependence of the code has been removed where possible. Thus, we expect that the code will also work in later versions. However, if for some reason the code does not work as expected, users are recommended to revert to version 3.10.4. If any bugs are found in this source code after its release, we may modify the code involving alterations to the present description.

10.2.2 Setup of an execution environment

The main body of Python provides only the minimum functions required by users. Extended functions other than the basic functions are provided as libraries, which can be called upon and used on demand. An extremely rich library collection is one feature that makes Python functional and appealing to users. However, the use of libraries also entails drawbacks. As the development of each library is independent of the main body of Python, functions of a library that were available in one version may not be available in a later version. Thus, in writing this source code, we had considered fully excluding the use of libraries. However, as this may increase the complexity and diminish the readability of the code, we decided to solely use NumPy, a library for efficient numerical computation. The simulation model was tested with NumPy 1.22.3.

The folder downloaded from the GitHub site contains a folder (data), nine Python files (cgprm.py, fire.py, g.py, gprm.py, init.py, main.py, ndat. py, rndm.py, and sprd.py), two bat files (exec.bat and setup.bat), and a txt file (ReadMe.txt). Among these files, 'setup.bat' is a batch file for setting up an execution environment on your PC, such as installing the NumPy libraries required for simulating the source code. Executing (double-clicking) 'setup.bat' opens a command window and executes the installation automatically. This creates two new folders (__pycache__ and venv) in the working folder. If the user is using a Windows machine and the double-click setup does not work due to the PC environment, launch 'Command Prompt' (cmd.exe), navigate to the folder where 'setup.bat' is saved, and then execute the file.

10.2.3 Execution of the code

To set up an execution environment, we simply had to run 'setup.bat' in the downloaded folder. Running the simulation is also simple. Execute (double-click) another batch file, 'exec.bat', in the same folder. The bat file executes the Python script (source codes) in the same folder and performs the fire spread simulation as per the conditions set in CSV files in the 'data' folder. We describe the details of the CSV files for setting simulation conditions in Section 10.4 with specific examples. The simulation results are saved as KML and CSV files in the 'data' folder. Running a KML file launches Google Earth, visualizing the burn state of fuel objects at a specified time. The simulation results on the dynamics of fire spread, such as the time evolution of the number of burning and burnt fuel objects, are saved in separate CSV files. The specific output to the CSV files is summarized in Table 10.1 in Subsection 10.3.3.

10.3 THEORETICAL FRAMEWORK OF THE SIMULATION MODEL

This section describes the theoretical framework of the simulation model along with its computational flow. As the phenomenological details of fire spread have already been described in the preceding chapters, this section focuses on integrating them into a prediction system.

10.3.1 Overall structure

Figure 10.2 describes the computational flow of the fire spread simulation model. The computational flow is simple, with a limited number of branches. First, the simulation conditions are loaded from external CSV files and converted into forms required for the computation of physical processes involved in fire spread. Next, the burning behavior of fuel objects is calculated, evaluating the hazard of fire spread to the surrounding fuel objects. This is continuously updated until the fuel object is burnt out through the iterational process of computation. Finally, the simulation results are saved in multiple KML and CSV files.

The source code consists of nine files (main.py, gprm.py, cprm.py, g.py, init.py, ndat.py, fire.py, sprd.py, and rndm.py). Among these files, 'main. py' is the main file that controls the overall computational flow. 'rndm.py' is responsible for generating random numbers for probabilistic computations.

The variables used in this simulation model are defined in three files, 'gprm.py', 'cprm.py', and 'g.py'. Variables that possess a hierarchical data

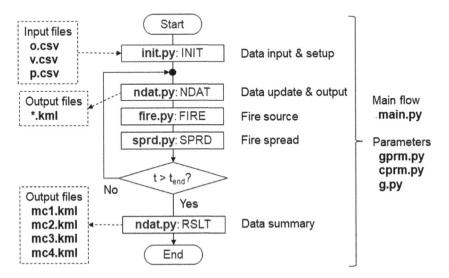

Figure 10.2 Computational flow of the fire spread simulation model.

structure, such as the coordinates of the vertices of each fuel object (object > vertex > coordinates), are defined in 'gprm.py'. Variables that do not possess a hierarchical structure are defined in the remaining two files. Variables updated during the computational process are defined in 'cprm.py', and the rest are defined in 'g.py'.

The four files actually responsible for the computation of fire spread between fuel objects are 'init.py', 'ndat.py', 'fire.py', and 'sprd.py'. Among these, 'init.py' loads simulation conditions from external files (o.csv, v.csv, and p.csv), and 'ndat.py' updates the status of some variables and exports the simulation results to external files (*.kml, mc1.csv, mc2.csv, mc3.csv, and mc4.csv). 'fire.py' is responsible for assessing the ignition of each fuel object when subject to external heating and calculating the mass loss rate (MLR) following the ignition. 'sprd.py' evaluates the hazard of fire spread between fuel objects considering the effects of radiative heating from flames tilted in a crosswind, temperature rise due to wind-blown fire plumes, and dispersal of firebrands.

The following describes the specific computations performed by each file and function included in this source code.

10.3.2 INIT: data input and setup

The 'INIT' file consists of four functions (INIT1, INIT2, INIT3, and INIT4) responsible for the data loading and initialization.

(1) INIT1: read from files

The function 'INIT1' loads the data from three external files (o.csv, v.csv, and p.csv). Among these files, 'o.csv' specifies the outline of the simulation, such as the total number of fuel objects and the weight of each fuel object. 'v.csv' defines the vertex coordinates composing each fuel object. 'p.csv' defines the vertex IDs composing the face polygons of each fuel object. The details of each external file are described with specific examples in Section 10.4.

(2) INIT2: objects

The function 'INIT2' computes necessary conditions to determine adjacency relations between fuel objects. The computed variables include the center of gravity of the face polygons composing each fuel object, the center of gravity of each object, the unit vector normal to each face polygon, the base area of each fuel object, and the mesh ID in which each fuel object is contained.

The mesh is a rectangular computational domain divided into a grid with an interval defined by a variable, dM. The computational domain is divided into nXX and nYY meshes in the north–south and east–west directions, respectively. The default mesh size is $dM = 10$ m, which is defined in

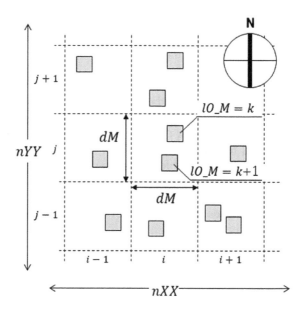

Figure 10.3 Determining corresponding mesh ID for each fuel object.

'cprm.py' together with nXX and nYY. The list, IO_M, is used to determine the corresponding mesh of each fuel object in advance of iterational computations to reduce the computational time. dM should be changed depending on the size of the fuel objects considered. As a guide, dM should be large enough to contain several fuel objects in a single mesh (Figure 10.3).

(3) INIT3: adjacency

In this model, radiation from the flame formed above a burning fuel object to the adjacent fuel objects is one of the causes of fire spread. As discussed in Chapter 3, the radiant heat flux from a flame to an adjacent object is determined by their geometric relationship. However, the geometric relationship changes over time as the flame shape changes along with the fuel combustion and ambient wind conditions (change in the ambient wind is not considered in the present case studies for simplicity). Numerical computation of the geometric relationship at every time step increases the computational load. Thus, in function 'INIT3', we list face polygons of fuel objects located in a position that could receive heating if a concerned fuel object starts to burn.

The list of the target face polygons is saved as a variable, ADJ. We used the computational procedure discussed in Chapter 9 to select the target face polygons. If the distance between two fuel objects is used as the sole criterion, we cannot exclude the case where another fuel object is in between them, shielding the radiation. Furthermore, even if two fuel objects are next to each other and no shielding exists between them, some face polygons may not be directly exposed to heating if oriented in opposite directions to the burning fuel object. Thus, we consider the visibility per face polygon in determining the adjacency list of face polygons that possess the possibility of heat exchanges.

(4) INIT4: from UTM coordinates to longitude and latitude

The model uses Google Earth for visualization of the simulation results. The simulated fire spread is represented by the color change of individual fuel objects arranged on the digital globe. In Google Earth, the location coordinates of an object need to be expressed in terms of longitude and latitude. In contrast, this simulation model uses a user-determined rectangular coordinate system in which the location and shape of the fuel objects are represented. The origin defined by the user is tied to the Universal Transverse Mercator (UTM) coordinate system. The UTM coordinates are further converted into the coordinates in latitude and longitude, enabling the representation of fuel objects at arbitrary locations on Earth.

The function 'INIT4' converts the vertex coordinates of fuel objects defined in the UTM coordinate system into longitude and latitude in a way manageable by Google Earth. Specific conversion procedures are beyond the scope of this book and are therefore omitted. Once the location of the computational domain is determined, the user needs to specify the longitude and latitude of the origin of the corresponding zone of the UTM coordinate system, LL_o and BB_o, respectively, and the displacement of the origin of the computational domain from it in the east-west and north-south directions, CC_o1 and CC_o2, respectively. These variables are defined in 'cprm.py' with the default values of $LL\_o = 141.0°$, $BB\_o = 0°$, $CC\_o1 = 416,130\,\text{m}$, and $CC\_o2 = 3,998,130\,\text{m}$. The default computational domain is in a cultivated field next to the National Institute for Land and Infrastructure Management (NILIM), the author's workplace, in the UTM zone '54N'.

10.3.3 NDAT: data update and output

The 'NDAT' file consists of three functions (NDAT, OUTP, and RSLT) responsible for updating the status of some variables and exporting the simulation results to external files.

(1) NDAT: data status

The function 'NDAT' identifies fuel objects that require an update of burn status at each time step based on their current burn status, whether they are currently burning, still unburnt but about to get ignited due to external heating, or the rest. The computational time can be shortened by restricting the number of fuel objects targeted for the computation of burning behavior. By default, an object is targeted for the computation if its surrounding temperature rise exceeds $50\,K$ or if the external heat flux on any of the face polygon exceeds $1\,kW \cdot m^{-2}$. Although these thresholds are empirical, the influence on simulation results can be avoided by adopting sufficiently low values.

(2) OUTP: data output

The function 'OUTP' exports the progress of the computation on the screen to external files (KML files) at regular time intervals. The elapsed time, the wind velocity, and the number of burning and burnt out fuel objects are displayed on the screen. A subordinate function 'KML' exports the data to external files in KML format. The time intervals can be adjusted using two variables in 'cprm.py': $nOUT1$ (default = 600) for the screen and $nOUT2$ (default = 600) for KML files. These variables represent the number of computational steps, $nSTP$, that are skipped without any data output. Thus, the time interval in the actual time scale depends on the time increment, $dTsec$. By default, the time increment is $dTsec = 1s$. Thus, if $nOUT1 = 600$, the computation results are displayed at an interval of $600\,s = 10\,min$.

(3) RSLT: data summary

The function 'RSLT' exports four files (mc1.csv, mc2.csv, mc3.csv, and mc4.csv) that summarize the results after all computations have been completed. The items included in each file are shown in Table 10.1. All of the output items are appended to the previous data in each file to maintain the computation records.

Table 10.1 Items included in each output file.

File name	Description	Items in order
mc1.csv	Outline	computation time (min), wind velocity $(m \cdot s^{-1})$, unit wind vector (x), unit wind vector (y), fire origin, number of burning objects, number of burnt objects
mc2.csv	Burn status	total number of objects, burn status for all objects (0, not ignited; 1, burning; 2, burnt out)
mc3.csv	Burning objects	the transition of the number of burning objects at an interval of one minute
mc4.csv	Burnt out objects	the transition of the number of burnt out objects at an interval of one minute

10.3.4 FIRE: burning behavior of objects

The 'FIRE' file consists of two functions (HRR and HTRF) responsible for predicting the burning behavior of fuel objects.

(1) HRR: heat release rate

For the calculation of radiant heat flux from a burning fuel object to the surrounding fuel objects, the flame geometry above the burning fuel object needs to be determined. It is dependent on several factors such as the fuel object geometry, the HRR (or MLR), and the crosswind velocity, as discussed in Chapters 5 and 6. In the present model, the fuel object geometry and the crosswind velocity are the given conditions. Thus, the HRR needs to be determined to enable evaluation of the flame geometry. The HRR is also required to evaluate the fire spread process due to firebrands, as discussed in Chapter 8.

This model assumes wood cribs as fuel objects. The users can adjust the wood crib condition to examine its effect on the fire spread behavior. Wood cribs have been regarded as surrogate fire sources for simulating various types of fire. Thus, an extensive number of experiments have been conducted to investigate its burning behavior. The combustion mode of wood cribs is controlled either by the exposed surface area of the combustible (fuel-controlled) or the rate of fresh air supply through the voids of the combustible (air-supply-controlled). The expected MLR in each mode of combustion can be represented by $\dot{m}_{B,S}$ and $\dot{m}_{B,P}$, respectively, the smaller of which can be assumed as the actual MLR of the burning wood crib, $\dot{m}_B$ (Babrauskas,2002a):

$$\dot{m}_B(t) = \min\{\dot{m}_{B,S}(t), \dot{m}_{B,P}\} \tag{10.1}$$

with $\dot{m}_{B,S}$ and $\dot{m}_{B,P}$ given as follows.

$$\text{Fuel-controlled:} \ \dot{m}_{B,S}(t) = \frac{4M_0 v_P}{D_C}\left(\frac{M(t)}{M_0}\right)^{1/2} \tag{10.2a}$$

$$\text{Air-supply-controlled:} \ \dot{m}_{B,P} = 4.4\times10^{-4}\left(\frac{S_C}{H_C}\right)\left(\frac{M_0}{D_C}\right) \tag{10.2b}$$

where M_0 is the initial mass of the wood crib, $M(t)$ is the mass at time t, D_C is the cross-sectional width of wood sticks, S_C is the spacing between wood sticks, H_C is the height of the wood crib, and $v_P\left(= 2.2\times10^{-6}\,D_C^{-0.6}\,(\text{m}\cdot\text{s}^{-1})\right)$ is the charring rate at the material surface (Babrauskas, 2002a). Figure 10.4 shows the results of sample calculations using Eq. (10.1). The assumed wood crib was a cube with a side of 3 m. The cross-sectional width of wood

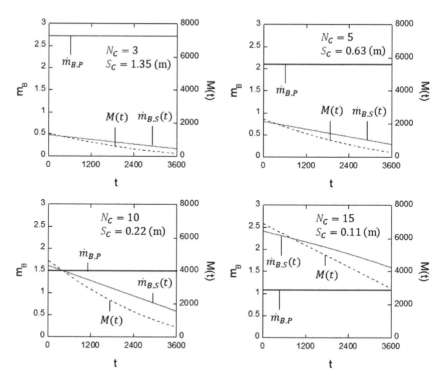

Figure 10.4 Calculated mass loss of a burning wood crib with $D_C = 0.1$ m and $H = W = 3$ m.

sticks composing the wood crib was $D_C = 0.1$ m. The number of wooden sticks aligned in each layer, N_C, was varied in four steps, 3, 5, 10, and 15, for comparison. According to Figure 10.4, the combustion mode was fuel-controlled when the N_C was smaller and the S_C was wider. Conversely, the combustion mode was air-supply-controlled when the N_C was larger and the S_C was narrower.

The $\dot{m}_{B,S}(t)$ in Eq. (10.2a) can also evaluate an attenuating MLR in the decay phase due to the charring of the material. However, according to this relationship, $M(t)$ never becomes zero, because $\dot{m}_B(t)$ also approaches zero as $M(t)$ becomes smaller. This is probably because Eq. (10.2a) was not intended to evaluate the burnout of combustible materials. Thus, in this model, we determine that a wood crib is burnt out when $M(t)$ falls below 20% of the initial mass, M_0. The value of 20% was determined with reference to the weight ratio of char formed by the combustion of wood.

With an abundant supply of fresh air, the HRR, $\dot{Q}$, can be evaluated by multiplying the MLR, $\dot{m}_B(t)$, by the heat of combustion, ΔH. However, $\dot{m}_B(t)$ in Eq. (10.1) is the burning rate at the stage when the entire wood crib is engulfed in flame. The time to reach the fully developed phase is a factor that affects the rate of fire spread between fuel objects. Assuming that the fire grows in proportion to the square of the time after ignition, t, the $\dot{Q}$ can be expressed as follows:

$$\dot{Q} = \min\left\{\alpha\left(t - t_{ig}\right)^2, \Delta H \dot{m}_B\left(t\right)\right\} \tag{10.3}$$

where α is the fire growth rate, which is empirically given by $\alpha = 0.01$ kW $\cdot$ s^{-2} as the default of the present model, t_{ig} is the time to ignition, and ΔH is the heat of combustion. In outdoor fires, multiple fuel objects located close together often burn simultaneously. In such a case, the burning behavior of fuel objects differs from that of a single fuel object due to mutual thermal interactions. However, such a difference is not considered in the present model for simplicity.

(2) HTRF: ignition due to external heating

The function 'HTRF' evaluates the ignition of a fuel object subject to the combined effect of radiative and convective heatings from fire plumes:

$$\dot{q}'' = \varepsilon\dot{q}''_R + \varepsilon\sigma\left(T^4 - T_S^4\right) + h\left(T - T_S\right) \tag{10.4}$$

where $\dot{q}''_R$ is the radiative heat flux from external fire sources, ε is the emissivity, h is the heat transfer coefficient assumed to be 0.02 kW $\cdot$ m^{-2} $\cdot$ K^{-1} in this model, T is the temperature around the fuel object, and T_S is the surface temperature. As discussed in Chapter 7, the surface temperature of a heated fuel object can be used as a criterion for ignition. However, this model does not predict changes in the surface temperature for simplicity. Alternatively, the ignition of a heated fuel object is determined based on the cumulative heat received on the center of each face polygon. In Chapter 7, we discussed that the time to ignition of a semi-infinite solid, t_{ig}, when subject to an incident heat flux, $\dot{q}''$, can be expressed as follows:

$$t_{ig} = \frac{\pi k\rho c}{4}\left(\frac{T_{ig} - T_0}{\dot{q}''}\right)^2 \quad \left(\dot{q}'' \geq \dot{q}''_{cr}\right) \tag{10.5}$$

where $k\rho c$ is the thermal inertia, T_{ig} is the ignition temperature, T_0 is the ambient temperature, and $\dot{q}''_{cr}$ is the critical heat flux. Eq. (10.5) assumes a constant incident heat flux, $\dot{q}''$. However, in actual fires, $\dot{q}''$ generally varies with time. Thus, Eq. (10.5) can be transformed as follows:

$$\int_0^{t_{ig}}\dot{q}''^2\,dt = \frac{\pi k\rho c}{4}\left(T_{ig} - T_0\right)^2 \quad \left(\dot{q}'' \geq \dot{q}''_{cr}\right) \tag{10.6}$$

This equation indicates that ignition of a heated fuel object occurs when the cumulative amount of $\dot{q}''^2$ exceeds a critical value on the right-hand side. The critical value on the right-hand side involves material-specific values such as $k\rho c$ and T_{ig}. Substituting the relevant values for Douglas fir in Table 7.1 in Chapter 7, the right-hand side of Eq. (10.6) can be approximated as $26,000 \ kW^2 \cdot m^{-4} \cdot s$. As an alternative to this approach, the *FTP* (flux time product) concept proposed by Shields and Silcock can be used to determine the ignition of a fuel object subject to time varying external heating (Shields et al., 1994; Silcock et al., 1995).

10.3.5 SPRD: fire spread behavior between objects

The 'SPRD' file consists of three functions (XRAD, FPLM, and BRND) responsible for evaluating the thermal interactions between fuel objects. Radiative and convective heat transfer from burning fuel objects, and firebrand dispersion in a crosswind are the causes of fire spread between fuel objects. In either mode of fire spread, the key issue is to identify the geometric relationship between the zone of influence formed around the burning fuel object and the other fuel objects.

(1) XRAD: flames

We need to identify the flame geometry to calculate the radiative heat transfer from the flame to the adjacent fuel object. For this, we consider two cases: one with no wind and the other with a crosswind.

Based on the discussion in Chapter 5, the following equation can be used to evaluate the flame height without wind, H_{fl} (Zukoski et al., 1981):

$$\frac{H_{fl}}{W} = \begin{cases} 3.3Q_W^{*2/3} & (Q_W^* < 1.0) \\ 3.3Q_W^{*2/5} & (Q_W^* \geq 1.0) \end{cases} \tag{10.7a}$$

$$Q_W^* = \frac{\dot{Q}}{\rho_\infty c_p T_\infty g^{1/2} W^{5/2}} \tag{10.7b}$$

where $\dot{Q}$ is the HRR, ρ_∞ is the ambient gas density, c_p is the specific heat of gas, T_∞ is the ambient temperature, g is the acceleration due to gravity, and W is the side length of a wood crib.

In windy conditions, the flame shape can be specified by determining the length along the central axis of the flame, L_{fl}, and the tilt angle, $\cos\theta$. Based on the discussion in Chapter 6, we use the following equation to evaluate L_{fl} (Thomas, 1963a; Thomas et al., 1963b, Thomas P H, 1967):

$$\frac{L_{fl}}{W} = 70\left(M_W^*\right)^{0.86}\left(U_W^*\right)^{-0.22} \tag{10.8a}$$

$$M_W^* = \frac{\dot{m}_B''}{\rho_\infty \sqrt{gW}} \tag{10.8b}$$

$$U_W^* = \frac{U_\infty}{\sqrt{gW}} \tag{10.8c}$$

where $\dot{m}_B''$ is the MLR per unit base area and U_∞ is the crosswind velocity. We use the following regression equation to calculate the $\cos\theta$ (Thomas, 1963a; Thomas et al., 1963b, Thomas PH, 1967):

$$\cos\theta = \begin{cases} 1 & (U_F^* < 1) \\ 0.7\left(U_F^*\right)^{-0.49} & (U_F^* \geq 1) \end{cases} \tag{10.9a}$$

$$U_F^* = \frac{U_\infty}{\left(\dot{m}_B'' gW / \rho_\infty\right)^{1/3}} \tag{10.9b}$$

The point source model, in which a single point represents the radiation source, is employed to calculate the radiative heat transfer from the flame to the surrounding fuel objects. We calculate the heat flux transferred to the center of gravity of each face polygon of wood cribs, G_A, within the visible range of the fire source. According to the discussion in Chapter 2 on radiative heat transfer, the incident heat flux, $\dot{q}_R''$, can be calculated by the following equation:

$$\dot{q}_R'' = \frac{\chi_R \dot{Q}}{4\pi r^2}\cos\phi \tag{10.10}$$

where χ_R is the radiative fraction of the HRR, r is the distance from the point heat source to the heat receiving point, and ϕ is the incidence angle.

There is arbitrariness in determining the representative point of radiation, G_B. The most straightforward approach is to use the center of the line segment connecting the base center of the burning wood crib and the flame tip. However, in this case, r in Eq. (10.10) includes the depth of the burning wood crib. With this representative point of radiation, $\dot{q}_R''$ attenuates at a rate proportional to r^{-2} even inside the flame, which is not necessarily true in reality. Such an error would be negligible if the width of the wood crib were short. Otherwise, the longer the width of the wood crib, the more $\dot{q}_R''$ is underestimated. Thus, we consider G_B to be at the center of a rectangular radiating surface with a width corresponding to the side length of the burning wood crib, W, and a height corresponding to the mean flame height, H_{fl}. The bottom edge is shifted for half the side length of the wood crib, $W/2$, toward the heat-receiving point, G_A, as described in Figure 10.5. Assume that points A and B are the bottom edge center of the heat-receiving

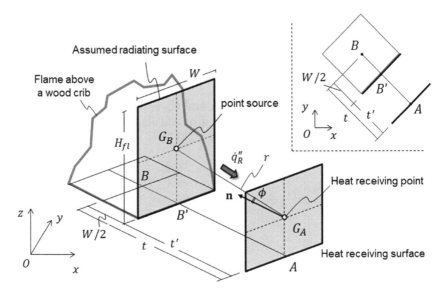

Figure 10.5 Thermal radiation from the flame formed above a wood crib to a heat receiving point on an adjacent wood crib.

surface and the base center of the burning wood crib, respectively. If we represent the intersection of the radiating surface and the line segment AB as point B', the position vector of G_B can be expressed as follows:

$$\overrightarrow{OG_B} = \overrightarrow{OB'} + \overrightarrow{B'G_B} \tag{10.11a}$$

$$\overrightarrow{OB'} = \frac{W/2}{t}\overrightarrow{OA} + \frac{t - W/2}{t}\overrightarrow{OB} \tag{10.11b}$$

$$\overrightarrow{B'G_B} = \begin{pmatrix} 0 \\ 0 \\ H_{fl}/2 \end{pmatrix} \tag{10.11c}$$

where t is the length of the line segment AB, which can be determined from the geometry of the wood cribs and their separations.

The above discussion is on no-wind conditions. Eqs. (10.11a) for $\overrightarrow{OG_B}$ and (10.11b) for $\overrightarrow{OB'}$ are still applicable in windy conditions when flames are tilted. However, Eq. (10.11c) for $\overrightarrow{B'G}$ needs to be modified for tilted flames:

$$\overrightarrow{B'G_B} = \begin{pmatrix} \left(L_{fl}\sin\theta/2\right)\times u_1 \\ \left(L_{fl}\sin\theta/2\right)\times u_2 \\ H_{fl}/2 \end{pmatrix} \tag{10.12}$$

where L_{fl} is the flame length, θ is the tilt angle, and u_1 and u_2 are the x – and y – components of the unit directional vector of the crosswind.

The incidence angle, $\cos\phi$, is nothing but the angle between the relative vector, $\mathbf{G} = \overrightarrow{G_A G_B}$,, and the unit vector normal to the heat receiving surface, $\mathbf{n}$. According to the discussion in Chapter 9, $\cos\phi$ can be calculated using the inner product equation of the two vectors as:

$$\cos\phi = \frac{\mathbf{G}\cdot\mathbf{n}}{|\mathbf{G}||\mathbf{n}|} = \frac{\mathbf{G}\cdot\mathbf{n}}{|\mathbf{G}|} \tag{10.13}$$

In the above discussion, we assumed that the heat-receiving point is in the center of the face polygon. However, it should be noted that the temperature in the center is not necessarily the maximum within the heat-receiving surface as it is not always the nearest point from the flame. Correspondingly, $\cos\phi$ taken at the center should also involve an error. Recall that in Worked example 9.4 in Chapter 9, the value of $\cos\phi$ was dependent on the position of the representative point. Thus, $\cos\phi$ given by Eq. (10.13) is likely to be an underestimation of the actual value. This is a limitation of the approximate computational procedure representing the heat source and heat-receiving surface by single points. Thus, in this model, we assume that the radiation is incident at a right angle to each face polygon of wood cribs, which is $\cos\phi = 1$, although short on rigor.

(2) FPLM: fire plumes

We refer to a fire plume as the non-reacting plume formed above a burning fuel object, tilted in a crosswind. Although the temperature rise of non-reacting plumes is not as high as that of reacting plumes, it can affect a wider area downwind. As discussed in Chapter 6, the temperature rise along the trajectory of fire plumes, ΔT_0, can be evaluated using the following equation (Himoto, 2019).

$$\frac{\Delta T_0}{T_\infty} = 2.08 Q_s^{*2/3} \tag{10.14a}$$

$$Q_s^* = \frac{\dot{Q}}{c_P \rho_\infty T_\infty g^{1/2} s^{5/2}} \tag{10.14b}$$

where T_∞ is the ambient temperature, $\dot{Q}$ is the HRR, c_p is the specific heat, ρ_∞ is the ambient gas density, g is the acceleration due to gravity, and s is the distance along the trajectory. Eq. (10.14) describes how ΔT_0 decreases with s due to mixing with the ambient air. However, as shown in Figure 10.6, the location of a fuel object generally deviates from the plume axis. The temperature rise at a distance, r, from the plume axis, $\Delta T(r)$, can be calculated by approximating the temperature profile with a normal distribution as follows:

$$\frac{\Delta T(r)}{\Delta T_0} = \exp\left[-\left(\frac{r}{b_T}\right)^2\right] \tag{10.15}$$

where b_T is the half-width. According to the results of wind tunnel experiments, b_T is proportional to the distance along the trajectory, s, which can be expressed as (Himoto, 2019):

$$b_T = 0.315s \tag{10.16}$$

The tilt of the plume axis changes with the fire source geometry. In the case of a square fire source, the default in this model, the plume trajectory is curved as shown in Figure 10.6 (Himoto, 2019):

$$\frac{z}{W} = 0.59\dot{Q}_W^{*1/3}U_W^{*-1}\left(\frac{x}{W}\right)^{2/3} \tag{10.17}$$

where $\dot{Q}_W^*$ is the dimensionless HRR defined in Eq. (10.7b) and U_W^* is the dimensionless wind velocity defined in Eq. (10.8c).

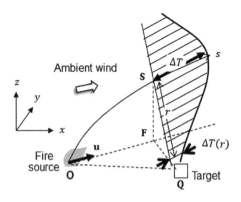

Figure 10.6 Temperature rise around a fuel object due to a wind-blown fire plume.

To calculate the temperature rise around a fuel object on the downwind side of the fire source, $\Delta T(r)$, the positional relationship between the fire source and the target fuel object needs to be identified. Assume that O and Q are the position vectors of the fire source and the target fuel object, respectively. F is the position vector of the foot of the perpendicular line from Q to the projection of the plume trajectory on the ground. S is the position vector of the intersection of the vertical line from F and the plume trajectory. The distance between the plume trajectory and the target fuel object, r, is nothing but the linear distance from S to Q, which can be calculated as:

$$r = \sqrt{|S - F|^2 + |Q - F|^2} \tag{10.18}$$

Among the two terms on the right-hand side of this equation, $|Q - F|$ can be calculated as the distance between a point and a line described in Chapter 9. Letting $t = |O - F|$ and denoting the unit direction vector of the wind as $\mathbf{u}$, we have $F = t\mathbf{u}$. As the two vectors, $t\mathbf{u}$ and $Q - F = Q - t\mathbf{u}$, are orthogonal, their inner product is zero. Thus, we obtain t as follows:

$$t = \mathbf{u} \cdot Q \tag{10.19}$$

As $\mathbf{u}$ and Q are given, we can readily obtain t from this equation. In addition, as the three position vectors, O, F, and Q, form a right triangle, we can compute $|Q - F|$ from the Pythagorean theorem. The other term in Eq. (10.17), $|S - F|$, is the height of the plume trajectory, z, which can be obtained by substituting $x = t$ into Eq. (10.17).

Note that a fuel object may be subject to the combined effect of multiple fire plumes in outdoor fires. This model calculates the temperature rise due to individual fire plumes separately. Based on a simple physical consideration on multiple fire plumes, they are composed using the following equation to evaluate the combined effect (Himoto et al., 2008):

$$\Delta T = \left(\sum_j \Delta T_j^{3/2} \right)^{2/3} \tag{10.20}$$

(3) SPOT: firebrands

As discussed in Chapter 8, the process of fire spread due to spot ignition of firebrands can be divided into three sub-processes: generation of firebrands from a burning fuel object, dispersal of firebrands in a crosswind, and ignition of recipient fuel object when firebrands have been deposited.

To calculate the total number of firebrands generated from a burning wood crib, we use the following equation obtained from a wind tunnel experiment (Miura et al., 2011):

$$\dot{N}''_{FB} = 3.11 \times 10^5 \dot{m}''^{3.57}_B \tag{10.21}$$

It should be noted that a substantial fraction of $\dot{N}''_{FB}$ will not actually contribute to causing spot ignitions due to a lack of thermal energy for heating fuel objects upon deposition. This is partly because most firebrands are fine fragments of burning wood sticks that cease combustion before falling on the ground. Presently, we do not have a simple method to determine the fraction of firebrands generated from a fire source that sustains combustion and thus possesses sufficient thermal energy to potentially cause the ignition of fuel objects upon deposition. To address this issue, we impose a reduction factor to Eq. (10.21), rFB, for estimating the effective number of firebrands sustaining combustion. In this model, we assume $rFB = 1.0 \times 10^{-4} U_\infty$, but this is empirical and should be updated in response to future research development. An alternative approach to introducing the reduction factor would be to straightforwardly track the mass loss and combustion status of each firebrand upon generation until deposition. However, we did not employ this approach for maintaining complications in the calculation procedure simple.

The projected area of firebrands, A_P, released from the fire source is assumed to follow the log-normal distribution:

$$p(A_P|\lambda, \xi) = \frac{1}{\sqrt{2\pi}\xi A_P} \exp\left\{-\frac{1}{2}\left(\frac{\ln A_P - \lambda}{\xi}\right)^2\right\} \tag{10.22a}$$

$$\lambda = \ln\left(\frac{\mu_A}{\sqrt{1 + \sigma_A^2/\mu_A^2}}\right) \tag{10.22b}$$

$$\xi = \sqrt{\ln\left(1 + \frac{\sigma_A^2}{\mu_A^2}\right)} \tag{10.22c}$$

where μ_A and σ_A are the mean and standard deviation of A_P, respectively. The μ_A and σ_A are evaluated using the following power-law model, which were obtained by regressing the results of wind tunnel and full-scale burn experiments (Hayashi et al., 2014; Himoto et al., 2022):

$$\left.\begin{aligned} \frac{\mu_A}{d^2} &= 0.445 W^{*1.07} \\ \frac{\sigma_A}{d^2} &= 0.0107 \end{aligned}\right\} \tag{10.23}$$

where W^* is the dimensionless parameter defined by (Himoto et al., 2021),

$$W^* = \frac{\rho_\infty}{\rho_P} \frac{W_0^2}{gd}$$ (10.24)

where d is the representative width of the fuel component, ρ_∞ is the ambient gas density, ρ_P is the firebrand density, W_0 is the initial ejection velocity of firebrands, and g is the acceleration due to gravity. Based on the discussion on turbulent diffusion flames in Chapter 5, we use the following equation for evaluating W_0 (Cox et al., 1980; Quintiere et al., 1998).

$$W_0 = \begin{cases} 6.83\dot{Q}^{1/5}\left(\dfrac{H}{\dot{Q}^{2/5}}\right)^{1/2} & \left(H \leq 0.08\dot{Q}^{2/5}\right) \\[2mm] 1.85\dot{Q}^{1/5} & \left(0.08\dot{Q}^{2/5} < H \leq 0.20\dot{Q}^{2/5}\right) \\[2mm] 1.08\dot{Q}^{1/5}\left(\dfrac{H}{\dot{Q}^{2/5}}\right)^{-1/3} & \left(0.20\dot{Q}^{2/5} < H\right) \end{cases} \; (m/s)$$ (10.25)

where H is the height of the fuel object and $\dot{Q}$ is the HRR.

As schematically shown in Figure 10.7, the transport distance of firebrands in the downwind direction (x-axis), x_P, and in the direction perpendicular to the wind direction (y-axis), y_P, are assumed to follow a

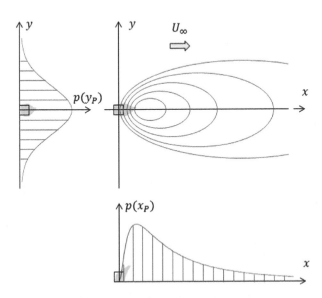

Figure 10.7 Probability distribution of firebrand deposition downwind of a fire source.

log-normal and a normal distribution, respectively (Hayashi et al., 2014; Himoto et al., 2022):

$$p(x_P|\lambda,\xi) = \frac{1}{\sqrt{2\pi}\xi x_P}\exp\left\{-\frac{1}{2}\left(\frac{\ln x_P - \lambda}{\xi}\right)^2\right\} \tag{10.26a}$$

$$p(y_P|\mu_y,\sigma_y) = \frac{1}{\sqrt{2\pi}\sigma_y}\exp\left\{-\frac{1}{2}\left(\frac{y_P - \mu_y}{\sigma_y}\right)^2\right\} \tag{10.26b}$$

where λ and ξ are the parameters of the log-normal distribution in Eq. (10.26a), which can be expressed as functions of the average, μ_x, and standard deviation, σ_x, of the transport distance as follows:

$$\lambda = \ln\left(\frac{\mu_x}{\sqrt{1 + \sigma_x^2/\mu_x^2}}\right) \tag{10.27a}$$

$$\xi = \sqrt{\ln\left(1 + \frac{\sigma_x^2}{\mu_x^2}\right)} \tag{10.27b}$$

Thus, the averages and the standard deviations of x_P and y_P are required to evaluate the dispersion of firebrands using Eq. (10.26). The following regression models are used to evaluate μ and σ (Hayashi et al., 2014; Himoto et al., 2022):

$$\left.\begin{aligned} \frac{\mu_x}{d} &= 41.0B^{*1.06} \\ \frac{\sigma_x}{d} &= 4.52 \end{aligned}\right\} \tag{10.28a}$$

$$\left.\begin{aligned} \mu_y &= 0 \\ \frac{\sigma_y}{H} &= 10 \end{aligned}\right\} \tag{10.28b}$$

where B^* is the governing dimensionless parameter defined by (Himoto et al., 2021):

$$B^* = \left(\frac{U_\infty W_0}{g\sqrt{d_P d}}\right)^2\left(\frac{\rho_\infty}{\rho_P}\right)\left(1 + \sqrt{1 + \frac{2gH}{W_0^2}}\right)^2 \tag{10.29}$$

The above procedure calculates the transport distance of a firebrand, x_P and y_P, relative to the burning fuel object with the x-axis corresponding to the wind direction. However, the position of the burning fuel object does not necessarily correspond to the origin of the reference coordinate system introduced in the computational domain. The relative coordinates, x_P and y_P, need to be converted to the reference coordinates, x'_P and y'_P, using the following coordinate transformation:

$$\begin{pmatrix} x'_P \\ y'_P \end{pmatrix} = \begin{pmatrix} x'_{P,0} \\ y'_{P,0} \end{pmatrix} + \begin{pmatrix} \cos\theta & \sin\theta \\ -\sin\theta & \cos\theta \end{pmatrix} \begin{pmatrix} x_P \\ y_P \end{pmatrix} \tag{10.30}$$

where $x'_{P,0}$ and $y'_{P,0}$ are the reference coordinates of the burning fuel object, and θ is the wind direction. As discussed in Section 10.3, the computational domain is divided into grid meshes (see Figure 10.3). We specify the recipient fuel object of a fallen firebrand based on the mesh inclusion relationship. Once the corresponding mesh is specified, we probabilistically determine one fuel object in this mesh according to the object-to-mesh area ratio.

The ignition of the recipient fuel object is probabilistically determined by the following logistic curve, as discussed in Chapter 8 (Urban et al., 2019b):

$$p_{SI} = \frac{1}{1 + \exp(-f)} \tag{10.31a}$$

$$f = 5.2591 - 1.186 \times 10^{-4} d_{fb}^{-2} - 1.5528 \times 10^{-1} FMC \tag{10.31b}$$

Eq. (10.30) was derived from an experiment using coastal redwoods sawdust as the recipient fuel (Urban et al., 2019b). As a recipient fuel, the difference between a sawdust bed and a wood crib is the contact area between deposited firebrands, affecting the rate of heat transfer. In addition, this equation determines the onset of smoldering combustion. However, it generally takes time to further progress to flaming combustion by which the heating to other fuel objects starts. Thus, it should be noted that the ignition probability by Eq. (10.30) yields an overestimate. The reduction factor, rFB, introduced earlier, also works as an adjustment for this overestimation.

10.4 CASE STUDY

Using the Python script that implements the aforementioned computational procedures, fire spread simulations were performed in a hypothetical arrangement of fuel objects.

10.4.1 Simulation conditions

Actual wildlands or urban areas are complex arrangements of various fuel objects with different burning characteristics and shapes. However,

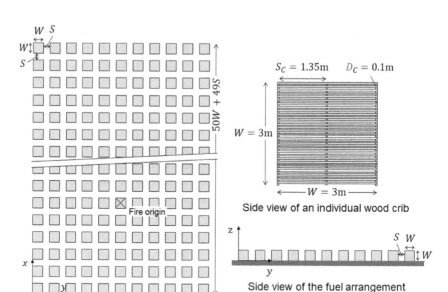

Figure 10.8 Equally spaced lattice arrangement of wood cribs on flat ground. Positive direction of x-axis corresponds to the wind direction.

in this case study, we considered an equally spaced lattice arrangement of cubic wood cribs with a side length of $W = 3$ m on flat ground (rectangular coordinate system xyz) as shown in Figure 10.8. The total number of arranged wood cribs was 550, with 50 and 11 in the x- and y-axis directions, respectively. The separation distance between the wood cribs, S(m), was varied. The fire was assumed to start from the fuel object marked with a cross in Figure 10.8. The process of successive fire spread from the fire origin to the other fuel objects was predicted under different wind direction and velocity conditions.

The wood crib was an assembly of Douglas fir sticks, often used as a fire source in various fire experiments. Wood cribs were selected as the fuel object in this case study because their burning behavior has been relatively well investigated. Their simple geometry was also desirable for understanding the computational flow of spatial data processing. The fuel combustion inside an enclosure, as expected in an urban fire, was not implemented to maintain the simplicity of the overall model structure.

The wood sticks comprising a wood crib had a cross-sectional length of $D_C = 0.1$ m and a gap spacing between sticks of $S_C = 1.35$ m. In this case, a wood crib consisted of 30 layers of wood sticks with three sticks in each

Table 10.2 Conditions of the assumed wood crib for fire spread simulation (refer to Chapters 3 and 7 for the thermo-physical properties of Douglas fir).

Species	Douglas fir
Size of a wood crib	3.0 m × 3.0 m ×3.0 m
Size of a component wood stick	0.1 m × 0.1 m ×3.0 m (D_c = 0.1 m)
Spacing between wood sticks S_c	1.35 m
Number of component wood sticks	90
Total weight	1,377 kg
Density ρ	510 kg·m^{-3}
Thermal inertia $k\rho c$	0.25 kJ^2·s^{-1}·m^{-4}·K^{-2}
Ignition temperature (piloted) T_{ig}	657 K
Critical ignition heat flux $\dot{q}''_{cr}$	16.0 kW·m^{-2}
Heat of combustion ΔH	16.4 kJ·g^{-1}

layer, which was 90 sticks in total. The porosity of this wood crib was 90%. Assuming the density of Douglas fir as 510 kg·m^{-3}, the weight of one wood crib was 1,377 kg. The other conditions of the assumed wood crib are listed in Table 10.2. Note that the transition in the MLR of the assumed wood crib corresponds to the one shown in the top left panel of Figure 10.4.

The simulation conditions were specified in three CSV files, 'o.csv', 'v.csv', and 'p.csv'. Among these, 'o.csv' specified the outline of the combustible space and the weather conditions, such as the ambient temperature, and the wind direction and velocity. 'v.csv' specified the vertex coordinates (x, y, z) of 550 fuel objects. 'p.csv' defined the face polygons that compose each fuel object by specifying the combination of the vertices defined in 'v.csv'. Specifications of each file are given below.

(1) Outline of the simulation: o.csv

The first 10 lines of 'o.csv', which specified the outline of the present simulation, are displayed in Figure 10.9.

The first line comprises two integers that represent the overall combustible space. The first and second items are the total number of fuel objects and the ID of the fire origin, respectively.

The second line is a real number representing the ambient temperature in K.

The third line comprises three real numbers on the ambient wind condition. The first item is the wind velocity in m·s^{-1}. The second and third items are the x- and y-components of the unit wind vector, respectively. For simplicity, we did not assume a change in the wind condition during the simulation.

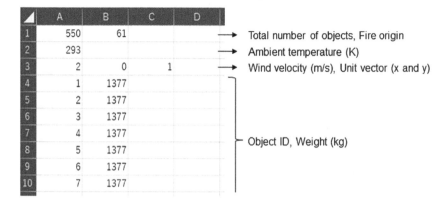

▲	A	B	C	D	
1	550	61			→ Total number of objects, Fire origin
2	293				→ Ambient temperature (K)
3	2	0	1		→ Wind velocity (m/s), Unit vector (x and y)
4	1	1377			⎤
5	2	1377			
6	3	1377			
7	4	1377			⎬ Object ID, Weight (kg)
8	5	1377			
9	6	1377			
10	7	1377			⎦

Figure 10.9 'o.csv' specified the outline of the present simulation (displays the first 10 lines).

The fourth and subsequent lines comprise data on individual fuel objects. The first item on each line is the object ID, and the second item is the weight (kg). The other object-related data are automatically calculated from the data specified in 'v.csv' and 'p.csv'.

(2) Vertices of the fuel objects: v.csv

The first 10 lines of 'v.csv', which specified the vertices of the fuel objects, are shown in Figure 10.10. As described above, each wood crib was composed of 90 wood sticks. However, we do not model the shape of individual sticks, but the shape of the enveloping cubic body. Thus, the number of vertices per object was eight.

'v.csv' specifies the coordinates of one vertex per line. The first through third items are the object ID, the number of vertices composing the object, and the serial number per object, respectively. The fourth through sixth items are the x-, y-, and z-coordinates, respectively. Although the sequence of vertices matters when defining face polygons in 'p.csv', 'v.csv' only defines the vertex coordinates. Thus, there is no particular rule for their sequence. However, in the sample data, the four vertices on the bottom of the cube came first, followed by the four vertices on the top. The four vertices on each horizontal face were aligned counterclockwise when the wood crib was viewed from above.

(3) Face polygons composing combustible objects: p.csv

The first 10 lines of 'p.csv', which specified the face polygons composing each fuel object, are shown in Figure 10.11. 'p.csv' defined the six face polygons using the vertex coordinates defined in 'v.csv'.

	A	B	C	D	E	F	G	
	Object ID	Total number of vertices	Serial #	Coordinate (x, y, and z) (m)				
1	1	8	1	0	0	0		⎤
2	1	8	2	3	0	0		
3	1	8	3	3	3	0		
4	1	8	4	0	3	0		ID=1
5	1	8	5	0	0	3		
6	1	8	6	3	0	3		
7	1	8	7	3	3	3		
8	1	8	8	0	3	3		⎦
9	2	8	1	6	0	0		⎤ ID=2
10	2	8	2	9	0	0		⎦

Figure 10.10 'v.csv' specified the vertices of the fuel objects (displays the first 10 lines).

	A	B	C	D	E	F	G	H	
	Object ID	Total number of polygons	Serial #	Exterior? (1: Yes, 0: No)	Vertex ID ("serial #" from "v.csv")				
1	1	6	1	1	1	2	6	5	⎤
2	1	6	2	1	2	3	7	6	
3	1	6	3	1	3	4	8	7	ID=1
4	1	6	4	1	4	1	5	8	
5	1	6	5	0	1	2	3	4	
6	1	6	6	1	5	6	7	8	⎦
7	2	6	1	1	1	2	6	5	⎤
8	2	6	2	1	2	3	7	6	
9	2	6	3	1	3	4	8	7	ID=2
10	2	6	4	1	4	1	5	8	

Figure 10.11 'p.csv' specified the face polygon data of each fuel object (displays the first 10 lines).

'p.csv' specifies data on one face polygon per line. Each line comprises eight numbers.

The first through third items are the object ID, the total number of face polygons that compose the object, and the serial number of the face polygon per object, respectively.

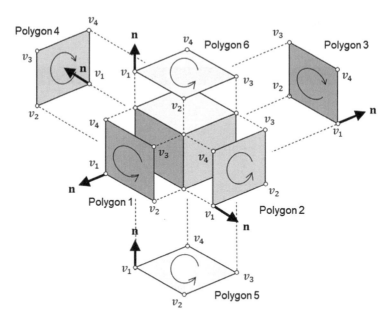

Figure 10.12 A cubic object with its component face polygons and vertices.

The fourth item is an identifier that takes '0' when the face polygon is the bottom face and '1' otherwise. This identifier determines whether the face polygon could be subject to external heating.

The fifth through eighth items are the four component vertices of the face polygon. The vertex IDs in 'p.csv' correspond to the serial numbers defined in 'v.csv'. In contrast to 'v.csv', the sequence of the four vertex IDs is meaningful. The vertex IDs composing each face polygon were assigned based on the discussion on the computational representation of the 3D shape in Chapter 9. When the face polygon was viewed from the outside, the vertices were aligned counterclockwise (Figure 10.12). In this way, the normal vectors of each face polygon required for judging the visibility from the other face polygons can be calculated systematically. The only exception was the bottom face, where the vertices were aligned clockwise when viewed from the outside. This is because the ground was not defined as a separate polygon in this case study. It was computationally more convenient to have the normal vector of the bottom face pointing toward the inside of the cube when judging the visibility from the other face polygons. It is noteworthy that although the number of vertices composing the face polygon in the sample data was four, an arbitrary number of vertices can be defined in the present simulation model.

Table 10.3 Cases tested in the fire spread simulation.

	Case A	Case B	Case C
Separation between objects S (m)	2	3	4
Wind velocity U_∞ (m·s^{-1})	0, 2, and 5		

10.4.2 Simulation results

In the fire spread simulation, the separation distance between fuel objects, S, was varied in three steps, 2, 3, and 4 m. We refer to these as Cases A, B, and C, respectively. In each case, the ambient wind velocity, U_∞, was varied in three steps, 0, 2, and 5 m·s^{-1}. Table 10.3 summarizes the simulation conditions considered.

The results of the fire spread simulations were visualized at 1-hour intervals in Figs. 10.13 through 10.15 for each tested case. Each figure also included the transition of the numbers of burning fuel objects (mc3.csv) and burnt out fuel objects (mc4.csv).

The rate of fire spread was the highest in Case A, where $S = 2$ m (Figure 10.13). When the wind velocity was $U_\infty = 0$ m·s^{-1}, the fire spread occurred symmetrically around the fire origin. The larger the wind velocity, U_∞, the larger the rate of fire spread in the downwind direction, and conversely, the smaller the rate of fire spread in the upwind direction. There was no significant difference in the transition of the number of burning fuel objects between the cases for $U_\infty = 0$ and $U_\infty = 2$ m·s^{-1}. This is because a decrease in the rate of fire spread in the upwind direction was canceled by an increase in the rate of fire spread in the downwind direction when $U_\infty = 2$ m·s^{-1}. The effect of spotting ignitions became prominent when $U_\infty = 5$ m·s^{-1}, exhibiting asymmetric fire spread in the downwind direction.

Due to a change in S from 2 to 3 m, the rate of fire spread in Case B was lower than in Case A (Figure 10.14). Although the overall patterns of fire spread were similar between the two cases, there was no fire spread in the upwind direction when $U_\infty = 5$ m·s^{-1}. This is due to a reduction in the radiative heating in the upwind direction by the tilt of the flames in a relatively strong wind.

In Case C (Figure 10.15), where $S = 4$ m, the fire spread from the fire origin to its adjacent fuel objects did not occur when $U_\infty = 0$ and 2 m·s^{-1}. Fire spread in the downwind direction occurred only when $U_\infty = 5$ m·s^{-1} due to spot ignitions. The initial fire spread was slow due to a relatively large separation distance between fuel objects. However, the number of burning fuel objects began to increase after 360 min from the start of the fire

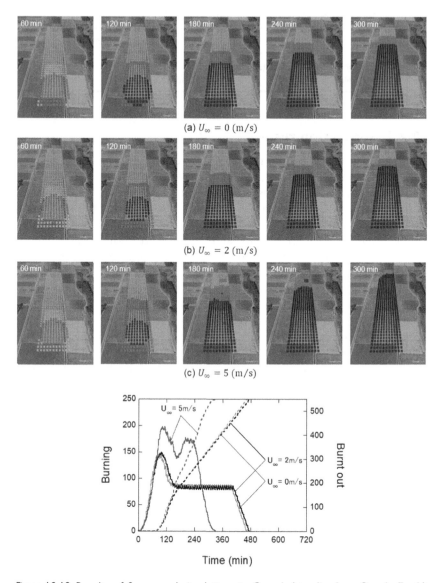

Figure 10.13 Results of fire spread simulations in Case A (visualized on Google Earth). The separation between fuel objects was $S = 2$ m.

due to successive occurrences of spot ignitions, though this is not visualized in Figure 10.15. There was no fire spread in the upwind direction, and a large number of fuel objects remained unburnt. The fuel objects on the upwind side were not free from radiative heating from the burning objects.

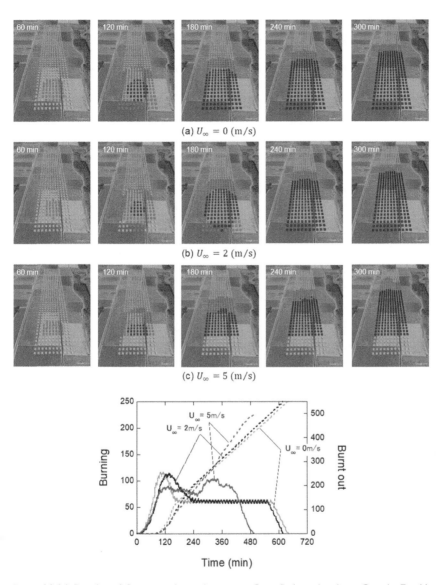

(a) $U_\infty = 0$ (m/s)

(b) $U_\infty = 2$ (m/s)

(c) $U_\infty = 5$ (m/s)

Figure 10.14 Results of fire spread simulations in Case B (visualized on Google Earth). The separation between fuel objects was $S = 3$ m.

However, the intensity was not high enough to cause ignitions due to the wind-induced tilt of the flames to the upwind side.

The simulation model was developed by integrating several independent models of physical processes involved in fire spread between fuel objects.

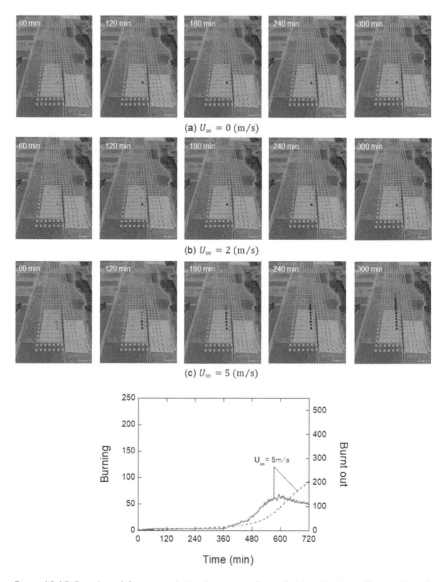

(a) $U_\infty = 0$ (m/s)

(b) $U_\infty = 2$ (m/s)

(c) $U_\infty = 5$ (m/s)

Figure 10.15 Results of fire spread simulations in Case C (visualized on Google Earth). The separation between fuel objects was $S = 4$ m.

The adopted models for each physical process were validated through comparison with relevant experimental data. However, the validity of the integrated model has not been fully examined. Comparing the results of the integrated model with those of appropriate experiments was desirable, but this was not accomplished. Despite such limitations, the results of the above case studies demonstrated that the integrated model could simulate the behavior of fire spread in a physically accountable manner. However, the assumptions employed in developing the integrated model are relatively simple among the available options. Users are encouraged to modify the model as necessary to improve the accuracy of predictions.

References

Aburano K, Yamanaka H, Ohmiya Y, Suzuki K, Tanaka T, Wakamatsu T. 1999. "Survey and analysis on surface area of fire load". *Fire Science and Technology* 19: 11–25. doi: 10.3210/fst.19.11

Adusumilli S, Chaplen JE, Blunck DL. 2021. "Firebrand generation rates at the source for trees and a shrub", *Frontiers in Mechanical Engineering* 7: 655593. doi: 10.3389/fmech.2021.655593

Akita K. 1959. "Studies on the mechanism of ignition of wood", *Report of Fire Research Institute of Japan*, Vol.9, No.1–2 (in Japanese).

Albini FA.1967. "A physical model for fire spread in brush", *Proc. Symposium (International) on Combustion* 11: 553–560. doi: 10.1016/S0082-0784(67)80180-2

Albini FA. 1976. "Estimating wildfire behavior and effects", *USDA Forest Service General Technical Report* INT-30, USDA Forest Service.

Albini FA. 1979. "Spot fire distance from burning trees – a predictive model", *USDS Forest Service General Technical Report* INT-56.

Albini FA. 1980. "Thermochemical properties of flame gases from fine wildland fuels", *USDA Forest Service Research Paper* INT-243, USDA Forest Service.

Albini FA. 1981. "A model for the wind-blown flame from a line fire", *Combustion and Flame* 43: 155–174. doi: 10.1016/0010-2180(81)90014-6

Albini FA. 1985. "A model for fire spread in wildland fuels by radiation", *Combustion Science and Technology* 42: 229–258. doi: 10.1080/00102208508960381

Albini FA, Alexander ME, Cruz MG. 2012. "A mathematical model for predicting the maximum potential spotting distance from a crown fire", *International Journal of Wildland Fire* 21: 609–627. doi: 10.1071/WF11020

Alexander ME, Stefner CN, Mason JA, Stocks BJ, Hartley GR, Maffey ME, Wotton BM, Taylor SW, Lavoie N, Dalrymple GN. 2004. "Characterizing the Jack pine – Black spruce fuel complex of the International Crown Fire Modelling Experiment (ICFME)", *Information Report*, NOR-X-393, Northern Forestry Centre, Canadian Forest Service.

Almeida M, Porto L, Viegas D. 2021. "Characterization of firebrands released from different burning tree species", *Frontiers in Mechanical Engineering* 7: 651135. doi: 10.3389/fmech.2021.651135

Anand C, Shotorban B, Mahalingam S. 2018. "Dispersion and deposition of firebrands in a turbulent boundary layer", *International Journal of Multiphase Flow* 109: 98–113. doi: 10.1016/j.ijmultiphaseflow.2018.07.012

Anderson DH, Catchpole EA, de Mestre NJ, Parkes T. 1982. "Modelling the spread of grass fires", *Journal of the Australian Mathematical Society* (B) 23: 451–466. doi: 10.1017/S0334270000000394

Anderson HE. 1969. "Heat transfer and fire spread", *Research Paper* INT-69, USDA Forest Service.

Anthenien RA, Tse SD, Fernandez-Pello AC. 2006. "On the trajectories of embers initially elevated or lofted by small scale ground fire plumes in high winds", *Fire Safety Journal* 41: 349–353. doi: 10.1016/j.firesaf.2006.01.005

Aoki Y. "Stochastic one-dimensional discrete model for urban fire spread", *Journal of Architecture and Planning* 381: 111–121. doi: 10.3130/aijax.381.0_111 (in Japanese).

Arya SP. 2001. *Introduction to Micrometeorology*, 2nd Ed., Academic Press.

Atallah S, Allan DS. "Safe separation distances from liquid fuel fires", *Fire Technology* 7: 47–56. doi: 10.1007/BF02588942

Atallah S, Raj P. 1974. "Radiation from LNG fires", *LNG Safety Program, Interim Report on Phase II Work*, Project IS-3-1, American Gas Association.

Babrauskas V. 2002a. "Heat release rates", in DiNenno PJ ed., *SFPE Handbook of Fire Protection Engineering, 3rd Ed.*, National Fire Protection Association.

Babrauskas V. 2002b. "Ignition of wood: A review of the state of the art", *Journal of Fire Protection Engineering* 12:163–188. doi: 10.1177/10423910260620482

Barr BW, Ezekoye OA. 2013. "Thermo-mechanical modeling of firebrand breakage on a fractal tree", *Proc. Combustion Institute* 34: 2649–2656. doi: 10.1016/j.proci.2012.07.066

Baum HR, McCaffrey BJ. 1989. "Fire induced flow field – theory and experiment", *Fire Safety Science* 2: 129–148. doi: 10.3801/IAFSS.FSS.2-129

Bearinger ED, Hodges JL, Yang F, Rippe CM, Lattimer BY. 2021. "Localized heat transfer from firebrands to surfaces", *Fire Safety Journal* 120: 103037. doi: 10.1016/j.firesaf.2020.103037

Beyler CL. 1986. "Fire plumes and ceiling jets", *Fire Safety Journal* 11: 53–75. doi: 10.1016/0379-7112(86)90052-4

Blinov VI, Khudyakov GN. 1961. "Diffusion burning of liquids", Russian Academy of Sciences Publication.

Bova AS, Mell WE, Hoffman CM. 2016. "A comparison of level set and marker method for the simulation of wildland fire front propagation", *International Journal of Wildland Fire* 25: 229–241. doi: 10.1071/WF13178

Box GEP, Jenkins GM, Reinsel GC, Ljung GM. 2015. *Time Series Analysis: Forecasting and Control*, 5th Ed. John Wiley & Sons.

Briggs GA. 1965. "A plume rise model compared with observations", *Journal of the Air Pollution Control Association*, 15(9): 433–438, doi: 10.1080/00022470.1965.10468404

Brown JK, Bevins CD. 1986. "Surface fuel loadings and predicted fire behavior for vegetation types in the northern Rocky Mountains", *USDA Research Note*, INT-358, USDA Forest Service.

Burgen RE, Rothermel RC. 1984. "BEHAVE: Fire behavior prediction and fuel modeling system FUEL subsystem", *General Technical Report* INT-167, USDA Forest Service.

Byram GM. 1959. "Combustion of forest fuels", in Davis KP ed., *Forest Fire: Control and Use*. McGraw-Hill.

CalFire. 2013–2021. Incident archive, California Department of Forestry and Fire Protection. (www.fire.ca.gov/incidents/, accessed March 2022)

Carslaw HS, Jaeger JC. 1959. *Conduction of heat in solids*, 2nd Ed., Oxford University Press.

Catchpole WR, Catchpole EA, Butler BW, Rothermel RC, Morris GA, Latham DJ. 1998. "Rate of spread of free-burning fires in woody fuels in a wind tunnel", *Combustion Science and Technology* 131: 1–37. doi: 10.1080/00102209 808935753

Caton SE, Hakes RSP, Gollner MJ, Gorham DJ, Zhou A. 2017. "Review of pathway for building fire spread in the wildland urban interface", *Fire Technology* 53: 429–473. doi: 10.1007/s10694-016-0589-z

Caton-Kerr SE, Tohidi A, Gollner MJ. 2019. "Firebrand generation from thermally-degraded cylindrical wooden dowels", *Frontiers in Mechanical Engineering* 5. doi: 10.3389/fmech.2019.00032

Cetegen BM, Ahmed TA. 1993. "Experiments on the periodic instability of buoyant plumes and pool fires", *Combustion and Flame* 93: 157–184. doi: 10.1016/ 0010-2180(93)90090-P

Chalk PL, Corotis RB. 1980. "Probability model for design live loads", *Journal of the Structural Division* 106(10): 2017–2033. doi: 10.1061/JSDEAG.0005542

Cheney NP, Gould JS, Catchpole WR. 1998. "Prediction of fire spread in grasslands", *International Journal of Wildland Fire* 8: 1–13. doi: 10.1071/ WF9980001

Church CR, Snow JT, Dessens J. 1980. "Intense atmospheric vortices associated with a 1000 MW fire", *Bulletin of the American Meteorological Society* 61: 682–694. doi: 10.1175/1520-0477(1980)061<0682:IAVAWA>2.0.CO;2

Churchil SW, Chu HHS. 1975. "Correlating equations for laminar and turbulent free convection from a vertical plate", *International Journal of Heat and Mass Transfer* 18: 1323–1329. doi: 10.1016/0017-9310(75)90243-4

Churchil SW, Bernstein M. 1977. "A correlating equation for forced convection from gases and liquids to a circular cylinder in crossflow", *Journal of Heat Transfer – Transactions of the ASME* 99: 300–306. doi: 10.1115/1.3450685

Clarke KC, Brass JA, Riggan PJ. 1994. "A cellular automaton model of wildfire propagation and extinction" *Photogrammetric Engineering and Remote Sensing* 60(11): 1355–1367.

Cohen MF, Wallace JR. 1993. *Radiosity and Realistic Image Synthesis*, Academic Press.

Cousins J, Heron D, Mazzoni S, Thomas G, LLoydd D, 2002. "Estimating risks from fire following earthquake", *Institute of Geological & Nuclear Sciences Client Report* 2002/60.

Counihan J. 1975. "Adiabatic atmospheric boundary layers: A review and analysis of data from the period 1880-1972", *Atmospheric Environment* 79: 871–905. doi: 10.1016/0004-6981(75)90088-8

Crowe CT, Schwarzkopf JD, Sommerfeld M, Tsuji Y. 2011. *Multiphase Flows with Droplets and Particles*, 2nd Ed. CRC Press.

Cox G, Chitty R. 1980. "A study of the deterministic properties of unbounded fire plumes", *Combustion and Flame* 39: 191–209. doi: 10.1016/0010-2180(80)90016-4

Cruz MG, Alexander ME, Wakimoto RH. 2003. "Assessing canopy fuel stratum characteristics in crown fire prone fuel types of western North America", *International Journal of Wildland Fire* 12: 39–50. doi: 10.1071/WF02024

Cruz MG, Gould JS, Alexander ME, Sullivan AL, McCaw WL, Mathews S. 2015. *A Guide to Rate of Fire Spread Models for Australian Vegetation*, CSIRO.

Csanady GT. 1971. "Bent-over vapor plumes", *Journal of Applied Meteorology* 10: 36–42. doi: 10.1175/1520-0450(1971)010<0036:BOVP>2.0.CO;2

Culver CG. 1978. "Characteristics of fire loads in office buildings", *Fire Technology* 14: 51–60. doi: 10.1007/BF01997261

Dakka SM, Jackson GS, Torero JL. 2002. "Mechanisms controlling the degradation of poly(methyl methacrylate) prior to piloted ignition", *Proc. Combustion Institute* 29: 281–287. doi: 10.1016/S1540-7489(02)80038-4

Davidson GA, Slawson PR. 1982. "Effective source flux parameters for use in analytical plume rise models", *Atmospheric Environment* 16: 223–227. doi: 10.1016/0004-6981(82)90436-X

Dayan A, Tien CL. 1974. "Radiant heating from a cylindrical fire column", *Combustion Science and Technology* 9: 41–47. doi: 10.1080/00102207408960335

de Dios Rivera J, Davies GM, Jahn W. 2012. "Flammability and the heat of combustion of natural fuels: A review", *Combustion Science and Technology* 184: 224–242. doi: 10.1080/00102202.2011.630332

Delichatsios MA. 1987. "Air entrainment into buoyant jet flames and pool fires", *Combustion and Flame* 70: 33–46. doi: 10.1016/0010-2180(87)90157-X

Delichatsios MA. 2007. "A correlation for the flame height in group fires", *Fire Science and Technology* 26: 1–8. doi: 10.3210/fst.26.1

de Mestre NJ, Catchpole EA, Anderson DH, Rothermel RC. 1989. "Uniform propagation of a planar fire front without wind", *Combustion Science and Technology* 65: 231–244. doi: 10.1080/00102208908924051

de Ris J. 1979. "Fire radiation – a review", *Proc. Symposium (International) Combustion* 17: 1003–1015. doi: 10.1016/S0082-0784(79)80097-1

DiDomizio MJ, Mulherin P, Weckman EJ. 2016. "Ignition of wood under time-varying radiant exposures", *Fire Safety Journal* 82: 131–144. doi: 10.1016/j.firesaf.2016.02.002

Dimopoulou M, Giannikos I. 2004. "Towards an integrated framework for forest fire control", *European Journal of Operational Research* 152: 476–486. doi: 10.1016/S0377-2217(03)00038-9

Dobson AJ, Barnett AG. 2008. *An Introduction to Generalized Linear Models*, 3rd Ed., CRC Press.

Drysdale D. 1998. *An Introduction to Fire Dynamics*, 2nd Edition. Wiley.

El Houssami M, Mueller E, Thomas JC, Simeoni A, Filkov A, Skowronski N, Gallagher MR, Clark K, Kremens R. 2016. "Experimental procedures characterizing firebrand generation in wildland fires", *Fire Technology* 52: 731–751. doi:10.1007/s10694-015-0492-z

Ellis PF. 2000. *The Aerodynamic and Combustion Characteristics of Eucalypt Bark – A Firebrand Study*, phD thesis, Australian National University.

Etlinger MG, Beall FC. 2004. "Development of a laboratory protocol for fire performance of landscape plants", *International Journal of Wildland Fire* 13: 479–488. doi: 10.1071/WF04039

Evans DD, Rehm RG, Baker ES. 2004. "Physics-based modeling for WUI fire spread – simplified model algorithm for ignition of structures by burning vegetation", *NISTIR* 7179, National Institute of Standards and Technology.

Fernandez-Pello AC. 1995. "The solid phase", in Cox G ed., *Combustion Fundamentals of Fire*, Academic Press.

Fernandez-Pello AC. 2017. "Wildland fire spot by sparks and firebrands", *Fire Safety Journal* 91: 2–10. doi: 10.1016/j.firesaf.2017.04.040

Ferreira-Leite F, Lourenço L, Bento-Gonçalves A. 2013. "Large forest fires in mainland Portugal, brief characterization", *Journal of Mediterranean Geography* 121: 53–65. doi: 10.4000/mediterranee.6863

Ferrero F, Muñoz M, Arnaldos J. 2007. "Effects of thin-layer boilover on flame geometry and dynamics in large hydrocarbon pool fires", *Fuel Processing Technology* 88(3): 227–235. doi: 10.1016/j.fuproc.2006.09.005

Ferziger JH, Perić M. 2002. *Computational Methods in Fluid Dynamics*, 3rd ed., Springer.

Filkov A, Prohanov S, Mueller E, Kasymov D, Martynov P, El Houssami M, Thomas J, Skowronski N, Butler B, Gallagher M, Clark K, Mell W, Kremens R, Hadden RM, Simeoni A. 2017. "Investigation of firebrand production during prescribed fires conducted in a pine forest", *Proc. of Combustion Institute* 36: 3263–3270. doi: 10.1016/j.proci.2016.06.125

Filippi JB, Bosseur F, Mari C, Lac C, Le Moigne P, Cuenot B, Veynante D, Cariolle D, Balbi JH. 2009. "Coupled atmosphere-wildland fire modelling", *Journal of Advances in Modeling Earth Systems* 1: #11. doi: 10.3894/JAMES.2009.1.11

Finney MA. 1998. "FARSITE: Fire ARea Simulator – model development and evaluation", *USDA Research Paper*, RMRS-RP-4, USDA Forest Service.

Fire Research Institute. 1976. "Reconnaissance report on the fire spread behavior of the Sakata Fire", *Technical Report of FRI*, No. 11 (in Japanese).

Fisher BEA, Metcalfe E, Vince I, Yates A. 2001. "Modelling plume rise and dispersion from pool fires", *Atmospheric Environment* 35(12): 2101–2110. doi: 10.1016/S1352-2310(00)00495-7

Flat Top Complex Wildfire Review Committee, 2012. *Flat Top Complex – Final Report*, Minister of Environment and Sustainable Resource Development

Fletcher CAJ. 1991. *Computational Techniques for Fluid Dynamics*, 2nd Ed., Springer.

Fons WL. "Analysis of fire spread in light forest fuels", *Journal of Agricultural Research* 72: 93–121.

Forestry Canada Fire Danger Group. 1992. *Development and Structure of the Canadian Forest Fire Behaviour Prediction System*, Canadian Department of Forestry. Inf Rep ST-X-3.

Fosberg MA. 1970. "Drying rates of heartwood below fiber saturation", *Forest Science* 16: 57–63. doi: 10.1093/forestscience/16.1.57

Frandsen WH. 1971. "Fire spread through porous fuels from the conservation of energy", *Combustion and Flame* 16: 9–16. doi: 10.1016/S0010-2180(71)80005-6

Frandsen WH, Andrews PL. 1979. "Fire behavior in nonuniform fuels", *USDA Forest Service Research Paper* INT-232, USDA.

Frankman D, Webb BW, Butler BW, Jimenez D, Forthofer JM, Sopko P, Shannon KS, Hiers JK, Ottmar RD. 2012. "Measurements of convective and radiative heating in wildland fires", *International Journal of Wildland Fire* 22(2): 157–167. doi: 10.1071/WF11097

Freitas SR, Longo KM, Trentmann J, Latham D. 2010. "Sensivity of 1-D smoke plume rise models to the inclusion of environmental wind drag", *Atmospheric Chemistry and Physics* 10: 585–594. doi: 10.5194/acp-10-585-2010

Fujita T. 1975. "Fire spread model and simulation", *Disaster Research* 8: 380–393 (in Japanese).

Fukuda Y, Kamikawa D, Hasemi Y, Kagiya K. 2004. "Flame characteristics of group fires", *Fire Science and Technology* 23: 164–169. doi: 10.3210/fst.23.164

Ganteaume A, Lampin-Maillet C, Guijarro M, Hernando C, Jappiot M, Fonturbel T, Pérez-Gorostiaga P, Vega JA. 2009. "Spot fires: fuel bed flammability and capability of firebrands to ignite fuel beds", *International Journal of Wildland Fire* 18(8) 951–969. doi: 10.1071/WF07111

Goldstein H, Poole CP, Safko JL. 2013. *Classical Mechanics*, Pearson New International ed. Pearson Education Limited.

Goodman TR. 1964. "Application of integral methods to transient nonlinear heat transfer", *Advances in Heat Transfer*, Vol. 1. Academic Press.

Gould JS, McCaw WL, Cheney NP, Ellis PF, Knight IK, Sullivan AL. 2008. *Project Vesta: Fire in Dry Eucalypt Forest: Fuel Structure, Fuel Dynamics and Fire Behaviour*, CSIRO Publishing.

Graham RT. 2003. "Hayman fire case study", *General Technical Report* RMRS-GTR-114, USDA Forest Service.

Grishin AM. 1996. "General mathematical model for forest fires and its applications", *Combustion, Explosion, and Shock Waves* 32: 503–519 doi: 10.1007/BF01998573

Gronli MG. 1996. *A Theoretical and Experimental Study of the Thermal Degradation of Biomass*, Doctoral thesis, The Norwegian University of Science and Technology.

Hadden RM, Scott S, Lautenberger C, Fernandez-Pello AC. 2011. "Ignition of combustible fuel beds by hot particles: An experimental and theoretical study", *Fire Technology* 47: 341–355. doi: 10.1007/s10694-010-0181-x

Hakes RSP, Salehizadeh H, Weston-Dawkes MJ, Gollner MJ. 2019. "Thermal characterization of firebrand piles", *Fire Safety Journal* 104: 34–42 doi: 10.1016/j.firesaf.2018.10.002

Hajilou M, Hu S, Roche T, Garg P, Gollner MJ. 2021. "A methodology for experimental quantification of firebrand generation from WUI fuels", *Fire Technology* 57: 2367–2385. doi: 10.1007/s10694-021-01119-9

Hamada M. 1951. "On the rate of fire spread", *Fire Research* I: 35–44, General Insurance Rating Organization of Japan (in Japanese).

Hamilton JD. 1994. *Time Series Analysis*, Princeton University Press.

Hamins A, Klassen M, Gore J, Kashiwagi T. 1991. "Estimate of flame radiance via a single location measurement in liquid pool fires", *Combustion and Flame* 86: 223–228. doi: 10.1016/0010-2180(91)90102-H

Hankinson G, Lowesmith BJ. 2012. "A consideration of methods of determining the radiative characteristics of jet fires", *Combustion and Flame* 159: 1165–1177. doi: 10.1016/j.combustflame.2011.09.004

Hanna SR, Briggs GA, Hosker RPJ. 1982. *Handbook on Atmospheric Diffusion*, U.S. Department of Energy DOE/TIC-11223.

Harada T, Hata T, Ishihara S. 1998. "Thermal constants of wood during the heating process measured with the laser flash method", *Journal of Wood Science* 44: 425–431. doi: 10.1007/BF00833405

Hasemi Y, Nishihata M. 1989. "Fuel shape effect on the deterministic properties of turbulent diffusion flames", *Fire Safety Science* 2: 275–284. doi: doi:10.3801/IAFSS.FSS.2-275

Hayashi Y, Hebiishi T, Izumi J, Naruse T, Itagaki N, Hashimoto T, Yasui N, Hasemi Y. 2014. "Firebrand deposition and measurements of collected firebrands generated and transported from a full-scale burn test using a large wooden building", *AIJ Journal of Technology and Design* 20: 605–610. doi: 10.3130/aijt.20.605 (in Japanese).

Hays RL. 1975. "The thermal conductivity of leaves", *Planta* 125: 281–287 doi: 10.1007/BF00385604

Hedayati F, Bahrani B, Zhou A, Quarles SL, Gorham DJ. 2019. "A framework to facilitate firebrand characterization", *Frontiers in Mechanical Engineering* 5. doi: 10.3389/fmech.2019.00043

Heskestad G. 1981. "Peak gas velocities and flame heights of buoyancy-controlled turbulent diffusion flames", *Proc. Combustion Institute* 18: 951–960. doi: 10.1016/S0082-0784(81)80099-9

Heskestad G. 1983a. "Luminous heights of turbulent diffusion flames", *Fire Safety Journal* 5: 103–108. doi: 10.1016/0379-7112(83)90002-4

Heskestad G. 1983b. "Virtual origins of fire plumes", *Fire Safety Journal* 5: 109–114. doi: 10.1016/0379-7112(83)90003-6

Heskestad G. 1984. "Engineering relations for fire plumes", *Fire Safety Journal* 7: 25–32. doi: 10.1016/0379-7112(84)90005-5

Himoto K. 2019. "Quantification of cross-wind effect on temperature elevation in the downwind region of fire sources", *Fire Safety Journal* 106: 114–123. doi: 10.1016/j.firesaf.2019.04.010

Himoto K, Tanaka T. 2002. "A physically-based model for urban fire spread", *Fire Safety Science* 7: 129–140. doi: 10.3801/IAFSS.FSS.7-129

Himoto K, Tanaka T. 2005. "Transport of disk-shaped firebrands in a turbulent boundary layer", *Fire Safety Science* 8: 433–444. doi: 10.3801/IAFSS.FSS.8-433

Himoto K, Tanaka T. 2008. "Development and validation of a physics-based urban fire spread model", *Fire Safety Journal* 43: 477–494. doi: 10.1016/j.firesaf.2007.12.008

Himoto K, Tsuchihashi T, Tanaka Y, Tanaka T. 2009a. "Modeling thermal behaviors of window flame ejected from a fire compartment", *Fire Safety Journal* 44: 230–240. doi: 10.1016/j.firesaf.2008.06.005

Himoto K, Tsuchihashi T, Tanaka Y, Tanaka T. 2009b. "Modeling the trajectory of window flames with regard to flow attachment to the adjacent wall", *Fire Safety Journal* 44: 250–258. doi: 10.1016/j.firesaf.2008.06.007

Himoto K, Deguchi Y. 2020. "Temperature elevation and trajectory in the downwind region of rectangular fire sources in cross-winds", *Fire Safety Journal* 116: 103183. doi: 10.1016/j.firesaf.2020.103183

Himoto K, Iwami T. 2021. "Generalization of probabilistically varying characteristics of the firebrand generation and transport from structural fire source", *Fire Safety Journal* 125: 103418. doi: 10.1016/j.firesaf.2021.103418

Himoto K, Hayashi Y. 2022. "Hierarchical Bayesian approach to developing probabilistic models for generation and transport of firebrands in large outdoor fires under limited data availability", *Fire Safety Journal* 134: 103679. doi: 10.1016/j.firesaf.2022.103679

Hirst E. 1971. *Analysis of Buoyant Jets Within the Zone of Flow Establishment*, Oak Ridge National Laboratory, ORNL-TM-3470.

Horiuchi S. 1972. *Fire Protection of Buildings*, Asakura Publishing (in Japanese).

Hottel HC. 1959. "Certain laws governing the diffusive burning of liquids – A review", *Fire Research Abstracts and Reviews* 1: 41–44.

Hottel HC, Williams GC, Steward FR. 1965. "The modeling of firespread through a fuel bed", *Proc. Symposium (International) on Combustion* 10: 997–1007. doi: 10.1016/S0082-0784(65)80242-9

Houf W, Schefer R. 2008. "Analytical and experimental investigation of small-scale unintended release of hydrogen", *International Journal of Hydrogen Energy* 33: 1435–1444. doi: 10.1016/j.ijhydene.2007.11.031

Hoult DP, Fay JA, Forney LJ. 1969. "A theory of plume rise compared with field observations", *Journal of Air Pollution Control Association* 19(8): 585–590. doi: 10.1080/00022470.1969.10466526

Hu L. 2017. "A review of physics and correlations of pool fire behavior in wind and future challenges", *Fire Safety Journal* 91: 41–55, doi: 10.1016/j.firesaf.2017.05.008

Hu L, Liu S, Xu Y, Li D. 2011 "A wind tunnel experimental study on burning rate enhancement behavior of gasoline pool fires by cross air flow", *Combustion and Flame* 158: 586–591. doi: 10.1016/j.combustflame.2010.10.013

Hu L, Liu S, de Ris JL, Wu L. 2013. "A new mathematical quantification of wind-blown flame tilt angle of hydrocarbon pool fires with a new global correlation model", *Fuel* 106: 730–736. doi: 10.1016/j.fuel.2012.10.075

Hu LH, Tang F, Delichatsios MA, Lu KH. 2013. "A mathematical model on lateral temperature profile of buoyant window plume from a compartment fire", *International Journal of Heat and Mass Transfer* 56: 447–453. doi: 10.1016/j.ijheatmasstransfer.2012.08.040

Huggett C. 1980. "Estimation of rate of heat release by means of oxygen consumption measurements", *Fire and Materials* 4: 61–65. doi: 10.1002/fam.810040202

Incropera FP, Dewitt DP, Bergman TL, Lavine AS. 2007. *Fundamentals of Heat and Mass Transfer*, 6th Ed. John Wiley & Sons.

ISO. 2021. *Large Outdoor Fires and the Built Environment — Global Overview of Different Approaches for Standardization*, ISO/DTR 24188.

Iwami T, Ohmiya Y, Hayashi Y. Kagiya K, Takahashi W, Naruse T. 2004. "Simulation of city fire", *Fire Science and Technology* 23: 132–140. doi: 10.3210/fst.23.132

Japan Society of Thermophysical Properties (JSTP). 2000. *Thermophysical Properties Handbook*, Yokendo (in Japanese).

Janssens M. 1991a. "A thermal model for piloted ignition of wood including variable thermophysical properties", *Fire Safety Science* 3: 167–176. doi: 10.3801/IAFSS.FSS.3-167

Janssens M. 1991b. "Piloted ignition of wood: A review", *Fire and Materials* 15: 151–167. doi: 10.1002/fam.810150402

Janssens M. 1994. "Thermo-physical properties for wood pyrolysis models", *Pacific Timber Engineering Conference 1994*: 607–618

Japan Association for Fire Science and Engineering (JAFSE). 1996. *Survey Report of Fires Following the 1995 Hyogo-ken Nanbu Earthquake* (in Japanese).

Jiang W, Wang F, Fang L, Zheng X, Qiao X, Li Z, Meng Q. 2021. "Modelling of wildland-urban interface fire spread within the heterogeneous cellular automata model", *Environmental Modelling and Software* 135: 104895. doi: 10.1016/j.envsoft.2020.104895

Johnson EA, Miyanishi K. ed. 2001. *Forest Fires – Behavior and Ecological Effects.* Academic Press.

Kalabokidis K, Athanasis N, Gagliardi F, Karayiannis F, Palaiologou S, Vasilakos C. 2013. "Virtual Fire: A web-based GIS platform for forest fire control", *Ecological Informatics* 16: 62–69. doi: 10.1016/j.ecoinf.2013.04.007

Karlsson B, Quintiere JG. 2000. *Enclosure Fire Dynamics*, CRC Press.

Kawagoe K. 1958. "Fire behaviour in rooms", *Report of the Building Research Institute*, No. 27.

Keane RE, Gray K, Bacciu V. 2012. "Spatial variability of wildland fuel characteristics in Northern Rocky Mountain ecosystems", *USDA Research Paper*, RMRS-RP-98,, USDA Forest Service.

Khorasani NE. 2014. "Fire load: Survey data, recent standards, and probabilistic models for office buildings", *Engineering Structures* 58: 152–165. doi: 10.1016/j.engstruct.2013.07.042

Kolb G, Torero JL, Most JM, Joulain P. 1997. "Cross flow effects on the flame height of an intermediate scale diffusion flame", *Proc. International Symposium on Fire Science and Technology* 169–177.

Kondo J. 2000. *Atmospheric Science near the Ground Surface*, University of Tokyo Press (in Japanese).

Koo E, Pagni PJ, Weise DR, Woycheese JP. 2010. "Firebrand and spotting ignition in large-scale fires", *International Journal of Wildland Fire* 19: 818–843. doi: 10.1071/WF07119

Koo E, Linn RR. Pagni PJ, Edminster CB. 2012. "Modeling firebrand transport in wildfires using HIGRAD/FIRETEC", *International Journal of Wildland Fire* 21: 396–417. doi: 10.1071/WF09146

Kortas S, Mindykowski P, Consalvi JL, Mhiri H, Porterie B. 2009. "Experimental validation of a numerical model for the transport of firebrands", *Fire Safety Journal* 44: 1095–1102. doi: 10.1016/j.firesaf.2009.08.001

Koseki H. 1989. "Combustion Properties of large liquid pool fires", *Fire Technology* 25: 241–255. doi: 10.1007/BF01039781

Koshimura S, Oie T, Yanagisawa H, Imamura F. 2009. "Developing fragility functions for tsunami damage estimation using numerical model and post-tsunami data from Banda Aceh, Indonesia", *Coastal Engineering Journal* 51: 243–273. doi: 10.1142/S0578563409002004

Kourtz SR, O'Regan WG. 1971. "A model for a small forest fire to simulate burned and burning areas for use in a detection model", *Forest Science* 17: 163–169. doi: 10.1093/forestscience/17.2.163

Küçük Ö, Bilgili E, Sağlam B. 2008. "Estimating crown fuel loading for Calabrian pine and Anatolian black pine", *International Journal of Wildland Fire* 2008: 17: 147–154. doi: 10.1071/WF06092

Lam CS, Weckman EJ. 2015. "Wind-blown pool fires, part II: Comparison of measured flame geometry with semi-empirical correlations", *Fire Safety Journal* 78: 130–141. doi: 10.1016/j.firesaf.2015.08.004

Lautenberger C. 2013. "Wildland fire modeling with an Eulerian level set method and automated calibration", *Fire Safety Journal* 62: 289–298. doi: 10.1016/j.firesaf.2013.08.014

Lautkaski R. 1992. "Validation of flame drag correlations with data from large pool fires", *Journal of Loss Prevention in the Process Industries* 5: 175–180. doi: 10.1016/0950-4230(92)80021-Y

Lawson D, Simms D. 1952. "The ignition of wood by radiation", *British Journal of Applied Physics* 3: 288–292 doi: 10.1088/0508-3443/3/9/305

Lee BS, Alexander ME, Hawkes BC, Lynham TJ, Stocks BJ, Englefield P. 2002. "Information systems in support of wildland fire management decision making in Canada", *Computers and Electronics in Agriculture* 37: 185–198. doi: 10.1016/S0168-1699(02)00120-5

Lee S, Davidson R, Ohnishi N, Scawthorn C. 2008. "Fire following earthquake—reviewing the state-of-the-art of modeling", *Earthquake Spectra* 24: 933–967. doi: 10.1193/1.2977493

Lee SL, Emmons HW. 1961. "A study of natural convection above a line fire", *Journal of Fluid Mechanics* 11(3): 353–368. doi: 10.1017/S0022112061000573

Lee SL, Hellman JM. 1969. "Study of firebrand trajectories in a turbulent swirling natural convection plume", *Combustion and Flame* 13: 645–655. doi: 10.1016/0010-2180(69)90072-8

Lee SL, Hellman JM. 1970. "Firebrand trajectory using an empirical velocity-dependent burning law", *Combustion and Flame* 15: 265–274. doi: 10.1016/0010-2180(70)90006-4

Lee SW, Davidson RA. 2010. "Physics-based simulation model of post-earthquake fire spread", *Journal of Earthquake Engineering* 14: 670–687. doi: 10.1080/13632460903336928

Lee YP, Delichatsios MA, Silcock GWH. 2007. "Heat fluxes and flame heights in façades from fires in enclosures of varying geometry", *Proc. Combustion Institute* 31: 2521–2528. doi: 10.1016/j.proci.2006.08.033

Lee YP, Delichatsios MA, Ohmiya Y. "The physics of the outflow from the opening of an enclosure fire and re-examination of Yokoi's correlation", *Fire Safety Journal* 49: 82–88. doi: 10.1016/j.firesaf.2012.01.001

Li J, Mahalingam S, Weise DR. 2017. "Experimental investigation of fire propagation in single live shrubs", *International Journal of Wildland Fire* 26: 58–70. doi: 10.1071/WF16042

Lin Y, Delichatsios MA, Zhang X, Hu L. 2019. "Experimental study and physical analysis of flame geometry in pool fires under relatively strong cross flows". *Combustion and Flame* 205: 422–433. doi: 10.1016/j.combustflame.2019.04.025

Linn RR. 1997. *A Transport Model for Prediction of Wildfire Behavior*, Ph.D dissertation, New Mexico State University.

Linn R, Reisner J, Colman JJ, Winterkamp J. 2002. "Studying wildfire behavior using FIRETEC", *International Journal of Wildland Fire* 11: 233–246. doi: 10.1071/WF02007

Lu X, Zeng X, Xu Zhen Guan H. 2017. "Physics-based simulation and high-fidelity visualization of fire following earthquake considering building seismic damage", *Journal of Earthquake Engineering* 23: 1173–1193. doi: 10.1080/13632469.2017.1351409

Ma TG, Quintiere JG. 2003. "Numerical simulation of axi-symmetric fire plumes: accuracy and limitations", *Fire Safety Journal* 38: 467–492. doi: 10.1016/ S0379-7112(02)00082-6

MacLean JD. 1941. "Thermal conductivity of wood", *Transactions American Society of Heating and Ventilating Engineers*, No. 1184, pp. 323–354.

Mandel J, Beezley JD, Kochanski AK. 2011. "Coupled atmosphere-wildland fire modeling with WRF 3.3 and SFIRE 2011", *Geoscientific Model Development* 4: 591–610. doi: 10.5194/gmd-4-591-2011

Manzello SL, Maranghides A, Shields JR, Mell W, Hayashi Y, Nii D. 2008. "Mass and size distribution of firebrands generated from burning Korean Pine (Pinus Koraiensis) trees", *Fire and Materials* 33: 21–31. doi: 10.1002/fam.977

Manzello SL, Foote EID. 2014. "Characterizing firebrand exposure from wildland-urban interface (WUI) fires: Results from the 2007 Angora fire", *Fire Technology* 50: 105–124. doi: 10.1007/s10694-012-0295-4

Manzello SL, Suzuki S, Gollner MJ, Fernandez-Pello AC. 2020. "Role of firebrand combustion in large outdoor fire spread", *Progress in Energy and Combustion Science* 76. doi: 10.1016/j.pecs.2019.100801

Markstein GH. 1977. "Scaling of radiative characteristics of turbulent diffusion flames", *Proc. Symp. (International) on Combustion* 16: 1407–1419. doi: 10.1016/ S0082-0784(77)80425-6

Marro M, Salozzoni P, Cierco FX, Korsakissok I, Danzi E, Soulhac. 2014. "Plume rise and spread in buoyant releases from elevated sources in the lower atmosphere", *Environmental Fluid Mechanics* 14: 201–219. doi: 10.1007/ s10652-013-9300-9

Matsuyama K, Fujita T, Kaneko H, Ohmiya Y, Tanaka T, Wakamatsu T. 1998. "A simple predictive method for room fire behavior", *Fire Science and Technology* 18: 23–32. doi: 10.3210/fst.18.23

McAllister S, Finney M. 2014. "Convection ignition of live forest fuels", *Fire Safety Science* 11: 1312–1325. doi: 10.3801/IAFSS.FSS.11-1312

McArthur AG. 1966. "Weather and grassland fire behaviour", *Australian Forestry and Timber Bureau Leaflet* 100, AFTB.

McCaffrey BJ. 1979. "Purely buoyant diffusion flames: Some experimental results", *NBSIR* 79-1910.

McCaffrey BJ, Quintiere JQ, Harkleroad MF. 1981. "Estimating room temperatures and the likelihood of flashover using fire test data correlations", *Fire Technology* 17: 98–119. doi: 10.1007/BF02479583

Mell W, Jenkins MA, Gould J, Cheney P. 2007. "A physics-based approach to modelling grassland fires", *International Journal of Wildland Fire* 16: 1–22. doi: 10.1071/WF06002

Mell W, Maranghides A, McDermott R, Manzello SL. 2009. "Numerical simulation and experiments of burning Douglas fir trees", *Combustion and Flame* 156: 2023–2041. doi: 10.1016/j.combustflame.2009.06.015

Mercer GN, Weber RO. 2001. "Fire plumes" in Johnson EA, Miyanishi K. ed. *Forest Fires – Behavior and Ecological Effects*. Academic Press.

Mitsopoulos ID, Dimitrakopoulos AP. 2007. "Canopy fuel characteristics and potential crown fire behavior in Aleppo pine (*Pinus halepensis* Mill.) forests", *Annals of Forest Science* 64: 287–299. doi: 10.1051/forest:2007006

Miura S, Hayashi Y, Ohmiya Y, Iwami T. 2011. "Experiments on firebrand generation", *Proc. Annual Meeting of Architectural Institute of Japan*, pp. 245–246 (in Japanese).

Mizutani Y. 1988. *Combustion Engineering*, 2nd. Ed., Morikita Publishing (in Japanese).

Modak S. 1977. "Thermal radiation from pool fires", *Combustion and Flame* 29: 177–192. doi: 10.1016/0010-2180(77)90106-7

Moghtaderi B, Novozhilov V, Fletcher DF, Kent JH. 1997. "A new correlation for bench-scale piloted ignition data of wood", *Fire Safety Journal* 29: 41–59. doi: 10.1016/S0379-7112(97)00004-0

Moorhouse J. 1982. "Scaling criteria for pool fires derived from large scale experiments", *The Assessment of Major Hazards*, The Institution of Chemical Engineers Symposium Series 71: 165–179.

Morandini F, Perez-Ramirez Y, Tihay V, Santoni PA, Barboni T. 2013. "Radiant, convective and heat release characterization of vegetation fire", *International Journal of Thermal Sciences* 70: 83–91. doi: 10.1016/j.ijthermalsci.2013.03.011

Morandini F, Santoni PA, Tramoni JB, Mell WE. 2019. "Experimental investigation of flammability and numerical study of combustion of shrub of rockrose under severe drought conditions", *Fire Safety Journal* 108: 102836. doi: 10.1016/j.firesaf.2019.102836

Morton BR, Taylor G, Turner JS. 1956. "Turbulent gravitational convection from maintained and instantaneous sources", *Proc. Royal Society* A 234: 1–23. doi: 10.1098/rspa.1956.0011

Morvan D, Dupuy JL. 2001. "Modeling of fire spread through a forest fuel bed using a multiphase formulation", *Combustion and Flame* 127: 1981–1994. doi: 10.1016/S0010-2180(01)00302-9

Morvan D, Méradji S, Accary G. 2009. "Physical modelling of fire spread in grasslands", *Fire Safety Journal* 44: 50–61. doi: 10.1016/j.firesaf.2008.03.004

Mudan KS. 1984. "Thermal radiation hazards from hydrocarbon fires", *Progress in Energy and Combustion Science* 10: 59–80. doi: 10.1016/0360-1285(84)90119-9

Muñoz M, Planas E, Ferrero F, Casal J. 2007. "Predicting the emissive power of hydrocarbon pool fires", *Journal of Hazardous Materials*, 144 (3): 725–729. doi: 10.1016/j.jhazmat.2007.01.121

Muraszew A, Fedele JB. 1976. "Statistical model for spot fire hazard", *Aerospace Report*, No. ATR-77(7588)-1.

Muraszew A, Fedele JB, Kuby WC. 1977. "Trajectory of firebrands in and out of fire whirls", *Combustion and Flame* 30: 321–324. doi: 10.1016/0010-2180(77)90081-5

Nakaya I, Tanaka T, Yoshida M, Steckler K. 1986. "Doorway flow induced by a propane fire", *Fire Safety Journal* 10: 185–195. doi: 10.1016/0379-7112(86)90015-9

Nelson RM, Jr. 2001. "Water relations of forest fuels", in Johnson EA, Miyakoshi K ed., *Forest Fires – Behavior and Ecological Effects*, Academic Press.

Noble IR, Gill AM, Bary GAV. 1980. "McArthur's fire-danger meters expressed as equations", *Australian Journal of Ecology* 5: 201–203. doi: 10.1111/j.1442-9993.1980.tb01243.x

Ohgai A, Gohnai Y, Watanabe K. 2007. "Cellular automata modeling of fire spread in built-up areas—A tool to aid community-based planning for disaster mitigation",

Computers, Environment and Urban Systems 31: 441–460. doi: 10.1016/j. compenvurbsys.2006.10.001

Ohkuma T, Kanda J, Tamura Y. 1996. *Wind-resistance Design of Buildings*, Kajima Institute Publishing (in Japanese).

Ohmiya Y, Tanaka T, Wakamatsu T. 1996. "Burning rate of fuels and generation limit of the external flames in compartment fire", *Fire Science and Technology* 16: 1–12. doi: 10.3210/fst.16.1

Ohmiya Y, Iwami T. 1996. "An investigation on the distribution of fire brands and spot fires due to a hotel fire", *Fire Science and Technology* 20: 27–35. doi: 10.3210/fst.20.27

Ohmiya Y, Hori Y. 2001. "Properties of external flame taking into consideration excess fuel gas ejected from fire compartment", *Journal of Architectural Planning and Environmental Engineering*, AIJ, 545: 1–8. doi: 10.3130/aija.66.1_7 (in Japanese).

Oka Y, Kurioka H, Satoh H, Sugawa O. 2000. "Modeling of unconfined flame tilt in cross-winds", *Fire Safety Science* 6: 1101–1112. doi: 10.3801/IAFSS.FSS.6-1101

Oke TR. 1987. *Boundary Layer Climates*, 2nd Ed., Routledge.

Oke TR. 1988. "Street design and urban canopy layer climate", *Energy and Buildings* 11: 103–113. doi: 10.1016/0378-7788(88)90026-6

Oran ES, Boris JP. 2001. *Numerical Simulation of Reactive Flow*, 2nd Ed., Cambridge University Press.

Pagni PJ. 1993. "Causes of the 20 October 1991 Oakland-hills conflagration", *Fire Safety Journal* 21: 331–339. doi: 10.1016/0379-7112(93)90020-Q

Pagni PJ, TG Peterson. 1973. "Flame spread through porous fuels", *Proc. Symposium (International) on Combustion* 14: 1099–1107. doi: 10.1016/S0082-0784(73)80099-2

Pastor E, Zárate L, Planas E, Arnaldos J. 2003. "Mathematical models and calculation systems for the study of wildland fire behaviour", *Progress in Energy and Combustion Science* 29: 139–153. doi: 10.1016/S0360-1285(03)00017-0

Penney G, Habibi D, Cattani M. 2020. *A Handbook of Wildfire Engineering*, Bushfire and Natural Hazards CRC.

Perry GLW. "Current approaches to modelling the spread of wildland fire: a review", *Progress in Physical Geography* 22: 222–245. doi: 10.1177/030913339802200204

Pitts WM. 1991. "Wind effects on fires", *Progress in Energy and Combustion Science*, 17: 83–134. doi: 10.1016/0360-1285(91)90017-H

Poinsot T, Veynante D. 2005. *Theoretical and Numerical Combustion*, 2nd Ed., Edwards.

Porter K, Kennedy R, Bachman R. 2007. "Creating fragility functions for performance-based earthquake engineering", *Earthquake Spectra* 23: 471–489. doi: 10.1193/1.2720892

Porterie B, Consalvi JL, Kaiss A, Loraud JC. 2005. "Predicting wildland fire behavior and emissions using a fine-scale physical model", *Numerical Heat Transfer, Part A: Applications*, 47: 571–591. doi: 10.1080/10407780590891362

Prahl J, Emmons HW. 1975. "Fire induced flow through an opening", *Combustion and Flame* 25: 369–385. doi: 10.1016/0010-2180(75)90109-1

Pyne SJ, Andrews PL, Laven RD. 1996. *Introduction to Wildland Fire*, 2nd Ed., John Wiley & Sons.

Quintiere J. 1981. "A simplified theory for generalizing results from a radiant panel rate of flame spread apparatus", *Fire and Materials* 5: 52–60. doi: 10.1002/fam.810050204

Quintiere JG. 1989. "Scaling applications in fire research", *Fire Safety Journal* 15: 3–29. doi: 10.1016/0379-7112(89)90045-3

Quintiere JG. 2006. *Fundamentals of Fire Phenomena*, John Wiley & Sons.

Quintiere JG, Harkleroad M. 1984. "New concepts for measuring flame spread properties", *NBSIR 84-2943*, National Bureau of Standards.

Quintiere JG, Grove BS. 1998. "A unified analysis for fire plumes", *Proc. Combustion Institute* 27: 2757–2766. doi: 10.1016/S0082-0784(98)80132-X

Raj PK. 2010. "A physical model and improved experimental data correlation for wind induced flame drag in pool fires", *Fire Technology* 46: 579–609. doi: 10.1007/s10694-009-0107-7

Rafi MM, Aziz T, Lodi SH. 2018. "A suggested model for mass fire spread", *Sustainable and Resilient Infrastructure* 5. 214–231. doi: 10.1080/23789689.2018.1519308

Ragland KW, Aerts DJ, Baker AJ. 1991. "Properties of wood for combustion analysis", *Bioresource Technology* 37: 161–168. doi: 10.1016/0960-8524(91)90205-X

Rehm R, Hamins A, Baum HR, McGrattan KB, Evans DD. 2002. "Community-scale fire spread", *NISTIR* 6891, National Institute of Standards and Technology.

Rehm R, McDermott RJ. 2009. "Fire-front propagation using the level set method", *NIST Technical Note* 1611, National Institute of Standards and Technology.

Rein G. 2013. Smouldering "Fires and natural fuels", in Belcher CM ed. *Fire Phenomena in the Earth System – An Interdisciplinary Approach to Fire Science*. Wiley and Sons. doi: 10.1002/9781118529539

Reinhardt E, Scott J, Gray K, Keane R. 2006. "Estimating canopy fuel characteristics in five conifer stands in the western United States using tree and stand measurements", *Canadian Journal of Forest Research* 36: 2803–2814. doi: 10.1139/x06-157

Ren AZ, Xie XY. 2004. "The simulation of post-earthquake fire-prone area based on GIS", *Journal of Fire Sciences* 22: 421–439. doi: 10.1177/0734904104042440

Riberiro LM, Oliveira L, Viegas DX. 2014. "The history of a large fire or how a series of events lead to 14000 hectares burned in 3 days", *Proc. International Conference on Forest Fire Research VII*. doi: 10.14195/978-989-26-0884-6_109

Rockett JA. 1976. "Fire induced flow in an enclosure", *Combustion Science and Technology* 12: 165–175. doi: 10.1080/00102207608946717

Rossetto T, Elnashai A. 2003. "Derivation of vulnerability functions for European-type RC structures based on observational data", *Engineering Structures* 25: 1241–1263. doi: 10.1016/S0141-0296(03)00060-9

Rothermel RC. 1972. "A mathematical model for predicting fire spread in wildland fuels", *USDA Forest Service Research Paper*, INT-115, USDA Forest Service.

Rouse H, Yih CS, Humphreys HW. 1952. "Gravitational convection from a boundary source", *Tellus* 4: 201–210. doi: 10.3402/tellusa.v4i3.8688

Saga T. 1990. "Temperature distribution of fire gas flow from a belt-shaped heat source under strong wind", *Journal of Structural and Construction Engineering* 408: 99–110. doi: 10.3130/aijsx.408.0_99 (in Japanese).

Sardoy N, Consalvi JL, Porterie B, Fernancez-Pello. 2007. "Modeling transport and combustion of firebrands from burning trees", *Combustion and Flame* 150: 151–169. doi: 10.1016/j.combustflame.2007.04.008

Sardoy N, Consalvi JL, Kaiss A, Fernancez-Pello, Porterie B. 2008. "Numerical study of ground-level distribution of firebrands generated by line fires", *Combustion and Flame* 154: 478–488. doi: 10.1016/j.combustflame.2008.05.006

Sarreshtehdari A, Khorasani NE. 2021. "Integrating the fire department response within a fire following earthquake framework for application in urban areas", *Fire Safety Journal* 124: 103397. doi: 10.1016/j.firesaf.2021.103397

Scott JH. 1999. NEXUS: "A system for assessing crown fire hazard", *Fire Management Notes* 59: 21–24.

Shen G, Zhou K, Wu F, Jiang J, Dou Z. 2019. "A model considering the flame volume for prediction of thermal radiation from pool fire", *Fire Technology* 55: 129–148. doi: 10.1007/s10694-018-0779-y

Shields TJ, Silcock GW, Murray JJ. 1994. "Evaluating ignition data using the flux time product", *Fire and Materials* 18: 243–254. doi: 10.1002/fam.810180407

Shinozuka M, Feng MQ, Lee J, Naganuma T. 2000. "Statistical analysis of fragility curves", *Journal of Engineering Mechanics* 126: 1224–1231. doi: 10.1061/(ASCE)0733-9399(2000)126:12(1224)

Shirley P, Marschner S. 2009. *Fundamentals of Computer Graphics*, 3rd. Ed., CRC Press.

Shokri M, Beyler CL. 1989. "Radiation from large pool fires", *Journal of Fire Protection Engineering* 1: 141–149. doi: 10. 1177/104239158900100404

Siegel R, Howell JR. 2002. *Thermal Radiation Heat Transfer*, 4th Ed., Taylor & Francis.

Silcock GWH, Shields TJ. 1995. "A protocol for analysis of time-to-ignition data from bench scale test", *Fire Safety Journal* 24: 75–95. doi: 10.1016/0379-7112(95)00003-C

Slawson PR, Csanady GT. 1967. "On the mean path of buoyant, bent-over chimney plumes", *Journal of Fluid Mechanics* 28(2): 311–322. doi: 10.1017/S0022112067002095

Song J, Liu N, Li H, Zhang L, Huang X. 2017. "The wind effect on the transport and burning of firebrands", *Fire Technology* 53: 1555–1568. doi: 10.1007/s10694-017-0647-1

Spearpoint MJ, Quintiere JG. 2001. "Predicting the piloted ignition of wood in the cone calorimeter using an integral model – Effect of species, grain orientation and heat flux", *Fire Safety Journal* 36: 391–415. doi: 10.1016/S0379-7112(00)00055-2

Steckler KD, Quintiere JG, Rinkinen WJ. 1982. "Flow induced by fire in a compartment", *NASA STI/Recon Technical Report N*, No. 83.

Steward FR. 1964. "Linear flame heights for various fuels", *Combustion and Flame* 8: 171–178. doi: 10.1016/0010-2180(64)90063-X

Stocks BJ, Alexander ME,Wotton BM, Stefner CN, Flannigan MD, Taylor SW, Lavoie N, Mason JA, Hartley GR, Maffey ME, Dalrymple GN, Blake TW, Cruz MG, Lanoville RA. 2004. "Crown fire behavior in a northern jack pine – black spruce forest", *Canadian Journal of Forest Research* 34: 1548–1560. doi: 10.1139/x04-054

Storey MA, Price OF, Sharples JJ, Bradstock RA. 2020. "Drivers of long-distance spotting during wildfires in south-eastern Australia", *International Journal of Wildland Fire* 29: 459–472. doi: doi.org/10.1071/WF19124

Stull RB. 1988. *An Introduction to Boundary Layer Meteorology*, Kluwer Academic Publishers.

Sugawa O, Satoh H, Oka Y. 1991. "Flame height from rectangular fire sources considering mixing factor", *Fire Safety Science* 3: 435–444. doi: 10.3801/IAFSS.FSS.3-435

Sugawa O, Takahashi W. 1993. "Flame height behavior from multi-fire sources", *Fire and Materials* 17: 111–117. doi: 10.1002/fam.810170303

Sullivan AL. 2009. "A review of wildland fire spread modelling, 1990-2007 1: Physical and quasi-physical models", *International Journal of Wildland Fire* 18: 349–368. doi: 10.1071/WF06143

Sullivan AL, McCaw WL, Cruz MG, Matthews S, Ellis PF. 2012. "Fuel, fire weather and fire behavior in Australian ecosystems", in Bradstock RA, Gill AM, Williams RJ ed. *Flammable Australia: Fire Regimes, Biodiversity and Ecosystems in a Changing World*, CSRIO Publishing.

Suzuki S, Brown A, Manzello S, Suzuki J, Hayashi Y. 2014. "Firebrands generated from a full-scale structure burning under well-controlled laboratory conditions", *Fire Safety Journal* 63: 43–51. doi: 10.1016/j.firesaf.2013.11.008

Suzuki S, Manzello SL. 2018. "Characteristics of firebrands collected from actual urban fires", *Fire Technology* 54: 1533–1546. doi: 10.1007/s10694-018-0751-x

Tajima S. 2020. *The Development of Radiation-based Vulnerability Functions for Building-to-Building Fire Spread Using 2016 Itoigawa Fire Data*, MS thesis, Tokyo University of Science.

Takeya S, Himoto K, Mizukami T, Kagiya K, Iwami T. 2017. "Report of the survey on the building damage by the large fire that occurred in Itoigawa city, Niigata Prefecture on December 22, 2016". *Technical Note of National Institute for Land and Infrastructure Management*, No. 980 (in Japanese).

Tanaka T. 1993. *Introduction to Fire Safety Engineering of Buildings*, Building Center of Japan (in Japanese).

Tanaka T, Sato M, Wakamatsu T. 1997. "Simple formula for ventilation controlled fire temperatures", *Fire Science and Technology* 17: 15–27. doi: 10.3210/fst.17.15

Tang F, Hu L, Delichatsios MA, Lu K, Zhu W. 2012. "Experimental study on flame height and temperature profile of buoyant window spill plume from an under-ventilated compartment fire", *International Journal of Heat and Mass Transfer* 55: 93–101. doi: 10.1016/j.ijheatmasstransfer.2011.08.045

Tang F, Hu L, Zhang X, Zhang X, Dong M. 2015. "Burning rate and flame tilt characteristics of radiation-controlled rectangular hydrocarbon pool fires with cross air flows in a reduced pressure", *Fuel* 139: 18–25. doi: 10.1016/j.fuel.2014.07.093

Tang T. 2017. "A physics-based approach to modeling wildland fire spread through porous fuel beds", *Theses and Dissertations – Mechanical Engineering* 84, University of Kentucky. doi: 10.13023/ETD.2017.027

Tao Z, Bathras B, Kwon B, Biallas B, Gollner MJ, Yang R. 2021. "Effect of firebrand size and geometry on heating from a smoldering pile under wind", *Fire Safety Journal*, 120: 103031, doi: 10.1016/j.firesaf.2020.103031

Tarifa CS, del Notario PP, Moreno FG. 1965. "On the flight paths and lifetimes of burning particles of wood". *Proc. Combustion Institute* 10: 1021–1037. doi: 10.1016/S0082-0784(65)80244-2

Tedim F, Leone V, McGee TK. 2020. *Extreme Wildfire Events and Disasters – Root Causes and New Management Strategies*, Elsevier.

Tennekes H, Lumley JL. 1972. *A First Course in Turbulence*, The MIT Press.

TenWolde A, McNatt JD, Krahn L. "Thermal properties of wood and wood panel products for use in buildings", *USDA Report* DOE/USDA-21697/1, USDA Forest Service.

Tewarson A. 2002. "Generation of heat and chemical compounds in fires", in DiNenno PJ ed., *SFPE Handbook of Fire Protection Engineering*, 3rd Ed., National Fire Protection Association.

Thekaekara MP. 1965. "Survey of the literature on the solar constant and the spectral distribution of solar radiant flux", *NASA SP-74*, National Aeronautics and Space Administration.

Thomas GC, Cousins WJ, Lloydd DA, Heron DW, Mazzoni S. 2002. "Post-earthquake fire spread between buildings estimating and costing extent in Wellington", *Fire Safety Science* 7: 691–702. doi: 10.3801/IAFSS.FSS.7-691

Thomas PH. 1963a. "The size of flames from natural fires", *Proc. Symposium (International) on Combustion* 9: 844–859. doi: 10.1016/S0082-0784(63)80091-0

Thomas PH. 1964. "The effect of wind on plumes from a line heat source", *Fire Research Note 572*.

Thomas PH. 1965. "Fire spread in wooden cribs: Part III The effect of wind", *Fire Research Note 600*, Fire Research Station.

Thomas PH. 1967. "Some aspects of the growth and spread of fire in the open", *Forestry* 40: 139–164. doi: 10.1093/forestry/40.2.139

Thomas PH. 1986. "Design guide: Structure fire safety CIB W14 Workshop report", *Fire Safety Journal* 10: 77–137. doi: 10.1016/0379-7112(86)90041-X

Thomas PH. 1995. "The growth of fire – ignition to full involvement", in Cox G. ed., *Combustion Fundamentals of Fire*, Academic Press.

Thomas PH, Webster CT, Raftery MM. 1961. "Some experiments on buoyant diffusion flames", *Combustion and Flame* 5: 459–367. doi: 10.1016/0010-2180(61)90117-1

Thomas PH, Pickard RW, Wraight HGH. 1963b. "On the size and orientation of buoyant diffusion flames and the effect of wind", *Fire Research Note 516*, Fire Research Station.

Thompson DK, Yip DA, Koo E, Linn R, Marshall G, Refai R, Schroeder D. 2022. "Quantifying firebrand production and transport using the acoustic analysis of in-fire cameras", *Fire Technology*. doi: 10.1007/s10694-021-01194-y

Thomson HE, Drysdale DD, Beyler CL. 1988. "An experimental evaluation of critical surface temperature as a criterion for piloted ignition of solid fuels", *Fire Safety Journal* 13: 185–196. doi: 10.1016/0379-7112(88)90014-8

Tien CL, Lee KY, Stretton AJ. 2002. "Radiation heat transfer", in DiNenno PJ ed., *SFPE Handbook of Fire Protection Engineering*, 3rd Ed., National Fire Protection Association.

Tohidi A, Kaye NB. 2016. "Highly buoyant bent-over plumes in a boundary layer", *Atmospheric Environment* 131: 97–114. doi: 10.1016/j.atmosenv.2016.01.046

Tohidi A, Kaye NB. 2017. "Stochastic modeling of firebrand shower scenarios", *Fire Safety Journal* 91: 91–102. doi: 10.1016/j.firesaf.2017.04.039

Tokyo Fire Department (TFD). 1997. "Analysis of causes and spread of the fires caused by an earthquake directly below Tokyo", *Report of the Fire Prevention Deliberation Council* (in Japanese).

Torero JL, Simeoni A. 2010. "Heat and mass transfer in fires: Scaling laws, ignition of solid fuels and application to forest fires", *The Open Thermodynamics Journal* 4: 145–155. doi: 10.2174/1874396X01004010145

Tran HC, White RH. 1992. "Burning rate of solid wood measured in a heat release rate calorimeter", *Fire and Materials* 16: 197–206. doi: 10.1002/fam.810160406

Tse SD, Fernandez-Pello AC. 1998. "On the flight paths of metal particles and embers generated by power lines in high winds – a potential source of wildland fires", *Fire Safety Journal* 30: 333–356. doi: 10.1016/S0379-7112(97)00050-7

Turns SR. 2011. *An Introduction to Combustion: Concepts and Applications*, 3rd. Ed., McGraw-Hill Science Engineering.

Urban JL, Vicariotto M, Dunn-Rankin D, Fernandez-Pello AC. 2019. "Temperature measurement of glowing embers with color pyrometry", *Fire Technology* 55: 1013–1026. doi: 10.1007/s10694-018-0810-3

Urban JL, Song J, Santamaria S, Fernandez-Pello C. 2019b. "Ignition of a spot smolder in a moist fuel bed by a firebrand", *Fire Safety Journal* 108. doi: 10.1016/j.firesaf.2019.102833

USDA Forest Service, Forest Product Laboratory. 2010. *Wood Handbook – Wood as an Engineering Material*, General Technical Report FPL-GTR-190.

Van Wagner CE. 1977. "Conditions for the start and spread of crown fire", *Canadian Journal of Forest Research* 7: 23–34. doi: 10.1139/x77-004

Vasconcelos MJ, Guertin DP. 1992. "FIREMAP-simulation of fire growth with a geographic information system", *International Journal of Wildland Fire* 2: 87–96. doi: 10.1071/WF9920087

Viegas DX. 2004. "A mathematical model for forest fires blowup", *Combustion Science and Technology* 177: 27–51. doi: 10.1080/00102200590883624

Viegas DX, Almeida M, Raposo J, Oliveira R, Viegas CX. 2014. "Ignition of Mediterranean fuel beds by several types of firebrands", *Fire Technology* 50: 61–77. doi: 10.1007/s10694-012-0267-8

Vince J. 2011. *Quaternions for Computer Graphics*, Springer.

Wadhwani R, Sutherland D, Ooi A, Moinuddin K, Thorpe G. 2017. "Verification of a Lagrangian particle model for short-range firebrand transport", *Fire Safety Journal* 91: 776–783. doi: 10.1016/j.firesaf.2017.03.019

Wang H. 2011. "Analysis on downwind distribution of firebrands sourced from a wildland fire", *Fire Technology* 47: 321–340. doi: 10.1007/s10694-009-0134-4

Wang S, Huang X, Chen H, Liu N. 2016. "Interaction between flaming and smouldering in hot-particle ignition of forest fuels and effects of moisture and wind", *International Journal of Wildland Fire* 26(1): 71–81. doi: 10.1071/WF16096

Weber RO. 1989. "Analytical models for fire spread due to radiation", *Combustion and Flame* 78: 398–408. doi: 10.1016/0010-2180(89)90027-8

Weber RO. 1991. "Modelling fire spread through fuel beds", *Progress in Energy and Combustion Science* 17: 67–82. doi: 10.1016/0360-1285(91)90003-6

Welker J, Sliepcevich C. 1966. "Bending of wind-blown flames from liquid pools", *Fire Technology* 2: 127–135. doi: 10.1007/BF02588541

Wieringa J. 1992. "Updating the Davenport roughness classification", *Journal of Wind Engineering and Industrial Aerodynamics* 41: 357–368. doi: 10.1016/0167-6105(92)90434-C

Wieringa J. 1993. "Representative roughness parameters for homogeneous terrain", *Boundary Layer Meteorology* 63: 323–363. doi: 10.1007/BF00705357

Williams FA. 1976. "Mechanisms of fire spread", *Proc. Combustion Institute* 16: 1281–1294. doi: 10.1016/S0082-0784(77)80415-3

Williams FA. 1982. "Urban and wildland fire phenomenology", *Progress in Energy and Combustion Science* 8: 317–354. doi: 10.1016/0360-1285(82)90004-1

Woycheese JP, Pagni PJ, Liepmann D. 1999. "Brand propagation from large-scale fires", *Journal of Fire Protection Engineering* 10: 32–44. doi: 10.1177/104239159901000203

Yamaguchi J, Tanaka T. 2005. "Temperature profiles of window jet plumes", *Fire Science and Technology* 24: 17–38. doi: 10.3210/fst.24.17

Yang G, Ning J, Shu L, Zhang J, Yu H, Di X. 2021. "Spotting ignition of larch (*Larix gmelinii*) duel bed by different firebrands", *Journal of Forestry Research*. doi: 10.1007/s11676-020-01282-9

Yeoh GH, Yuen KK. 2009. *Computational Fluid Dynamics in Fire Engineering – Theory, Modelling and Practice*, Academic Press.

Yin P, Liu N, Chen H, Lozano JS, Shan Y. 2014. "New correlation between ignition time and moisture content for pine needles attacked by firebrands", *Fire Technology* 50: 79–91. doi: 10.1007/s10694-012-0272-y

Yokoi S. 1960. "Study on the prevention of fire spread caused by hot upward current", *Building Research Report* 34, Building Research Institute, Ministry of Construction.

Yokoi S. 1965. "Temperature distribution downwind of the line heat source", *Bulletin of Japan Association for Fire Science and Engineering* 13: 49–55. doi: 10.11196/kasai.13.49

Yoshioka H, Hayashi Y, Masuda H, Noguchi T. 2004. "Real-scale fire wind tunnel experiment on generation of firebrands from a house on fire", *Fire Science and Technology* 23: 142–150. doi: 10.3210/fst.23.142.

Yuan LM, Cox G. 1996. "An experimental study of some line fires", *Fire Safety Journal* 27: 123–139. doi: 10.1016/S0379-7112(96)00047-1

Yuen WW, Tien CL. 1977. "A simple calculation scheme for the luminous-flame emissivity", *Proc. Symposium (International) Combustion* 16: 1481–1487. doi: 10.1016/S0082-0784(77)80430-X

Zalok E, Eduful J. 2013. "Assessment of fuel load survey methodologies and its impact on fire load data", *Fire Safety Journal* 62: 299–310. doi: 10.1016/j.firesaf.2013.08.011

Zhang S, Liu N, Lei J, Xie X, Jiao Y, Tu R. 2018. "Experimental study on flame characteristics of propane fire array", *International Journal of Thermal Sciences* 129: 171–180. doi: 10.1016/j.ijthermalsci.2018.02.024

Zhao S. 2011. "Simulation of mass fire-spread simulation in urban densely built areas based on irregular coarse cellular automata", *Fire Technology* 47: 721–749. doi: 10.1007/s10694-010-0187-4

Zhou K, Jia J, Zhu J. 2018. "Experimental research on the burning behavior of dragon juniper tree", *Fire and Materials* 42: 173–182. doi: 10.1002/fam.2469

Zhou L, Zeng D, Li D, Chaos M. 2017. "Total radiative heat loss and radiation distribution of liquid pool fire flames", *Fire Safety Journal* 89: 16–21. doi: 10.1016/j.firesaf.2017.02.004

Zukoski EE. 1995. "Properties of fire plumes" in Cox G ed., *Combustion Fundamentals of Fire*, Academic Press.

Zukoski EE, Kubota T, Cetegen B. 1980/81. "Entrainment in fire plumes", *Fire Safety Journal* 3: 107–121. doi: 10.1016/0379-7112(81)90037-0

Index

For Product Safety Concerns and Information please contact our
EU representative GPSR@taylorandfrancis.com Taylor & Francis
Verlag GmbH, Kaufingerstraße 24, 80331 München, Germany